Warriors
of the
Rainbow

Greenpeace 40th Anniversary Edition

This edition published in 2011 by
Greenpeace International
Ottho Heldringstraat 5
1066 AZ Amsterdam
The Netherlands
www.greenpeace.org

in association with

Fremantle Press
25 Quarry Street
Fremantle Western Australia 6160
www.fremantlepress.com.au

Designer Tracey Gibbs
Printed by Everbest Printing Company, China
Production: This book is printed on 100% recycled paper; print
Ceritifcation ISO 14001; inks are vegetable based; water-based varnish
on the cover.

National Library of Australia Cataloguing-in-Publication entry:

Hunter, Robert, 1941-2005
Warriors of the Rainbow : a chronicle of the Greenpeace
movement from 1971-1979 / Robert Hunter.
2nd ed.
ISBN: 978 1921 888 80 9 (pbk.)
ISBN: 978 1921 888 81 6 (ebook)
Includes index.
Greenpeace Foundation--History.
    Environmental protection--History.
    Nature conservation--History.
333.78216

Permission to use the lyrics from the song "We Are Whales" by
Melville and Sybille Gregory has been granted by the authors for
the first edition. Permission to reprint excerpts from "A Labrador
Journal" by Ms. Brigitte Bardot granted by the author and *Greenpeace
Chronicles* for use in the connection with the work of Greenpeace.

*For further information on Greenpeace activities and membership, see
www.greenpeace.org*

Warriors
of the
Rainbow

**A Chronicle of the Greenpeace Movement
from 1971 to 1979**

by Robert Hunter

GREENPEACE
in association with
FREMANTLE PRESS
fine independent publishing

For Mother Gaia

Foreword

It is June and summer time in Greenland. Not that I can really tell since it is freezing as a Zodiac inflatable carries me and fellow activist Ulvar Arnkvaern from the Greenpeace ship *Esperanza* toward the massive oil rig *Liev Erikson*. Baffin Bay is not where you would expect to find an African from Durban, South Africa—even in "summer"—but there is a multi-national oil company intent on opening up the Arctic to deep sea oil drilling here, and so Greenpeace is also here to protest, bear witness and take action. We have no choice but to do so.

The story you hold in your hands is too powerful to ignore. It is the story of the birth and early years of Greenpeace, the most important environmental activist organization to come out of North America. It is a vivid and often hilarious firsthand account of how a group of journalists, anti-nuclear campaigners, mystics and mechanics bickered their way through the 1970s, of how their direct tactics inspired popular movements to save the whales and seals, stop French nuclear weapons tests and the live capture of marine mammals and more.

This account was written by Bob Hunter, who for many of those early years was the recognised leader of Greenpeace. Hunter was a reporter and a writer before Greenpeace, and during the years this book describes he ventured into a state of mystical, hallucination-induced hippie vision—one where humans are not at the centre of the universe. By his own admission, Bob was not always easy to work with, in part because of the vision that drove him, but he never ceased being an affable, talented communicator and, in a manner peculiar to the times, an effective and generous leader.

It was under Bob Hunter's tenure that Greenpeace achieved its global reputation for daring, media-savvy campaigns. Using the most cutting edge technology of the day, such as short-wave radio, radio direction finders, Zodiacs and 35mm film, Greenpeace voyages sought to expose and bear witness to environmental malfeasance. They stopped nuclear weapon tests in places where news organizations had not dared to venture and endeavoured to live up to a Cree tale about the Warriors of the Rainbow who literally chased rainbows and the moon across the sea.

The name of this book is taken from the narrative described by a Cree elder and grandmother called Eyes of Fire. She foretold of a time when the Indigenous people would lose their spirit, and the Settlers' greed would destroy Mother Earth. Over time, some of the Native peoples would regain their spirit, and teach the Settlers to have reverence for the Earth. Together, using the rainbow as a symbol, all the Peoples of the world would join together as Warriors of the Rainbow and work with one another to end the destruction and desecration of the planet. Hunter found this story inspirational and shared it with his companions, the idea stuck and a new legend was born.

Most organizations have mission statements. Thanks to Bob, and to the Cree tale that inspired him, Greenpeace has a prophecy.

Bob Hunter's narrative describes the early Greenpeacers sailing around the Pacific, consuming excessive amounts of recreational drugs, bickering over internal politics, risking their lives, developing arguments about whales being sentient beings and staving off bankruptcy. Somehow they managed to start a global movement, take on both super powers of the time, save seals and whales, protest oil tankers in coastal waters and drive nuclear testing underground. No author of fiction would have dared propose a tale this amazing.

Along the way they were denounced by movie stars like John Wayne, feted by movie stars like Brigitte Bardot, and supported by 1970s musical icons such as Joni Mitchell, Phil Ochs, Country Joe McDonald, James Taylor and Gordon Lightfoot.

Several of the figures mentioned in the book went on to have a great influence in the environmental movement, and their more recent history is worth mentioning.

Paul Watson, prominent in early accounts of Greenpeace campaigns, was ejected from the Greenpeace Board in 1977 and later went on to found the Sea Shepherd Conservation Society. Together, Hunter and Watson were named among the men of the twentieth century by *Time* magazine. Hunter went on to advise Sea Shepherd and accompanied Watson on numerous campaign journeys and the two remained friends until Hunter's untimely death in 2005.

David McTaggart was in Australia, the Pacific Islands, or in France for much of that period, pursuing his own campaigns against nuclear testing. Later McTaggart would carry on the work Bob Hunter had started in driving Greenpeace's expansion around the world.

Paul Spong, who introduced Bob Hunter and Greenpeace to the plight of the whales, continues his important research on cetaceans to this day.

Rex Weyler, the photographer on many of the early voyages, would become an important Greenpeace historian with the publication of his book *Greenpeace*.

Bobbi Hunter, first treasurer, first fundraiser, first woman to put herself between a whale and a harpoon, Bob's second wife and a stalwart for both he and Greenpeace.

And then there was Patrick Moore, a key figure in the early years who would become a vociferous critic of Greenpeace and other environmental organizations. Moore has opposed our work from outside Greenpeace for longer than he supported it from within.

After ending his formal role with Greenpeace and spending several years as a freelance writer, Hunter settled in Toronto where he worked for the upstart local television station CITY-TV. By the mid 1990s, he had his sights set on the goal of stopping catastrophic climate change. Environment ministers would cringe when CITY-TV's Eco-Specialist attended their news conferences. Environmental activists were delighted when their campaigns were brought to air, and to life, by a man who always had time to encourage activists. He covered numerous Greenpeace actions in Canada and the U.S., travelled to the 1997 Kyoto Conference, and wrote a paper for Greenpeace International on the threat of climate change, which was transformed into the last of his thirteen books, *Thermaggedon, Countdown to 2030*.

In preparing this edition only a few sentences have been altered from the original, changes that Bob himself had agreed should be made to any future editions, at the request of a crew member from the first voyage. We are pleased to be able to honour that now.

Greenpeace carries with it to this day the Spirit of Bob Hunter, as do many others. Among them is Bob's youngest daughter Emily, who has followed in her father's footsteps as an eco-activist and journalist. We continue to value life more than material things, we continue to sail ships with rainbows painted on the bows, and we still make it our business to voyage to the frontiers, to bear witness to environmental destruction and, where we find it, we take action to stop it.

There have been many books and memoirs written by participants of these events. All are worth reading. The first, the most poignant, and the best is *Warriors of the Rainbow*. It is also, in my humble opinion, the most hilarious.

Kumi Naidoo
International Executive Director, Greenpeace
2011

Author's note

We all bear witness: What plays nightly on the newscast if not Twilight of the Gods?

Deserts on the move. Oceans turned black and lifeless. Birds vanishing from the sky. Jungles burning. Smog tightening its carbon dioxide grip around city after city, while the countryside turns to sand. More Northern forests aflame than ever. The sky itself rotting before our computerized censors while we do nothing to save ourselves from perishing in a few generations from unbearable extra ultraviolet radiation.

A storm called Greenhouse Effect.

Like a new kind of all-at-once global tsunami of savage climatic gyrations. A steady percolating-to-the-boiling-point inside the goldfish-bowl atmosphere, followed, as some computer projections foresee, by a swift and terrible plunge in temperature leading to a million-year Ice Age. Super-hurricanes that sweep the soil away across backed-up empty lakebeds.

The nightmare this very day, you can truthfully agree, is no longer a future prediction. It is upon us.

Not alone. Not striving in a vacuum. An aspect of nature.

Such a notion smacks of legendary old-world magicians who could reportedly make nature do their bidding. They worked their miracles, it is said, through a profound understanding of the natural world, enough so that a grand and fantastic co-operative effort between man and nature could occur.

Maybe it is a time for magicianship, if such a thing is still possible.

Certainly, it is a time for ecologism, not just environmentalism. It is a time to face up to the likelihood that if ecological conditions are allowed to deteriorate to the point where an entire generation finds itself engulfed in a burning, radiation-lashed hell on Earth, it will be too late.

I cannot avoid the nagging feeling that change—good, solid, environmental reformist type of change—is just not coming fast enough.

If we are as close to the edge as I fear we are, we should be leaping for our lives. I think when the real Greenhouse Effect hits, we'll wake up en masse in a hurry. But we might be up against something that has burned too high and can't be stopped.

Strongly suspecting this scenario to be the one that really happens, isn't it my duty to start resisting disaster immediately? Not tomorrow. Not next New Year's Eve. The disaster we are talking about would leave our children—certainly our grandchildren—nowhere to run. It would leave our countries in rubble. It would devastate—has already devastated to a terrifying degree—our very planet.

Everybody must consider that in the absence of a settlement of the struggle over the remaining commons—the seas, the rivers, the atmosphere, Antarctica, orbital space—a "war of the millennium" may indeed fall quite possibly too late upon the shoulders of today's young and not-so-young—and by then conditions will be enormously worse.

The generation of the Eco-Holocaust will look back on the present moment and history as an hallucination—so briefly enjoyed, the opulent wealth so quickly squandered, then gone forever, leaving only ruin.

And they will ask: Why, for God's sake, didn't somebody fight for us?

Didn't they know we were coming? Didn't they care? Were they crazy or simply too dazed collectively by lead poisoning, ozone brain damage, carbon dioxide overdose, low-level ultraviolet radiation injuries, soaring dioxin concentrations, and even electromagnetically-induced cellular disruption? Maybe it was the caffeine or the alcohol or the trihalomethanes in the water. Maybe there was an invisible, corresponding cancer of the mind or spirits going on that no one recognized because they were all affected by it. Maybe they were technically all insane, an entire generation, a whole world full of mad people who threw the planet and the future away. Ate it alive, as it were.

It was out of this context the Greenpeace Foundation, as we originally called it, came into being. I thought of it from the very beginning as some kind of natural phenomenon. All of us were involved because we had to be. We were part of a reflex, rather than being the reflex itself. We had been summoned into action by a collective impulse that came up out of the bottoms of our brains, with our spines acting as cables touching the ground, and somehow being commanded from there.

For me there was little doubt: The marching orders came from the Earth. If awareness is the accumulated amount of subconscious material that is rendered conscious, the ultimate root of my subconscious was my planet, and not just the psychological Collective Unconscious that Carl Jung talked about, or even just the sacred Noosphere envisioned by Teilhard de Chardin, but the physical supramother, Mother Earth, whom the Indigenous Peoples persist, correctly, in worshipping.

No matter how big a movement grows, its blueprint is to be found in its seed. By the moment of its birth, the basic imprinting is done. It remains for environment and circumstances to shape what emerges.

Greenpeace emerged out of the tail end of the 1960s in Vancouver, Canada, which really says it all: a unique place at a unique moment in history. There was, at that time, no other place on Earth quite so perfectly suited to give birth to an environmental revolution. The city's sons and daughters weren't involved in a war. We were fabulously rich by the world's standards. We lived in a democracy. We could afford the luxury of giving ourselves—for a while, anyway—to a cause, and still somehow survive. Most importantly, our valleys and mountains were still mainly green, beaches swimmable, tap water delicious and untainted. We had plenty to lose.

Freedom and wealth and knowing. We had one thing more: access to the highly-advanced Canadian electronic communications grid. We could plug into an enormous media machine that was fully integrated with the American system, a system which not only connected to the international news and entertainment networks, but largely controlled them. We could not have done what we did from, say, Nairobi. When we sent out our major mind-bombs, they were launched from Vancouver, but boosted into orbit from New York and San Francisco.

What caught the attention of the public were the dramatic confrontations we staged, very much with an eye to the camera, since my theory was that the camera was mightier than the harpoon, and the global media was the only force capable of staying The Bomb. But we did have a philosophy. Patrick Moore and I sat down and wrote it in the autumn of 1976. And since it is not included in this book, I believe it is worth quoting. It was titled The Greenpeace Declaration of Interdependence. It is as valid today, I believe, as it was then.

> We have arrived at a place in history where decisive action must be taken to avoid a generational environmental disaster. With nuclear reactors proliferating and over nine hundred species on the endangered list, there can be no further delay or our children will be denied their future.
>
> The Greenpeace Foundation hopes to stimulate practical, intelligent actions to stem the tide of planetary destruction. We are "rainbow people" representing every race, every nation, every living creature. We are patriots, not of any one nation, state or military alliance, but of the entire Earth.

It must be understood that the innocent word "ecology" contains a concept that is as revolutionary as anything since the Copernican breakthrough, when it was discovered that the Earth was not the center of the entire universe. Through ecology, science has embarked on a quest for the great systems of order that underlay the complex flow of life on our planet. This quest has taken us far beyond the realm of traditional scientific thought. Like religion, ecology seeks to understand the infinite mysteries of life itself. Harnessing the tools of logic, deduction, analysis and empiricism, ecology may prove to be the first true science-religion.

As surely as Copernicus taught us that the Earth is not the center of the universe, ecology teaches us that humanity is not the center of this planet. Each species has its function in the scheme of life. Each has a role, however obscure that role may be.

Ecology has taught us that the entire Earth is part of our body and that we must learn to respect it as we respect ourselves. As we love ourselves, we must also love all forms of life in the planetary system—the whales, the seals, the forests and the seas. The tremendous beauty of ecological thought is that it shows us a pathway back to an understanding of the natural world—an understanding that is imperative if we are to avoid a total collapse of the global ecosystem.

Ecology has provided us with many insights. These may be grouped into three basic "Laws of Ecology," which hold true for all forms of life—fish, plants, insects, plankton, whales and humanity. These laws may be stated as follows:

THE FIRST LAW OF ECOLOGY states that all forms of life are interdependent. The prey is as dependent on the predator for the control of its population as the predator is on the prey for a supply of food.

Example: Humans, in their self-interest, often lay plans for the extermination of species that are viewed as 'undesirable.' There would be few objections raised to a programme of mosquito eradications. It would be more difficult, however, to gain acceptance of a programme to eradicate swallows, so beautiful as they flit about chasing insects through the sky. But wait, the swallows are eating mosquitoes. Before the

12

mosquitoes are eliminated it would be wise to consider the number of birds that will starve to death as a result.

THE SECOND LAW OF ECOLOGY states that the stability (unity, security, harmony, togetherness) of ecosystems is dependent on their diversity (complexity). An ecosystem that contains one hundred different species is more stable than an ecosystem that has only three species. Thus the complex tropical rainforest is more stable than the fragile arctic tundra.

Example: Consider a natural ecosystem such as a forest which contains a dozen different species of trees. Each species is susceptible to specific disease that can kill individual trees. If there are many tree species it is unlikely that they would all be attacked at once, and since the trees of any one species are spaced apart, with other species between them, there is less chance of an epidemic that would wipe out every tree of that species. Enter humanity—the forest is clear-cut for lumber and pulpwood, and is replanted with seedling trees of one species, a species that suits humanity's needs better than any of the original dozen. Now if a disease that is specific to the new species strikes the forest, all the trees are susceptible and an epidemic is far more likely.

THE THIRD LAW OF ECOLOGY states that all resources (food, water, air, minerals, energy) are finite and there are limits to the growth of all living systems. These limits are finally dictated by the finite size of the Earth and the finite input of energy from the sun.

Example: There are so many examples of our inability to recognize this law that no instance would explain the severity of the problem. Let it suffice to say that we are now coming up against the limits of many natural resources, including agricultural lands, the fisheries, whales, petroleum, minerals, water and forests. In the process we are creating a desperate situation for other species who also depend on many of these resources for food and energy.

If we ignore the logical implications of these Laws of Ecology we will be guilty of crimes against the Earth. We will not be judged by other people for these crimes, but with a justice meted out by the Earth itself. The destruction of the Earth will lead, inevitably, to the destruction of ourselves.

So let us work together to put an end to the destruction of the Earth by the forces of human greed and ignorance. Through an understanding of the principles of ecology we must find new directions for the evolution of human values and human institutions. Short-term economics must be replaced with actions based on the need for conservation and preservation of the entire global ecosystem. We must learn to live in harmony, not only with our fellow humans, but with all the beautiful creatures on this planet.

This is the story of the first seven years of Greenpeace. In seven years, physical body sheds all its cells and grows new ones—except for those of the brain, which have their own slower decaying mechanism. In the first seven years Greenpeace found its essential nature, which was, as it turned out, an expression of nature.

I don't think for a minute that as a phenomenon Greenpeace was at all "autonomous." It was something that had to happen, that was in some profoundly unknown way made to happen.

With the launching of Greenpeace on the West Coast of Canada, a trace of long-dormant shamanistic magic wafted over the continent and ocean that Europeans had long ago swarmed across. This dew of magic moved backward over the checkerboard plains where today only a handful of buffalo survive. It passed the nuclear reactors leaking into the Great Lakes. It skudded through the carbon dioxide clouds and above the acid rain-eaten forests, over Love Canal and across the Sea of Slaughter in the St. Lawrence, whispering over the bloody ice packs of Labrador. The magic crossed the lonely North Atlantic where you can voyage for a week now without seeing a single whale, and reached at last the shores of Europe, whence the carnage and the poisoning began, like the ghost of the Red Man holding up his hand and saying: No more!

At its deepest level—even though this will sound crazy to some—I believe it was a part of the awakening of the planet into a state of self-consciousness.

I should also say, as a result of my experiences with Greenpeace, I firmly believe in miracles. I define a miracle this way: If a coincidence occurs, fine, that's a coincidence. If two coincidences occur simultaneously, that's rather interesting but nothing to get especially worked up about.

If three coincidences occur simultaneously, however, the hair ought perhaps to stand up a bit on the back of your neck. If four occur, all at once, you would surely pause to take a second look. If five or more happen at the same moment in time, you have begun to move into the realm of small "m" miracle. Draw the line where you will, according to your own sense of reality, but there is a point where the occurrence of too many multiple coincidences, too many times in a row, leave even the most determinedly rationalistic mind facing the distinct possibility that it has just encountered a true miracle.

During some of the Greenpeace campaign, miracles were not only commonplace—we got quite a few of them on film and on tape.

The events described here cover a "medicine period" in the history of the ecological struggle. Among the Plains Indians, when disaster threatened, the medicine men were sent forth to deal with it. Only when they failed to solve the problem were the warriors brought out. So this is not really so much an account of the early activities of the Warriors of the Rainbow, whose coming was prophesized by a Cree Indian grandmother named Eyes of Fire some two hundred years ago, as it is the tale of the Medicine Men and Women of the Rainbow.

If the green and peaceful—and indeed loving—way fails, it must be war next. The trouble with waiting for the meek to inherit the Earth is that there won't be anything left for them to inherit.

Robert Hunter
May 1990

This text was found in 2011 among the papers left by Robert Hunter. It was written in 1990, more than a decade after the first edition, and reflects the author's prophetic view of the future of the planet.
Robert Hunter died May 2, 2005.

I'm not a Red, I'm a Green.

—ROD MARINING

Introduction

This is the story of the first seven years of a movement that attempted to fulfill an ancient North American Indian prophecy of an age when the different races and nationalities would band together to defend the earth from her enemies.

Adopting the peace and ecology symbols, its members fought—we fought—an unequal battle against American and French nuclear weapons makers; Russian, Japanese, and Australian whalers; Norwegian and Canadian seal hunters; multinational oil consortiums and pesticide manufacturers; cynical politicians; angry workers; and, again and again, ourselves. The people involved were men and women, young and old, not all of them brave or wise, who found themselves face-to-face with the fullest ecological horrors of the century—the mass slaughter of infant harp seals on the heaving stage of the Labrador Ice Front, fleets of metal ships armed with explosive harpoons that pursue the last of the great whales into the depths of the ocean, and the incomparable fury of the hydrogen bomb itself.

At its heart, this movement sought to give political form to an awareness that predates Buddhism but is at the same time as new as the science of interdisciplinary ecology. It grew out of a flickering awareness that *all* our relationships are political, and that the crucial political relationships with which we must concern ourselves now have almost nothing to do

17

with man's relation to man, but with man's relation to the earth itself. It is our relationship to our planetary environment which is the most important issue of all. All human structures inevitably rest upon it.

The time has arrived when we must begin to examine the underlying realities of our relationship to all life around us, the life that we are just beginning to appreciate as the true medium of our innermost identity. We need to move neither further to the Left nor further to the Right— rather, we must seriously begin to inquire into the rights of rabbits and turnips, the rights of soil and swamp, the rights of the atmosphere, and, ultimately, the rights of the planet. For these in the end are the containers of our entire future evolution, and everything rests upon whether or not we come to terms with the politics of earth and sky, evolution and transformation, God and nature. Otherwise, in our lifetimes, we shall suffer the enactment of the saga of Genesis: our expulsion from paradise and the fall of nature itself.

Machiavellianism and mysticism alike played their parts in the shaping of the consciousness this movement expressed. It embodied at times a religious fervor, at other times, a ruthlessness that bordered on savagery; it was born out of desperation, out of fear, out of a love that transcended not only the boundaries of race or creed but even mere humanness. Corruption and greatness both played their parts and both took their tolls.

1

From Amchitka to Mururoa

A flower is your brother.

—PATRICK MOORE

Not since the War of 1812 had the border between Canada and the United States—the world's longest undefended border—been closed. So there was a sense of history in the air as six thousand Canadians and roughly one thousand American draft dodgers and deserters converged on the Douglas Border Crossing between British Columbia and Washington State to plug the main highway with our bodies. It was October 1, 1969. A hard wind was blowing from the Pacific Ocean, crackling the banners and placards that said: DON'T MAKE A WAVE. IT'S YOUR FAULT IF OUR FAULT GOES. AMERICA IS DEATH.

The Alma Mater Society of the University of British Columbia had chartered several dozen buses to bring students from Vancouver. The rest of us had traveled the fifty miles to the border in vans and cars. There were street freaks and Marxists and Maoists and Trotskyites and Yippies and members of the radical Vancouver Liberation Front in the crowd that now spilled across the large landscaped bowl of the Peace Arch Park, but most of the faces were clean-shaven, sheepskin jackets were the style, and there were textbooks under most of the protesters' arms.

The attorney general of the province had declared that morning: "persons actually taking part in the obstruction of traffic which normally flows between Canada and the U.S. lay themselves open to the most serious charges under the Criminal Code." Rumors swirled through the crowd to the effect that busloads of Royal Canadian Mounted Police and their dogs were parked just out of sight beyond a hill to the north and that several units of National Guardsmen had been flown into the nearby American town of Blaine. To the south lay an open sweep of land and the low brick buildings of Customs and Immigration. Cars and trucks were still crawling along the highway, maneuvering between the mob that was pressing on both sides of the road and waiting for the signal to step onto the pavement. A great white monolith called the Peace Arch towered above the scene, emblazoned with the words: MAY THESE GATES NEVER BE CLOSED.

While the war in Vietnam was an issue in the back of everyone's mind, and fears were growing that the blue coastal waters sparkling beyond the bulk of the Peace Arch might soon be fouled by oil spills from supertankers

20

on their way down from Alaska, the focus of the crowd's concern and anger this day was on Amchitka Island, a remote outcropping of rock some twenty-four hundred miles away near the tail end of the Aleutian Islands. Until recently, few people in British Columbia had ever heard of the place, but now, thanks to a sudden blitz of attention in the media, its name was a household word on the West Coast.

As a columnist for the Vancouver *Sun*, the province's largest daily newspaper, I had contributed as much as anyone to the sudden rush of anxiety that had driven seven thousand souls down to the border. A week before, September 24, I had written:

> Beginning at midnight tonight, the United States will begin to play a game of Russian roulette with a nuclear pistol pressed against the head of the world. As of midnight, a blockade will be thrown around remote Amchitka Island near the tail end of the Aleutians. Sometime between tomorrow and Oct. 15, a 1.2 megaton atomic bomb will be triggered at the bottom of a 4,000-foot hole on the island.

> No one knows what the consequences will be, but scientists in Canada, the U.S., Japan and Hawaii have warned that there is a distinct danger that the test might set in motion earthquakes and tidal waves which could sweep from one end of the Pacific to the other. ... By setting off its underground nuclear test in the Aleutians—one of the most earthquake-prone areas in the world—the U.S. Atomic Energy Commission is taking a chance on triggering a chain reaction of earthquakes and tidal waves which would slam the lips of the Pacific Rim like a series of karate chops.

The West Coast had been battered only five years before by a tidal wave. A deep fault line that ran within a few miles of Amchitka passed offshore of Vancouver Island before winding its way inland and emerging in California as the San Andreas Fault.

Information and culture tended to flow on a north-south axis along the West Coast, bringing British Columbia in many ways closer to the mood of California than that of eastern Canada. Since the beginning of the war in Vietnam, Vancouver itself had become the main refuge for fleeing Americans. Amchitka played on the worst fears of British Columbians and Californians alike. It was perhaps no surprise that the combined threat of tidal waves and earthquakes was enough to stimulate this rush for the gates of the Douglas Border Crossing. Certainly, in the previous century and a half, no issue had generated that kind of response on the part of the Canadians.

Moments before the scheduled closing of the border—at 1:00 P.M.— Gene Kiniski, a famous wrestler, came barreling out of Customs and Immigration. Stopping in the midst of the crowd, he leapt onto the hood of his car where he was greeted by jeers and laughter. "I'll break you punks in half!" he roared, then got back in the car and drove through fast enough to send students leaping for the curb. The last car to cross the border was a white Cadillac with massive fins and a California license plate. As it left the bay at U.S. Customs, howls burst through the air, and the crowd began surging onto the highway. Then a black arm snaked out of the window, and a black clenched fist stabbed at the sky. Half the crowd was still howling, but those close to the car who could see the black man began to cheer. Confusion. The Cadillac slipped through. Suddenly, a police siren started up somewhere on the American side. The crowd jerked, and there was a convulsive rush for the road. Everybody seemed to want to be in the arena when the fight started.

The border stayed blocked for an hour. Neither the RCMP nor the National Guardsmen made any move. One by one, the speakers climbed up on the edge of the Peace Arch to make their speeches denouncing the tests at Amchitka, the Cold War, Vietnam, and America in general. When my turn came at the microphone, I tossed away a written speech and simply reacted to the mood. It was the first time I had ever addressed so large and volatile a mob. They didn't want reason, they wanted to be made to roar. My own will seemed to vanish. The words that came out made hardly any sense at all, except that they were called forth by the excitement and tension in the crowd. Watching the demonstration on television that night, I was stunned to discover that one of the demagogues up there ranting and raging was myself.

After an hour, the organizers of the blockade announced that the point "has been made," and ordered everyone back off the road.

Two days later, a one-megaton bomb went off at Amchitka. It neither leaked radiation nor caused any tidal waves or earthquakes, although a previous eighty-kiloton test in 1965 had reportedly leaked slightly.

But in the minds of many of the people who had been in the crowd, the reaction to the test at Amchitka had signaled a turning point of some kind in the otherwise normally docile attitude of British Columbians. No issue with environmental and antiwar overtones had yet roused more than a couple of hundred people at once. Whatever else Amchitka might be, it was clearly something around which a massive amount of energy could be mobilized.

Back in Vancouver, other minds were at work trying to think of some way to carry the momentum of the border blockade further, but without resorting to an outright invasion of the United States or expecting too much from the "revolutionary mood" of the Canadian people at large.

One of those minds was that of Jim Bohlen, a forty-three-year-old American who had fled to Vancouver with his second wife when their son became eligible for the draft. Bohlen was a composite-materials researcher. At one time—around 1955—he had worked with Buckminster Fuller and had been one of perhaps one hundred people on Earth at that time who understood Fuller's theories about synergistic effects. A veteran of the Second World War, Bohlen had once been involved in the design of a missile-rocket motor system known as Sprint, which was later to be incorporated in the Advanced Ballistic Missile System—the same system whose warheads were now being tested at Amchitka.

Another individual opposed to the test at Amchitka was a slightly older former Philadelphia lawyer named Irving Stowe, a Jew who had joined the Quaker religion, and who, like Bohlen, had fled to Vancouver when his own son came of age to be drafted. Both men had soon become involved in the Committee to Aid American War Objectors, providing bed and board and job counseling for hundreds of draft dodgers and deserters. Both of them had also been involved in groups like the Sierra Club back in the U.S., and had noticed, shortly after arriving in B.C., that Vancouver lacked any kind of organized environmental group at all. It didn't take them long to set up a British Columbia chapter of the Sierra Club.

Shortly after the Douglas Border Crossing blockade, Bohlen and Stowe, along with a young Canadian lawyer named Paul Cote, decided to set up a mechanism for keeping the feeling against Amchitka alive. Borrowing one of the slogans that had appeared at the blockade, they incorporated themselves as the Don't Make a Wave Committee. Primarily, it was made up of members of the B.C. Sierra Club.

Their first few meetings led nowhere. Irving Stowe tended to talk for hours nonstop if given a chance, and while everyone on the committee agreed that Amchitka was a potent symbol of war craziness and environmental degradation wrapped up into one, the precise means of turning it into a major ongoing issue eluded them. It wasn't until the morning after one of the committee's long-drawn-out meetings, at the breakfast table, that Bohlen's wife, Marie, said: "Why the hell doesn't somebody just sail a boat up there and park right next to the bomb? That's something everybody can understand."

Bohlen was still mulling that one over in his head a few minutes later when the phone rang. It was a reporter from a television station wanting to know what the Sierra Club's plans were in relation to Amchitka. Were any further demonstrations in the works?

Shooting a quick glance at Marie, Bohlen took the plunge: "Well, as a matter of fact, we're going to sail a ship up to Amchitka."

For the next week, the phone rang constantly. Calls came in from news services all over North America. Group after group rang up to pledge their support. Mail began to pile up until it overflowed the mailbox. Reporters and cameramen arrived one after the other at the door. Jim Bohlen talked until he began to lose his voice. At a hastily called meeting, the Don't Make a Wave Committee endorsed the idea. And at the end of the meeting, as everyone was leaving, Irving Stowe waved his usual V sign and said, "Peace." The youngest member of the committee, a twenty-three-year-old Canadian named Bill Darnell, said: "Make it a *green* peace."

Everyone paused. That had a certain ring to it. Not just peace, but an ecological peace.

"That sounds good," said Jim Bohlen. "If we ever find a boat, that's what we'll call it. *Greenpeace*."

Paul Cote was in his mid-twenties, an intense young lawyer barely out of the university, so uncomfortable speaking in public that he blushed from the time he cleared his throat until the time he sat down. He and Bill Darnell were the only Canadians involved in the immediate group. Darnell seldom spoke. If anyone had ever seemed to personify the myth of the strong, silent Canadian, it was Darnell. Yet when I confessed this to his wife, she laughed hysterically.

Bohlen and Stowe were the patriarchs. Both had beards and wives and children. Marie Bohlen was a strongly built woman with iron will. Her graying hair was styled severely. She radiated power. She and Dorothy Stowe, a plump, grandmotherly woman with a kind of Bronx accent, were the matriarchs. They were also mothers on the run with their children from America of the 1960s.

24

Terry Simmons, twenty-five, a cultural geographer from the West Coast of America, had joined the group early and had been one of the cofounders of the Sierra Club of B.C. Simmons had a swift, academic mind.

This was the core group out of which the Greenpeace Foundation would eventually emerge.

The idea of sending a boat into a bomb test zone was not quite new. The Quakers had tried it twice before with the *Phoenix* and the *Golden Rule*, but in both cases the boats had been seized long before reaching their goals. The weakness had been that the owners were Americans setting out to challenge their own government and could therefore be grabbed on any pretext. But if a Canadian boat were to sail into a test zone, so long as it stayed outside the actual territorial limit, then there would be nothing the American authorities could do, unless they were willing to try to seize a foreign vessel in international waters—an outright act of piracy.

Shortly after the Douglas Border Crossing blockade, the U.S. Atomic Energy Commission announced that its next test at Amchitka would be carried out in two years' time: early autumn, 1971. The 1969 test, code-named "Milrow," had been 1.2 megatons. The next, to be code-named "Cannikin," could be in the neighborhood of five megatons, which was 250 times the force of the blast that hit Hiroshima. These were big toys the Americans were playing with. No one outside the AEC knew exactly how many tests were planned, but there were reports of as many as seven in the works, each one larger than the last.

In the two years between the closing of the border and the launching of the *Greenpeace*, the Don't Make a Wave Committee held something like fifty meetings, most of them in a Unitarian church in South Vancouver. They ranged in size from get-togethers of the "triumvirate"—Bohlen, Stowe, and Cote—to sessions where as many as forty people would all sit down on wooden folding chairs and talk aimlessly for six or seven hours at a time.

The committee by this time had been joined by a young ecology student, Patrick Moore, who was working on his Ph.D. in ecology at the University of British Columbia. He had a reputation in local environmental circles as a "heavy radical" because he had done a study of biological monitoring.

Single-handedly, he had stood up to a multinational corporation and forced it to modify an outrageous scheme to dump some nine million gallons of effluent a day for the next twenty-five years into an otherwise-pristine coastal inlet. Moore was quickly accepted into the inner circle on the basis of his scientific background, his reputation, and his ability to inject practical, no-nonsense insights into the discussions. He had been raised on the coast in a logging camp, and he knew more about the weather and sea conditions involved than anyone else on the committee. He had hair like candy floss and looked more like a rock star than a scientist.

But the committee was attracting more than scientists. Bohlen and Stowe had for some time been deliberately trying to enlist the media, bombarding reporters and broadcasters with information about the environmental crisis. In a province like British Columbia, with a population of only two million, it was a relatively simple matter to contact the handful of journalists, program directors, hot-liners, columnists, and editorial writers who were the shapers of public opinion. By 1970, British Columbia was already known throughout Canada as the most "environmentally minded" province, thanks in no small degree to the background work done by the Sierra Club.

British Columbians had been picking up on the sudden flurry of attention to ecology that was taking place in the U.S., but more than that, they had the advantage of living in a lush, rich, under- populated region where the natural environment had not yet been altered beyond recognition or repair. It was this characteristic of B.C. that had attracted so many refugees from Philadelphia, Pittsburgh, New York, and other urban American "sinks." Unlike the natives, who took British Columbia's natural state for granted, the newly arrived Americans were determined to preserve "Beautiful B.C." from the kind of industrialization that had already destroyed so much of their homeland.

Accordingly, it came as no surprise when Irving Stowe asked me if I would take part in the voyage to Amchitka. My column at the *Sun* had evolved from a counterculture forum—I was the paper's "token hippie"—to an increasingly environmental platform. Although I had a bad reputation for having written too much too openly about drugs, I had also devoted more space to ecology than any other columnist in the province. Moreover, I had been attacking Amchitka since early 1969, and my speech at the Peace Arch had been one of the more provocative.

Two other writers were invited. One was a Canadian Broadcasting Corporation free-lancer, Ben Metcalfe, a veteran of the bad old days of western Canadian journalism, who not only served as the CBC's West Coast theater critic, but who also at one time ran his own public-relations firm. Another was young Bob Cummings, a former private detective now working for the local underground newspaper, whose function was to represent the counterculture. The idea was to hit the establishment press, the underground press, and the airwaves all at once. Irving Stowe himself had by this time taken to writing a column in an underground paper, the *Georgia Straight*, called "*Greenpeace* Is Beautiful." So, from the beginning, there was a major departure from the spirit of the earlier Quaker voyages against bomb tests. Whereas the Quakers had been content to try to "bear witness," *Greenpeace* would try to make *everybody* bear witness—through news dispatches, voice reports, press releases, columns, and, of course, photographs. A photographer was added to the list of possible crew members very early in the planning. The candidate who emerged was a tall long-haired chemistry student, Bob Keziere, who was also a jazz musician.

The committee had by this time attracted a number of other people, not all of them considered desirable by Bohlen, Stowe, and Cote. Among them were two in particular whom Bohlen considered "too radical." Both were in their early twenties. Both were dropouts. Both had been in trouble with the police, although neither had ever been convicted of anything. The first was Rod Marining, the self-styled "nonleader" of the Northern Lunatic Fringe of Yippie! He had joined Jerry Rubin during an occupation of the Faculty Club at the University of B.C., and had been involved in several attempts to liberate park areas destined for development. He usually carried a tape recorder slung over his shoulder, had an immense mane of chestnut hair, one blue eye, one dark brown eye, and a habit of talking in McLuhanisms. Bohlen had him figured for an acidhead.

The other was Paul Watson, a muscular seaman who had knocked about the world on freighters, and who would have been a perfect candidate for the voyage, except that he had North Vietnamese flags stitched to his army jacket, wore Red Power buttons, Black Power buttons, and just about any kind of antiestablishment button that could be imagined. He, too, wrote for the underground press, and had been busted in at least a dozen demonstrations, some of them staged by the Vancouver Liberation Front.

During the two years it took to get ready for the voyage, the heaviest task fell to Jim Bohlen and Paul Cote. Theirs was the responsibility for finding a boat.

While they wandered the docks in pouring Vancouver rain, Irving Stowe took charge of finding the money. Using his contacts in the U.S., he elicited donations from various Quaker groups, such as the Palo Alto Meeting of Friends, the Eugene Meeting of Friends, the Eldrige Foundation, and the Sierra Club. Finally, he organized a benefit concert featuring Joni Mitchell, James Taylor, Phil Ochs, and a B.C. group called Chilliwack, which netted seventeen thousand dollars. Stowe was a tightfisted, efficient treasurer. He would not allow so much as a penny to go unaccounted for, even if it meant someone having to deliver the dimes from their last button sales directly to his house, however late at night. But he did wonders with money, and very little time got wasted wondering where it had gone. Everybody knew that Irving Stowe *knew*, and he was incorruptible.

Meanwhile, Bohlen and Cote prowled the docks.

It was no easy task, trying to find someone with a boat big enough to get to Amchitka, who was willing to take the thing up there and park next to a nuclear explosion, in defiance of the combined forces of the U.S. Navy, U.S. Coast Guard, and probably the U.S. Air Force thrown in for good measure. Such a boat would represent an investment of at least half a million dollars. Insurance would probably not be available. Heavy fines, seizure, and blacklisting awaited the skipper who would try.

And it was not just the political situation. The next test at Amchitka would probably occur in early October, close to the autumn equinox, when the waters around the Aleutian Islands become so dangerous no sane fishing boat will go near them and the military itself often remains pinned to the ground. As the water sluices back and forth between the vast bodies of the Gulf of Alaska and the Bering Sea, caught between the weight of the earth and the weight of the moon, incredible riptides occur. Sometimes fifty feet in height, these tides have been known to tear steel boats in half. The furious waters around the Rat Islands—of which Amchitka is a member—is home to huge Sargasso Sea-like blooms of seaweed thick enough to tangle a prop. Winds off the Bering Sea at that time of year have been known to reach 170 miles per hour. Finally, there was something else to deal with: a weird meteorological phenomenon known as a williwaw, a hurricane like wind that leaps unpredictably out from behind the Aleutian Islands and pounces on passing boats like a dervish, literally shredding them.

It was not until late November 1970, thirteen months after he had announced that a ship would be sent to Amchitka, that Jim Bohlen finally stumbled across Captain John C. Cormack.

Bohlen had gone down to look at a boat called the *Sir Thomas Crosby*. It was owned by the United Church, and there had been discussions about the possibility of her being used for the voyage. In the end, the United Church backed out, but the discussions had served their purpose. When Bohlen arrived in his raincoat down at the gray dock where the boat sat, he found himself talking to an old man in grease-stained work clothes who seemed to be acting as the janitor. The old man talked slowly and cautiously about the *Sir Thomas Crosby*, frequently rubbing his jaw with a huge hand that was more like a paw because it was missing two fingers. Most of his teeth also seemed to be missing. One eye drooped. When he removed his peaked work cap to scratch his head, he revealed a bald skull with only a wisp of white around the edges.

"Wal," said John Cormack, after he had listened to Bohlen for a while, "I've got a boat m'self."

At first, Bohlen didn't take the old man seriously. Heavily built, Cormack showed signs of having long since started to sag everywhere. With his U.S. Navy background, Bohlen had a hard time relating to this ancient grease monkey as a *captain*. Yet it was true. John Cormack owned an eighty-foot halibut seiner, *Phyllis Cormack*, named after his wife. The *Cormack* was some thirty years old and had a twenty-year-old Atlas engine. She could only do nine knots, but he had taken her up more than once into Aleutian waters—although, of course, never that late in the year. Cormack owed a lot of money, he was facing foreclosure, and he had experienced several terrible halibut seasons in a row. He was desperate.

Not that Cormack admitted any of this at the time. It only came out later, once we were at sea.

Bohlen's first glimpse a few days later of the *Phyllis Cormack* left him with the urge to laugh and cry. During the war he had served on the U.S.S. *Recovery*, and he remembered the waters off Attu. He remembered the cloudy skies and the heaving gray flanks of sea. The idea of approaching such waters in this rusting, weathered old clunker of a fish boat first of all was surely not serious, and, second, not worth thinking about. Yet, despite his despair, there was the memory of the last year and all those long wet walks he had taken along dock after dock after dock, with no captain willing to do much more than laugh at him. Here was an actual living sea captain with at least the semblance of a boat, who was willing to talk about it. Bohlen had privately been close to giving up, although he never admitted it at the meetings. Now, absurd and maybe as irresponsible as it seemed, here at least was one old junk heap that could probably get away from the dock, and that was a start.

I had been drinking at the Press Club most of the afternoon before driving out to the first meeting Bohlen called on board the *Phyllis Cormack* herself. Arriving at the crumbling wharf in the Fraser River where the boat was tied up, I wandered past the old girl no less than half a dozen times, searching for a ship coming even close to the size of the one I imagined we needed to get to Amchitka. My first overwhelming impression of the *Phyllis Cormack* was one of shock. Paint peeling and damp, with ropes hanging like mossy vines from her rigging, she looked too dilapidated to start up, let alone cross the Gulf of Alaska. It was only my boozy, cheerful mood that kept me from walking away on the spot. I concluded the whole thing must be a joke.

It was smoky and crowded in the white-painted galley of the *Phyllis Cormack* as it wobbled mildly in the Fraser currents. Everyone seemed to be laughing hysterically at the slightest quip. I still couldn't believe that we were seriously planning to travel over the sea in this cozy little floating farmhouse. When I got home that night, I was laughing harshly. "Forget it," I told my wife. "There isn't gonna be any trip."

For a long time afterward, the committee continued to look for a boat. Not a single one of the people who had been down to look at the *Phyllis Cormack* believed that such a junk bucket could make it to Amchitka. We wanted something bigger, faster, safer.

Speed became an important issue when it was learned that the tides around Amchitka ran at twelve knots. John Cormack's old halibut seiner could only do nine. We could all see the boat being carried backward from the island, not being able to hold its position. The fact that it was a fishing boat in itself posed a whole new set of problems. Fishing might not be legal within twelve miles, which meant that we might not be technically capable of pushing up to the edge of the three-mile territorial limit. Could a fish boat entering a twelve-mile fish-management zone be seized? Certainly it could. We had by this time also learned more about the huge blooms of seaweed around the island.

"We've discovered," Paul Cote said morosely, "that it's impossible to get within even four miles of Amchitka on the Gulf of Alaska side, because, well, the island is famous as a sea-otter sanctuary, and the reason the sea otters go there, logically enough, is because of the kelp beds. These kelp beds are apparently something else. The problem is they extend outward from the south side of the island as much as four miles, sometimes five. It depends on the weather. And they're so thick that any boat that, say, gets blown into them can't move. The kelp gets tangled around the propellers, and it's thick enough to break the propellers."

Cote was blushing furiously as he spoke, as though the kelp beds were his own private shame. He wore black horn-rimmed glasses, he had black carefully combed hair, and dressed impeccably. He had been struggling for over a year to make this voyage happen, but now things were beginning to look hopeless. His energy was running low. As he faced about twenty of us at a meeting shortly after the visit to the *Phyllis Cormack*, talking about the kelp beds and the twelve-knot tides, it was easy to see he was nearly ready to say: "I give up."

The sight of the *Phyllis Cormack* had jolted us all. We knew that she was sitting there, that she existed, that it would be entirely possible to make a charter deal with John Cormack, and we would be off! If no other boat turned up, we would climb on that creaking rust bucket and put to sea. The *Phyllis Cormack*, aged and filthy as she might be, was the vehicle for transforming fantasy into reality. The problem was, now that we saw the reality beginning to unfold, we could no longer avoid seeing more clearly the *rest* of the reality: the kelp beds, tides, riptides, williwaws, red tape, legalities, gales, and nothing between us and the icy seas but this mossy antique wooden tub. And *then* the Bomb.

Suddenly, I was very frightened.

I was not the only one. After one of the meetings—which were now happening every week—I caught up with Patrick Moore, the radical ecology student, and we went off for a few drinks. Until now, I had viewed him from a distance, aware of his reputation, and had been pleased by his calm, solid way of dealing with problems. I had always secretly admired scientists for their discipline. When Moore stood up in a meeting to deliver an observation or make a point, he radiated certainty and a soft sense of humor. He was a bit psychedelic-looking for a scientist, but lab coat or not, when he spoke it sounded very definitely that he knew what he was talking about. I had already begun to count on him for sound judgment.

Now, after a few drinks, he looked at me glassily and said: "Shit, man, this is crazy! We're gonna *die!*"

Whatever was going to happen, most of us were taking the voyage much more seriously than we had at the beginning. Five uncomfortable months passed. Finally, in May 1971, Bohlen announced that he had been unable to find any other boat, and so the committee was recommending that the *Phyllis Cormack* be chartered for six weeks for twelve thousand dollars. She would depart in mid-September. There would be room for twelve people in the crew. And that was that. We were stuck with it.

Within a week of chartering the boat, we ran into our first political crisis.

Paul Cote got up at the beginning of the meeting. His face was completely red. He looked ready to start wrecking the walls. "This is incredible," he said, holding up a piece of paper. "We've just been notified that John Cormack will *not* be able to get government insurance for the trip."

It took a moment for that to set in. Plans to get insurance from some company like Lloyd's of London had collapsed because of a requirement to post a forty-thousand-dollar bond. The next step had been to simply maintain Cormack's regular fishing coverage. The usual coverage was on a year-round basis. But it happened: suddenly, an exception was being made; John Cormack would be the only working fisherman on the West Coast that autumn whose insurance would not be maintained.

The insurance came from a Crown corporation, the Industrial Development Bank, which was owned by the Canadian government. The decision to refuse Cormack's insurance had been made by no less a person than the federal cabinet minister in charge of the environment, Jack Davis. Mr. Davis had written to Paul Cote stating flatly that the government would not continue Cormack's coverage if he took his boat to Amchitka.

It was the Don't Make a Wave Committee's first chance to test its political muscle in the media. Ben Metcalfe rushed back to his tape recorder to prepare a CBC broadcast accusing the Canadian government of sabotaging the "one serious effort being made to stop Amchitka." Bob Cummings and I leapt for our respective underground and establishment typewriters, spitting out diatribes.

Jack Davis happened to have his constituency in West Vancouver. The night after the news hit that he had personally scuttled the anti-Amchitka voyage—which was tantamount to coming out in favor of the nuclear tests themselves—his phone began to ring and telegrams began to pour into his office. And they were not just random shots—they were coming en masse from his own constituents. Within a week, Mr. Davis did a complete flip-flop and announced that he would take a "closer look" at coverage for John Cormack. In the end, he agreed to maintain the insurance with the provision that the boat "must be fishing."

The last week before the boat left was madness. Thirty-five people had applied to be on the crew. With spaces for only twelve, that left the triumvirate of Bohlen, Stowe, and Cote with some hard decisions to make. Stowe himself would not be going, because "I get seasick easily," nor would Cote be going because he was scheduled to take part in an Olympic sailing race as a member of the Canadian team. Apart from Jim Bohlen, Captain Cormack, and an engineer, that left nine bunks. Furious lobbying and arguing ensued. There were many women in the group, all of whom were alert for any signs of sexism in the crew selection.

When the crew list was finally announced, the crew was entirely male. Marie Bohlen had passed up on her option to go. A lady named Lou Hogan, who had been selling GREENPEACE buttons for months, was so enraged at being rejected that she left the meeting in tears. Privately, she plotted with her boyfriend, Bob Keziere, the photographer, to stow away in the hold. Rod Marining, the Yippie, was also planning to stow away. The final crew included three Americans: Bohlen, Terry Simmons, and a political science teacher from the University of Alaska, named Richard Fineberg. The rest were Canadians: Cormack, an engineer named Dave Birmingham, Bob Cummings, Patrick Moore, Bob Keziere, Bill Darnell, Ben Metcalfe, Dr. Lyle Thurston, and myself.

Irving Stowe was in full flight during that whole last week. Cameras were everywhere. Reporters were starting to call up from the United States. Radio shows couldn't get enough news about the upcoming voyage. Stowe's reasons for leaving America were, of course, Vietnam, refusal to pay taxes that went into the war effort, anger over corruption of the political system, and "creeping fascism." He tended to let all his personal views spill out in his interviews. His public denunciations of American imperialism and atrocities were beginning to embarrass other members of the committee. There were fears that he was stirring up anti-Americanism for its own sake, not because of the specific issue of Amchitka. In fact, Irving Stowe's attacks on America were to leave such a lasting impression in Vancouver that for years afterward, Greenpeace would be viewed as a tool of Peking or the Kremlin, a reputation that was not helped much a few years later when Stowe traveled to China and came back singing its praises.

While Stowe handled the media end of things, the speeches and interviews, Bohlen worked furiously away in the hold of the *Phyllis Cormack* with a handful of people like Bill Darnell and Terry Simmons, building the racks and braces for the extra fuel tanks. By the time the day came to leave, he was hollow-eyed and close to speechless from exhaustion.

Everybody else was laughing and yelling, strung out, hyped, dizzy on swoops of emotion and tiredness. For days, forklifts had been trundling back and forth with supplies. Cartons, crates, and bags were being handed down from the dock to the deck. A rock band had come down to serenade the boat. They danced about, barefoot, getting in the way of the forklift. People had been surging up and down the ladders and ramps, scrubbing the walls and floorboards, packing, unloading, stashing things away. Camera crews from the American Broadcasting Company had arrived. United Press International, CBC, Canadian Television, and Canadian Press had all showed up, clamoring for interviews. The National Film Board had announced it would make a movie about the voyage, and its crew was everywhere, trailing cables all over the decks. Reporters appeared from Montreal, Toronto, and even New York. There were yet other reporters from local papers, weeklies, conservation magazines; there were handfuls of politicians, officials from unions, churches, antiwar groups, environmentalists, pacifists, anarchists, poets, sculptors, artists, and dozens of ordinary hippies with their beads and cowboy hats and "old ladies" and liberated little children running everywhere.

At the last minute, Customs and Immigration officials refused to give Cormack clearance to sail unless he had halibut fishing gear with him, to comply with the environment minister's insistence that the boat "must be fishing." He and Bohlen and Stowe had to rush off to his locker to load a car up with gear, rush it back, and throw it on the deck. In the middle of the confusion, Bob Cummings confessed that he had been made temporary president of Georgia Straight Publishing Company, because the underground paper's editor was expecting an obscenity bust and had reasoned that the cops wouldn't be able to serve a warrant if the president of the company was out at sea. That added to the anxiety nicely. Any minute, the Vancouver police might arrive to grab Cummings, and possibly the rest of us as accomplices.

The boat had to move from its loading dock to Burrard Inlet to fuel. But the light was failing, and the television crews were getting panic-stricken. If the boat did not leave until after dark, they wouldn't be able to get any footage of its departure.

To oblige them, the boat "left" at 4:00 P.M. Several people jumped on board for the ride around to Burrard Inlet. The sail—a green canvas triangle with the peace and ecology symbols and the word *Greenpeace* in the middle—was hoisted. As the old boat shuddered away from the dock in reverse, then turned, with the skyline of Vancouver in the background, the sun caught her glossy white hull just perfectly. Whereas she had looked like a junk heap the first time—I could see that her lines were actually graceful, like a swept-out gull wing. Cormack tooted the horn:

FAWOOOOOOOOOOOOOOOOOO!

Several of us who were avoiding the hard labor of the last-minute loading and fueling slipped away to a pub before rendezvousing with the boat at the fuel dock. We were quite plastered by the time the six-o'clock news came on, and there, in living color, was the *Greenpeace* sailing off from Vancouver for Amchitka. The media image had been launched. The Battle of Amchitka had begun—and there we were, comfortably belting back beer in our favorite pub.

"If we can keep this up, it won't be so bad after all," commented Bob Cummings.

The other item on the news that grabbed our attention was a report that John Wayne had just pulled into the nearby harbor of Victoria, on board his privately owned minesweeper. Asked about the Amchitka protest boat, he snapped: "They're a bunch of Commies. Canadians should mind their own business."

There was a delightful irony to that. One of Dr. Thurston's patients was Chief Dan George, probably the world's most famous Indian movie actor. Chief Dan had come out wholeheartedly in favor of what we were doing, so we now had a situation where the world's most famous cowboy actor was against us, and the most famous Indian actor was on our side. Of course, thanks to John Wayne's remark, almost every Canadian was now automatically with us.

We got lost on the way to the boat and ended up—about six of us—having to scramble down a steep dirt hill and stumble in darkness across a train yard, dodging rumbling freight cars. By the time we found our way to the fuel dock, the hoses had been uncoupled and the lines were ready to go. Someone had tipped Jim Bohlen that the Yippie would try to stow away, so he was firmly shaking Rod Marining's hand and ushering him down the gangplank to the dock. Lou Hogan had abandoned her own stowaway plan. I looked around desperately for my wife, who was trapped several hundred yards away behind a barbwire fence. I couldn't leave the boat to join her because we might be departing any minute, making a dash for it before any more officials came down with new excuses to delay us, and, of course, there was the danger of the Vancouver cops coming any moment to serve a warrant on Bob Cummings.

Then, suddenly, the engines were chugging and the boat was moving. The dock was dwindling, loaded with people who were waving and whooping and whistling. The lights of downtown Vancouver dropped swiftly away. The dark band of the underside of the Lion's Gate Bridge swung briefly over us, obscuring the great wash of stars. The blackness of the harbor was all around, and the smell of hissing salt water.

Captain Cormack and his engineer were up in the wheelhouse. The remaining ten of us gathered in the galley and looked at each other. It was the first time we had ever been together—all of us—in a single room. It was like a subway at rush hour, except that Metcalfe had broken open one of the huge bottles of cheap B.C. wine—Cold Duck, it was called—that had been donated at the last minute. *Pop! Pop! Pop!* Dr. Thurston plunked down a case of Guinness. Soon we were all singing and cheering.

It was crowded but warm in the galley, with a huge steel stove blazing away. The boat rocked slightly, but not so much that anyone was bothered by it yet. We were as boisterous and noisy as men on a troop train. Only one note was sounded to spoil the mood. The engineer, Dave Birmingham, an aging man with wire-rim spectacles, who looked very much like a country parson, climbed into the galley and stood frowning at us for several moments. When someone offered him a drink, he stepped back angrily. In a loud voice, he announced: "I had expected the crew of the *Greenpeace* to be men of religion." With that, he stalked out on deck.

The party went on until close to midnight, when John Cormack clumped into the galley, still clad in his grease-covered work clothes. He sniffed the smoke-filled air, looked pointedly at the ashtrays overflowing with butts and at the empty wine and beer bottles, grunted, shook his head, squinted around the room, looking at us one at a time, and then said: "Huh."

"Would you have a drink, Captain?"

"Nope. Never drink on the job." Then he paused and rubbed his chin.

"Speakin' of the *job*, which of you fellas is gonna take the first watch?"

Cummings and I elected to follow the skipper up to the wheel-house. It was like the conductor's booth on an old wooden tram, wide enough for four men to stand shoulder-to-shoulder. The varnish on the walls was tattered like peeling skin. There was a small glow from the night-light—otherwise everything was blackness. Later, we would learn that the barometer was corroded, the depth-sounder needed to be hammered with a fist to make it work, the clock would have fetched a good price in an antique shop, and the wooden doors had to be kicked hard to make them open.

But now—darkness! And the horror of the wheel.

Once the boat began to swing in one direction, it moved like an avalanche. There was a lag before even the most desperate thrashing at the wheel would force it to respond. Then, ever so slowly, the bow would begin to come about. The light at Sisters Isle was approaching, moving like a satellite back and forth across our field of vision. Neither Cummings nor I had ever steered any kind of boat before, let alone an eighty-foot halibut seiner. It took so long for her to respond that I would panic and begin spinning the wheel the other way. Soon I was sweating and shaking with confusion, afraid to call the captain, but afraid that if we didn't, we would soon crash.

Cummings was bent over the radar, his face jammed against the rubber viewing nozzle, staring down into phosphorescent ripples of light and fuzzy glimmering clouds. "Faaar out," he whispered.

The compass needle was swinging like a pendulum, sweeping over the line representing our course, then swinging back, never remaining where it should be for even a second.

The Sisters Isle lighthouse finally swung by on our—right. Captain Cormack emerged briefly from his bunk room and stood in the darkness, not saying anything.

After a while, the wild swinging of the bow settled down and the *Phyllis Cormack* began to slough contentedly. I wandered out on the upper deck. There were dark bulks of mountains all around. The wind was like a breath of crystal queens. The old engine pulsed and throbbed like a living creature, a creature named Phyllis. Pipes and tubes ran over her decks in intricate patterns. The wood was rough and seasoned. There was a feeling of being somehow out of phase with time, of having passed beyond the reach of the century. We seemed to be out in a Viking age or a time of frigates and galleons.

Trapped in the funky old wheelhouse with the dark wood and rusting instruments, we sank down into the vessel's rhythm, our bodies swaying slightly. The torture rack of the wheel evolved slowly into a mandala. Already, we were in love with the *Phyllis Cormack*.

Our course took us past Maude Island, Cape Lazo, and Cape Mudge, all the time hugging the east coast of Vancouver Island until the Strait of Georgia began to break up like pieces of a jigsaw puzzle. We steered into the mouth of Discovery Passage, with Quadra Island on our right and the town of Campbell River coming up on our left in the early morning light. Whether it was because we had arrived too early or the locals had slept in, an expected flotilla of support boats never showed up. We hung around for half an hour, then gave up and plunged northward into the passage.

Our first day out was like a blessing.

Noon found everybody except the skipper and the engineer loafing on the decks with their shirts off, enjoying the sweetest Indian summer to have happened on the coast in years. The sun sent legions of three-dimensional pellets of light down on blue water. Eye-paining bright white arrowheads of peaks broke above the walls of forest on both sides of the boat. Rocks drifted by with seams like ancient faces.

Down through the narrow passages were blowing millions of feathery seeds, glinting like snowflakes against purple cliff shadows.

The northern British Columbia coast could not have been a more perfect place through which to pass on our way to confront the bomb they were building at Amchitka, for this was already the graveyard of a civilization that had been destroyed. Abandoned canneries stood along the shores, patchworks of rusting corrugated metal sheets with broken windows. Gutted wooden sheds perched at odd angles on barnacled pilings. Bull kelp floated among fallen wharves. Jellyfish drifted past half-sunk barges.

A silence hung over the coast like the hush of broken worlds. Streams gurgled across tide flats past the wrecks of old fishing boats. The herring populations were dying off. The halibut were dwindling. Whales were rare. One by one, the canneries were closing down. As the fish disappeared, the government responded by cutting down on the number of fishing licenses. The canneries and wrecked boats were the symbols of the retreat of the White Man from a land he had devastated, leaving only the Indians to live among the ruins. Chugging up through the lonely stretches of coast, there were moments when one shuddered, for the silence was like a forewarning of what was to come. Western civilization had already begun its great fall.

Dr. Thurston had brought along a tape recorder. Over and over again he played Beethoven's *Fifth*, and a Moody Blues tape called *On The Threshold of a Dream*.

It was such a relief to have escaped from the vortex of energy that had seized us all in the final days before the launching, that everyone's reaction this first day seemed to be to lie back and smile, float with the music, and stare in awe at the scenery.

It wasn't until the second day that the bickering began.

Patrick Moore opened a can of butter upside down. The captain flew into a rage. "Don't you thirty-three pounders know *anything*? That's bad luck!" Then Bill Darnell hung a china cup on a hook facing inward instead of outward. The captain roared, "That's worse luck! Now you've done it! We're in for it now!" A few of us started to laugh, but Cormack wasn't kidding. He was furious.

Bob Keziere made the mistake of trying to get down into the engine room to take pictures. Cormack chased him out. The reason was a two-by-four jammed between the huge old Atlas engine and the hull—evidently placed there to stop the engine from falling over in bad weather. The captain was as defensive as a mother bear about his boat. He didn't want anyone knowing—or showing—that there was anything wrong with it at all. He also didn't like "newspaper men always makin' mountains out of molehills" and he had nothing but scorn for "professors" who didn't know which way to hang a cup. There were five people on board who either already had their degrees or were going to university, and there were three newsmen. That put the captain on a collision course with most of his crew to begin with.

Cormack was a shrewd old warrior who had been at sea for forty-five years. Shorter than most of us, he weighed well over two hundred pounds, and had once been an amateur wrestling champion. Although close to sixty years old, he could still have easily taken on any four of us and beaten us to death. He was used to spending months at a time out on the water, squinting at the clouds and currents, sniffing the air. He knew only the rudiments of navigation. His theory was that you could never get lost at sea because "all yah gotta do is turn right and keep going til yah bump into the beach." He couldn't stand "thirty-three pounders" or "mattress lovers." When he noticed that Metcalfe and I had attached curtains to our bunks, he sneered: "Ha! Jerk-off curtains!"

Cormack's bad moods quickly became the centerpiece of our existence. Each time we turned around in this new environment, it seemed that we were breaking some rule that Cormack had taken for granted for decades but which none of the rest of us had ever dreamed existed. For his part, it was certain he had never put to sea with a crew such as this one. He was uncomfortable, out of his element, wary, suspicious—and on top of it all, worried about what was going to happen when we got to Amchitka. The rest of us were merely risking our lives. He was risking his boat. Every time Keziere attempted to get a picture of him, the old man snarled, ashamed to be photographed in his grease-covered work clothes.

The engine conked out briefly the second day. While Cormack and Birmingham were below repairing it, Metcalfe made the mistake of sending a radio message to Vancouver reporting what had happened. When Cormack heard on the six-o'clock news that his boat had "broken down," he flew into a rage that made his previous outbursts seem like minor tremors. Thurston and Bill Darnell had been in the wheelhouse with him at the time. Now they came flying out the door. "That man's dangerous," said Thurston. For hours, no one dared go back up there. Birmingham reported that the captain was so angry he was talking about putting the whole bunch of us ashore at the next port. Metcalfe finally patched things up by apologizing and putting out a report praising the *Phyllis Cormack* as "one of the finest fishing boats on the West Coast." Down below, we all nervously scanned the mugs and cups to make sure they were being hung facing outward; I tore up a column I had been composing about the breakdown; and when Cummings opened a can the wrong way, he was pounced on by half a dozen people, and the can was quickly thrown overboard before the skipper found out about it. It was agreed that any future mechanical problems would be described simply as "routine maintenance."

Cormack's outbursts were not only irritating—they were fearsome. It occurred to many of us that he could kill us with his bare hands. And it was impossible to predict what might anger him next.

In the galley he would wait until everyone had squeezed around the table, then he would use his elbow to chop out a space for himself. Very quickly, we turned into a unique little floating subculture that followed very strange customs. We hung mugs facing outward. We were fussy about the ways we opened cans. We had a taboo about taking pictures in the engine room. We performed elaborate seating rituals before meals.

The second most important person on a boat is the cook, and if the captain is unhappy with the cook, then you have trouble. Our cook, unfortunately, had never done much cooking before. This meant that Bill Darnell took the brunt of Cormack's black moods.

"Yah call this shit food?" the captain snorted.

Darnell was soft-spoken, polite, and intelligent. He was working for the Company of Young Canadians, a Canuck version of the Peace Corps. He was strongly built, although not of Cormack's girth. He had been pleased with his own work so far, justifiably thinking he had done everything that could reasonably be expected to provide decent meals, and his only reward from the captain was to be bumped heavily aside as the old man grabbed for the peanut butter and bread, grumbling that he had "never tasted such shit" in his life.

The figure of Cormack loomed over everything we did. Bohlen, the leader during the previous two years, now seemed so exhausted by his efforts that he was content to leave any more decisions that had to be made in the hands of "consensus." But it quickly became evident that getting this crew to agree on anything at all was going to be tricky. It was not just relations with the captain that were strained. By the end of the second day out, disturbing signs of conflict had begun to appear on several fronts. There was, for openers, the division of the crew into "straights" and "freaks." Then there were other divisions between the "academics" and the "propagandists," the establishment and underground press, the middle-aged men and the "youngsters," and, inevitably, the Canadians and the Yanks.

Most young Americans that you met in Vancouver in those days were on the run. More than a difference of culture, there was a difference in situation. Canada had not been at war since Korea, and no young Canadian had had to face the draft. None of us could think of anybody in our generation who had been wounded or killed in battle. Vietnam had touched none of us directly. Americans like Richard Fineberg were far more personally and emotionally involved in the war issue. Fineberg and Simmons were also professional academics—something that grated on the nerves of Ben Metcalfe, who had come up through the ranks in journalism and who remembered being so poor that "the only thing you got to play with Christmas morning was a hard-on." He was verbally swift. He dressed impeccably, even on board the old workboat, and relished spinning out stories late into the night.

Cummings and I had taken the two forward bunks on what we were now coming to understand to be the starboard side of the boat. We had an odd situation in relation to each other. Technically, he was the "radical" writer, whereas I wrote for a family newspaper. It developed that just as much as I privately wanted to be thought of as a "radical writer," Cummings really wanted to write for the establishment press. Fate had placed us in opposite positions from the roles we wanted to play, like mirror images of each other.

Dave Birmingham proved to be an eccentric genius. Although he dressed in the working-class style of the Depression era, he carried a smart, locked attache case around with him, which he opened and shut with businesslike clicks. Inside were the plans for a one-man submarine he had been building in his backyard for several years. When pressed as to why he had come along, he said cheerfully that his wife and children had made him do it. His wife was president of a passivist women's organization and his son "turned out to be one of them danged hippies."

In theory we were all equals, except for Cormack, who was "hired." And while we all knew he was to be feared and handled as carefully as possible, he was still considered basically to be our "employee," rather like the driver of a large wooden taxi. None of us had yet liberated ourselves from class consciousness. Cormack and Birmingham were viewed as the working class, and the rest of us fancied ourselves as managers, or at least coleaders, of a fast-growing little empire.

The coleaders had unequal status in the community out of which we had emerged, and our social standing came with us. Bohlen was respected as a model of integrity and guts. Metcalfe was a famous radio personality, a veteran newspaperman, and a backroom figure in Canadian politics, believed to be moderately wealthy. Darnell, while serving with Gandhian humility as cook, was actually a director of the Sierra Club of B.C. and a prime Don't Make a Wave Committee mover. Terry Simmons had similar official status and seemed to be Bohlen's main advisor. Doc Thurston had his automatic medical-officer's authority. I had my columnist status, which made me a bit of a local celebrity. Bob Keziere was known as a dedicated and intelligent conservationist. He and I had been assigned by the committee to chronicle the events of the voyage, so we had important jobs. Cummings was on board mainly due to Irving Stowe's passionate insistence that "the liberated underground news media of the world must be represented as well as the capitalist press, which only serves the war machine that is right now pouring napalm on women and children in the Democratic Republic of Vietnam." Dave Birmingham was viewed as an upright, religious man, and until we finally realized that he was a secret genius, we all tried to be condescending toward him. Our last crew addition and only U.S. resident, Richard Fineberg had scored a couple of articles about our voyage in U.S. newspapers (something we were having difficulty doing). This unknown outsider quickly became the low man on our social totem.

It quickly became apparent that no two of us agreed on what it was we were doing in the first place. For one man, the voyage's real power lay in the symbols it created in the mass mind. For another, it was useful only as a battering ram in a larger American battle: the struggle to overthrow Richard Nixon. The Canadians were excited about it, especially the younger ones, because we were finally getting into action. Until then, we had been thoroughly frustrated by our inability to directly engage the Yankee war effort. This was the closest thing to battle that we had ever experienced.

It was tough for the Americans to avoid being patronizing. The Canadians, in turn, especially Metcalfe, found it difficult to avoid betraying a sense of moral superiority because Canada had had the good sense to stay out of Asia, and Canada had no nuclear weapons herself. Tribal differences were with us from the beginning, coupled with traditional Canadian resentment, because the country is, after all, an American economic, cultural, and political satellite.

The stakes *seemed* to be high. There could be no doubt that the voyage had already served its first purpose. It had "drawn attention" to the issue. It had captured public imagination. Politicians were having to deal with it. The U.S. military itself was known to be "making preparations." We were being treated seriously. The wire services were covering us. There were reports of border blockades being organized across Canada.

Three nights out, we got word that the prime minister of Canada had tried to call us on the radiophone to announce that he was asking the government of the United States to halt the test. That cheered Cormack out of his foul mood for a short while. His boat had never been called by a prime minister before, he had to admit. One newspaper editorial described us as "the unconventional symbol of all Canada's revulsion at the Amchitka test." None of us had ever had our hands on this much power before.

*

I had on board a copy of a well-worn pamphlet containing a collection of North American Indian prophecies and myths. It had been given to me, rather mysteriously, by a Jewish dulcimer maker who described himself as a gypsy and predicted that the book would reveal a "path" that would affect my life. It contained one particular prophecy made some two hundred years ago by an old Cree grandmother named Eyes of Fire, who saw a time coming when the birds would fall out of the skies, the fish would be poisoned in their streams, the deer would drop in their tracks in the forest, and the seas would be "blackened"—all thanks to the White Man's greed and technology. At that time, the Indian people would have all but completely lost their spirit. They would find it again, and they would begin to teach the White Man how to have reverence for Mother Earth. Together, using the symbol of the rainbow, all the races of the world would band together to spread the great Indian teaching and go forth—Warriors of the Rainbow—to bring an end to the destruction and desecration of sacred Earth.

As we were approaching the Kwakiutl Indian village of Alert Bay on Cormorant Island, we got a radio message inviting us ashore to accept a "blessing" and a gift of coho salmon. When we tied up at the dock, a group of Indians climbed on board, including two radiant women, Lucy and Daisy Sewid, the daughters of the chief. Eloquently, they explained that the good wishes of all the Indian people on the West Coast went with us, for they understood as much as anybody the dangers of nuclear testing, and particularly the tests at Amchitka. The Kwakiutl were busy carving the world's longest totem pole. They hoped to have it erected shortly. We were invited to stop in at Alert Bay on our return voyage, when our names would be carved on the totems—no small honor. Coupled with the support of Chief Dan George and the antagonism of John Wayne, this seemed to cinch the vague affinity most of us already felt with the Indians. John Cormack, it turned out, had helped the Kwakiutl out several times in the past and was much respected by them. The *Phyllis Cormack* was a familiar sight in the waters around Alert Bay.

We left Cormorant Island in high spirits, laughing and shaking our heads. I hauled out my copy of *The Warriors of the Rainbow* and passed it around. Predictably, the older men were less impressed than the youngsters. But rainbows *did* appear several times the following day and it all *did* seem somehow magical as we chugged through a maze

44

of inlets and channels and sounds and bays. Emerging from Johnstone Strait, we swung briefly into the open mouth of Queen Charlotte Strait. Immediately, the slow-motion caboose movement of the boat changed into long glides as the swells lifted us and set us down. Then, as suddenly as they had come, the swells were gone, and we were moving behind the shelter of Calvert Island.

The boat's boiler sprang a leak, forcing us to put into the tiny Kitasoo Indian fishing village of Klemtu, some 350 miles north of Vancouver. We knew better than to send any stories back talking about mechanical breakdowns. The problem was simply "routine maintenance." While Cormack and Birmingham welded the boiler plates, the rest of us wandered through the village, to the delight of the local children. There were at least four of us who looked like nothing so much as hippies, and no figure from white society is quite so beloved by Indian children. In addition, there was the fact that they had seen the departure of the *Greenpeace* from Vancouver on television—the 4:00 P.M. launching of the media image, which we, too, had watched on TV. We were the first television image they had ever seen come to life in their own little bay. It was a bit as though Captain Kangaroo had materialized out of the forest. They swarmed over the decks and hung around for the five or six hours we were there, babbling excitedly. As we were getting ready to leave, the kids gathered on the dock and began singing nursery rhymes, urging us to come out on deck and do whatever magic tricks we must be able to do in order to get ourselves on television.

While the kids were singing, Metcalfe clambered onto the upper deck with his tape recorder. The rest of us, one by one, straggled out in a fine mood after finishing off the coho salmon from Alert Bay and also a few more bottles of Cold Duck wine. Having gone through a series of nursery rhymes and even the national anthem—to which Keziere responded: "Yuks, can't you do better than that, you kids?"—they struck upon a catchy tune called "We Love You, Conrad!" from the Broadway musical *Bye Bye Birdie*. After a couple of choruses, the kids started yelling at us to tell them our names. One by one, Metcalfe called out the names, and the kids responded by singing:

> We love you Uncle Ben
> Oh yes we do
> Oh Uncle Ben
> We love you

Or "Uncle Bob" or "Uncle Bill" or "Uncle Lyle." When it came time to sing to Cormack, however, it was not "Uncle John," it was *Captain John.* We were in a tremendous mood now.

Bill Darnell shouted:

"What do you think of *Greenpeace*?"

And the kids sang with gusto:

> We love you Greenpeace
> Oh yes we do
> Oh Greenpeace
> We love you

With that, the boat took off, the kids tearing off headbands and plastic peace symbols, throwing them to us, and continuing to sing and whoop and cheer, while we assembled on the deck, except for the skipper, and sang back at them:

> We love you Klemtu
> Oh yes we do ...

And from then on, whenever we were feeling affectionate or comradely, we would address each other as "Uncle."

From Finlayson Channel, we passed into the long trough of Princess Royal Channel, into Wright Sound, then up the long avenue of Renville Channel.

I had switched with someone to take the afternoon watch and was at the wheel when Cummings wandered in from the upper deck. He seemed nervous. Lighting a cigarette, he stood by the radar, staring ahead up the channel. Finally he said:

"What's your opinion about Dick Fineberg?"

"He's beautiful," I said, trying to avoid whatever was coming.

"No, seriously," said Cummings, "hear me out. I'm not kidding around. Have you noticed all the notes he's been taking? Nobody really knows him very well. We didn't check him out, you know."

"You know what? I don't give a shit."

Cummings persisted:

"I'm beginning to think he's an agent for the CIA."

"That's ridiculous, man. Absolute crap. You have to be putting me on."

He was talking about the guy who broke news stories in the Anchorage Daily News about U.S. Army germ warfare testing in Alaska U.S., the guy who was invited aboard the *Greenpeace* because of another exposé he wrote, about thousands of tons of poison gas canisters that were dumped into the water not far from Amchitka—canisters that might be smashed by the blast, releasing all sorts of deadly shit into the environment.

"No I'm not. I've been watching him."

"Crap, man," I said. "Utter crap. Underground press paranoia."

The trouble with you," he said as he left, "is that you've never been busted. I have."

I went back to grooving on the slow sweep of cliff faces as the boat chugged peacefully northward. Then Doc Thurston wandered in, camera slung over his shoulder just like a tourist. He stood around for a while, fingers tugging at the point of his beard. I was just about to share the joke of Cummings's paranoia about Fineberg when he leaned toward me and half whispered:

"Listen, maybe I'm losing my mind—that's entirely possible—but there's a rumor going around that Dick Fineberg is working for the CIA."

"Oh, Christ, not you too, Doc?"

"Well, why not? Anything's possible. Put it this way: do you believe a grand piano could come flying into the wheelhouse? Think about it."

"Well, sure, if you mean anything's possible, I'd have to be crazy to say it couldn't happen."

"Right on," said Thurston. "There you are. I don't think Dick's a CIA agent, but what the hell? Grand pianos, man. That's where it's at. This *boat* is a grand piano. The States is a crazy place. They do things like that, you know."

We changed the topic and talked about metaphysics and psychosis for a while. Then Thurston left. A few minutes later, Terry Simmons came in. He had a string tied around the back of his head to keep his glasses from falling off in the heavy seas to come.

"Have you been getting wind of the rumors about Fineberg?"

"Yep."

"Well, they're a bunch of crap, you know. *Somebody's* trying to discredit Fineberg. I have my suspicions about who it is, but I'm not saying yet."

Simmons departed. Shortly Metcalfe appeared with a book in his hand, *The Strange Last Voyage of Donald Crowhurst*—a book about a lone yachtsman whose mind came apart as he sent back false radio reports making it look like he was the round-the-world race leader while all the time he hadn't left the Atlantic.

"This is a hell of a book to be reading on a trip like this," he said. "You know, I don't know what it is, but I can't get it out of my head: there's something about Fineberg that bothers me. Have you noticed it? Nothing's come into focus yet. It's more of a gut feeling than anything else, but ... something."

Just as my turn at the wheel was ending, Bohlen sauntered in, looking elfishly happy. "Great day, eh?"

"Sure. Listen, Jim, have you been picking up on all these rumors about Fineberg?"

"What rumors?"

"Oh, about him being a CIA agent and all that shit."

"Oh oh, don't go spreading rumors like that," said Bohlen. "Hah! Listen, I lived in the States too long not to be able to spot a CIA. Believe me, our boy Dick isn't the type."

"Well, you'd better do something to calm everybody down."

"Me? *You* do something. I want to look at the scenery. Come on, quit hogging the wheel."

Out on deck, Fineberg was complaining bitterly to Keziere that no one would listen to his logic. He had studied and reported on immigrant labor issues, and he thought we were naively blasé about border protocols. When he wasn't busy behind his camera, the quiet and thoughtful Keziere would usually listen sympathetically and offer feedback when Fineberg ventilated. Not this time. Keziere said: "Look, this may not be the most logically oriented crew you could ask for, but it happens to be the *only* crew on a boat on its way to Amchitka right now, man. I mean, either you can dig it or you can't."

"I see. Just keep in my place, hmmmmmm?"

"No, that's not what I meant."

Poor Fineberg. He was the last man to join the crew, and as the only official American on board, he is a Big People (as in Lord of the Rings); the rest of us, including the expatriate Bohlen, think of ourselves as Hobbits. It is totally unfair, ethnocentric, racist—the whole sorry bundle. Fineberg bore up quite gracefully, all things considered. But when the boat finally cleared Dixon Entrance for the open sea, the mood on board had gone sour.

In fading wintry light, we stood in the battenclaim watching the faint silver outlines of the mountains gradually being engulfed by rising swells. The wind was cold. The sheltered waters of the Inside Passage were behind us now, and we had launched ourselves on a great-circle route for the Aleutian Islands. Up in the radio room, Bohlen fiddled with the loran—the long-range-navigation equipment—which Cormack himself did not know how to use. Metcalfe worked the dials of the single-side-band radio, pumping a steady stream of propaganda back to shore. I retreated to the galley to write to the Vancouver *Sun*. Keziere and Darnell, immediately made sick by the new motion of the boat, crashed into their bunks. Doc Thurston handed out Gravol pills by the handful, but for the two main seasickness cases it was too late, and the effect of the pills on the rest of us was to make us so drowsy we could hardly think. Cormack, of course, sneered with disgust when the thirty-three-pounders started resorting to pill popping to cope with the new sensations of movement.

The boat rolled like a drum. The wake was heaved out in a pathway of backing-up drains: one moment the water fell down from the stern as we came up over a swell, the next moment the cold gray boil of the wake wagged over our heads like a serpent of phosphorescence, dragging our stomachs into a slow pulpy collision with our lungs. The horizon would only be yards away, then it would drop out from under us. Somehow, instead of falling, it felt like we were being pushed. Each upward haul of the bow was like a last gasp. Downward we surfed slow-motion into a sea canyon, only to wallow in sloshes of water gone suddenly still as though totally disengaged from gravity. Fumbling through the troughs, we clambered over the tops of water hills that immediately collapsed beneath us.

Down in his bunk, Keziere looked like a squashed grape. He was quite unable to eat. But after the first day out in the open, it didn't matter, because the cook was collapsed into the bunk above him, his face blank as though in a trance, and those of us who were still functioning had to feed ourselves.

The passage across the Gulf of Alaska turned into an endurance contest. Our language got down to a series of grunts. Everyone was tight-jawed, stiff, sore from being tossed against the walls and bunks. There was no way to move without holding on to something and hauling yourself along hand over hand. Sleeping became a matter of hanging on to nearby pipes, curled into a fetal ball with one's buttocks pressed against the rim of the bunk, knees jammed against the wall, expecting any minute to be flipped onto the floor.

The Gulf was the color of a basilica—granite and limestone—with foam traces of fossils. Trilobite impressions of waves swept by. There was a hint of marble archways. It was like an immense cathedral with all its walls thrown down. The mood, to me, was like that of the High Masses I had attended as a child.

The huge anchors stowed in the battenclaim had been gnawed and sculpted by the rust. Even ordinary railings and ladders had been transformed into spidery grillworks of pop art. Peanut-butter jars were used to encase the deck lights, and years of salt water had done something to the rust—turning it *green*.

Our internal divisions and arguments melted away. There was no time now for dickering. Those of us who were still standing had to take regular four-hour watches, and these sometimes had to be extended several hours to make up for the seasickness cases. As we got more and more tired, we started to feel like children at the entrance of an infinite arch, with the swells coming up like organ notes and hymns. And far down the aisle between the rolling, sloshing pews stood John C. Cormack, alert to strange invisible presences and forces, exactly like a sorcerer. We began to refer to him as the Lord of the Piston Rings.

A feeling of awe grew stronger every day. As we looked out at the sea, whether curled in our bunks or up in the wheelhouse, the awe got into the middle of our heads and stayed there. The ocean was a giant across whose flanks we were skittering like insects, holding our breath lest the giant roll over in its slumber and crush us. With the swells coming in two-hundred-foot-long strides, like a canyonland getting up and walking, we were soon at the point where *nobody* opened a can the wrong way, and we ran our eyes compulsively over the mugs on the hooks to make absolutely certain they were safely facing outward. All John Cormack's superstitions became our own.

The skipper dismissed the swells over which we were riding as being "nuthin'."

"That's nuthin'," he said repeatedly. "Out here the waves get as high as treetops." He took obvious pleasure from our nervousness.

The sunsets were the color of cataracts, and the furthest the light moved up the spectrum was toward the lower edges of brass.

Life had become distinctly uncomfortable. Thurston's porthole leaked so that icy water kept sloshing over his feet while his head was wedged against the chimney from the engine room, which was usually about 120 degrees Fahrenheit. There was a transom in the floor of the radio room, directly above the bunk room, which meant that all the noises from the nightly broadcasts were blasted directly down into the ears of anyone trying to sleep. From the radio came the most awful sounds any of us had ever heard, like filtered stereo recordings of lizards being hatched at the dawn of time.

From below came the sounds of the engine, like an underground train passing along an endless track. In comparison, the sound of a power lawn mower would have been a hum. There were explosions of gasoline and air mixtures, whirring metal pieces reciprocating at blurred-out speeds, spent gases detonating and fan belts screaming, explosions within explosions, cylinders and pistons moving in regimented imitations of masturbation. Then there was the clattering and crashing of pots and pans in the galley, china dishes smashing on the floor. The sea itself made noises so weird that exhausted imaginations could easily picture monsters from the Black Lagoon flopping along the deck at night.

Our biggest problem, of course, was "amateurs at the wheel." By night, the wheelhouse was a sensory deprivation tank, because the captain insisted that only the tiny compass light could be used, "otherwise yuh can't see a damned thing outside."

"But John," someone had replied, "I can't see a damned thing anyway."

"Goddamn weak-eyed landlubbers: can't even see at night."

One night, Ben Metcalfe got caught unaware by an exceptionally big swell. There was a sudden lurch, and the chinaware threw itself all over the galley. The boat gave a sickening sideways plunge, like a shot buffalo, and Metcalfe was thrown to the deck, hanging on to the wheel, whirling the whole boat around. By the time he could get to his feet, the boat, which had been going due west, was traveling due east. Four or five of us had been thrown halfway out of our bunks. Up in the radio room, the rim of Dick Fineberg's bunk had broken under the strain, and he went sailing out into the air, landing with a horrible thud that made us all think that the whole wheelhouse was breaking apart. "Oh, fuck *around*," yelled Thurston. "Shape up, Metcalfe, or you're fired! God damn, what am I *doing* here?"

"Isn't it awful?" asked Moore from the bunk below. Darnell said, "Bhleeee." Keziere went, "Yuks." He had just begun to get over the worst of his seasickness, and now his stomach was swiftly, powerlessly plunging back into the stew pits. With a spastic movement, he was up and stumbling out through the galley door to throw up, but no sooner had he pushed open the door than a gigantic screaming black wave engulfed him, almost washing him away. Crawling back, dripping, to his bunk, he muttered: "Christ, a guy can't even have a decent puke around here."

One afternoon, Doc Thurston and I were on watch in the wheelhouse—or the Penthouse, as Cormack called it. Thurston had placed his tape recorder beside the compass, and we were listening to Beethoven and the Moody Blues. Everything was in motion, inside our heads and out. The doctor was conducting an imaginary orchestra, seeming to call forth the waves and flatten them, raise them up again to accompaniment of cymbals and harpsichords, pulling the clouds across the sky with a gesture, causing the boat to surge and pause in its gropings across the swells. Like tram conductors in a surreal electric train we sloughed across the water while the music soared through the Penthouse, eventually luring up the younger members of the crew. Soon, Patrick Moore was climbing through the hatchway. Fineberg was wedged between the radar and the door, with a string of beads around his neck. Darnell had recovered enough to leave his bunk. He was sporting an engineer's cap and flaming yellow suspenders. Keziere still looked sick and weak, but he was grooving on the beautiful synchronistic meshings of sea and sound. Terry Simmons was slouched beneath the depth sounder, red beard starting to burst like prickles all over his chin, hands tucked in the pockets of his blue nylon jacket so that he looked somehow amputated. Cummings puffed on a pipe, filling the Penthouse with heavy clouds of smoke. Thurston continued to conduct the ocean. And I clung dreamily to the wheel.

The radio-room door opened and Jim Bohlen squeezed himself into the packed Penthouse, coughing on the smoke. He took one amused look at us all and said: "Well, well, the Fabulous Furry Freak Brothers!" That drew a good chuckle. Then he spotted the tape recorder next to the compass. "Hey! Isn't that going to throw the magnet off?"

"Nah," somebody reassured him.

"Are you sure?"

"Sure."

Actually, none of us was sure, but the scene was so delightful nobody wanted to mess it up.

Thus began and ended the first naval engagement of the Battle of Amchitka. For, in fact, the tape recorder *did* affect the compass, and while we were grooving on the Moody Blues, the boat wandered ninety miles off course. The day was September 21, still well ahead of the expected test date of October 2, and so we lost nothing by straying slightly from our destination. But, in the process, we cost the American taxpayer something like twenty-two thousand dollars. The Seventeenth District of the U.S. Coast Guard had dispatched a Hercules HC-130 from Kodiak, Alaska, that morning to run out over the Gulf and get some pictures of us for identification purposes later. We had been making no effort to cover our tracks. Metcalfe had, in fact, been radioing our position back regularly to his wife, Dorothy. And Dorothy had gotten a request from the Coast Guard commander in Juneau to keep him "posted," just in case we ran into trouble and needed to be rescued. It was all very paternalistic and friendly. So, armed with specific data about our position, speed and course, the Hercules had swept out across the Gulf.

A notice had gone up on the flight-schedule board at the Kodiak Air Station: SEARCH CAN VES GREEN PEACE. The cost of one hour's flying time for a Hercules is eleven hundred dollars. In all, the aircraft spent ten hours searching for us September 21, and another ten hours the following day, but we weren't where we were supposed to be, courtesy of Thurston's tape recorder. Late that night, the Hercules swung back into Kodiak, having found no trace of its prey. From that moment on, a mood of paranoia began to develop in the headquarters of the Seventeenth District. The only conclusion the Coast Guard could draw was that we had deliberately deceived them, and if we could elude them in the open sea in our thirty-year-old fishing boat, capable of doing no more than nine knots, what tricks might we have up our sleeves when we got to Amchitka?

The next morning, the same notice was up on the Kodiak flight-schedule board, and that day the Hercules came gliding like a silver steel albatross across the waves at us, four long fingers of smoke trailing from its wings. It roared overhead, banked, and took another run, exactly like a strafing run, except that as the side doors came open they revealed no weapon more fearsome than a camera.

Then the blow landed that wiped us out.

Moore and Darnell and I were in the galley when Cormack came tromping in from the poop deck and said: "Test's been delayed."

We let out three simultaneous groans. It was like sustaining a terrific body punch. We were eight days out from Vancouver. It was Thursday, September 23. We were within one day's sailing of the Aleutian Islands, and once we'd reached the Aleutians it was just a matter of seven hundred more miles to Amchitka. The delay, we learned, would set the test back until early November.

Immediately we had a meeting in the galley. There were several questions, and no answers. Was the report accurate? Was it just a trick cooked up by the AEC to throw us off the track? The problem was that we didn't really know whether our mission mattered or not. It might well be that we had already succeeded in creating a bad situation for the weaponsmakers. The opposition building up to the test back home had reached considerable proportions, and Canadian-American relations were already strained. So it was possible that the test had been delayed just to shake us off. Possible. But Russia's Premier Kosygin was also due to arrive in Canada for a cross-country tour soon, and it might simply be bad diplomacy for the Americans to blow off the bomb while he was on tour. And, shortly, Japan's Emperor Hirohito was due to arrive in Alaska, where Richard Nixon would shake his hand. It would be even worse diplomacy to blow off the bomb virtually under the emperor's nose, especially since there had been huge demonstrations against the test in Tokyo, the nearest major city to Amchitka. Apart from all this international intrigue, the real question was whether we could hang on for another whole month. We had fuel and supplies for only six weeks, and a November test would leave us only a few hundred miles from the Russian mainland with nothing to eat or drink, probably not enough fuel to get back to Vancouver, and with the winter storms from the Bering Sea coming down on top of us.

Bohlen summed up the situation:

"It seems to me we have three choices at this stage. Either we can turn around and go back to Vancouver and wait until a date is announced for the test. But they probably won't announce the date. Or at least Nixon only has to give something like seven days warning. And of course we can't possibly make it back up here in that time. Or else we can head into Kodiak, park the boat, fly home, wait, then fly back up and jump on the boat and race for the island. But I've been talking to John about this, and he says you can't count on the weather this time of year. If we get bad weather, it might take as much as seventeen days to get from Kodiak to Amchitka. The problem there is that if we do pull into an American port, they might be able to pin us down in red tape. Our third option is to keep going, try to find some place in the Aleutians where we can restock our supplies and kill time until we know one way or another when the damned thing is going to go off."

We debated for hours. My own feeling was that at least it was true that the test had been delayed, which meant we probably did have them on the run, and so we should keep up the pressure, or, better yet, increase it somehow. "I think we should go straight up to Amchitka, go into orbit, and go on a hunger strike until the damned thing is canceled," I argued.

The hunger strike idea didn't exactly catch fire. In the end, when it became apparent that we couldn't get a consensus on what to do, we resorted to a vote. Cormack and Birmingham were excluded. The vote came down seven to two in favor of heading for Kodiak, with Bohlen and Metcalfe opposing, and Cummings asleep in his bunk.

But that night, while the rest of us slept, thinking we were on our way to Kodiak, Bohlen and Metcalfe got on the radio to Vancouver, heard another newspaper report to the effect that the bomb might actually go off as planned in early October, and told Cormack to get us back on course for the Aleutians. There were mutterings and grumblings in the morning and questions about how come the vote got reversed without another meeting, but nothing much came of it. Bohlen was still "the representative" of the committee, and it had been Metcalfe who had pulled the political strings that resulted in the prime minister coming out against the test. Their combined authority, coupled with the fact they they were both Second World War veterans, was formidable. There was resentment, but it was muted because there really didn't seem to be any other course, given the uncertainty of the situation. I wrote in my notebook that night: "This trip is going to be at least one hundred times as heavy as I thought. Possibility of a freak-out must now seriously be considered."

Thus divided, hung up, with the seeds of a mutiny already sown, late in the afternoon of September 24, we came within view of the Aleutian Islands. Back in Canada, some five thousand protesters had blocked over two dozen border crossings from the Atlantic to the Pacific, preventing any traffic from moving to or from the United States for most of the day. The initial report of a delay in the test had given everybody back home a feeling that we might just win this one, and Nixon was widely rumored to be vacillating, torn between the usual Cold Warrior mentality, and the fear of badly damaging relations with Canada.

The Aleutian Islands came up out of the mists—snag-toothed and bleak-looking, like the wrecks of giant grape-stained wine urns. We could see froth on white-gummed beaches and hear the distant flush and whoosh of breakers. With cracked chunks of cliff coming into focus like a photo developing on hard-grain paper, the feeling emerged that we had actually been traveling in a Time Machine back to the beginning of the planet, when volcanoes were still chucking up blood flows of lava. Dankness and cold came from invisible fleets of icebergs in the Bering Sea. The mood could only be described as spooky. We were, after all, entering the most geologically unstable area on earth, epicenter of the Great Alaska Earthquake of 1964. And, of course, an atomic bomb named Cannikin was soon to roar among the volcanoes, sending its vibrations out among the smaller earthquakes that occur almost daily.

We went through Unimak Pass in the dark. Cormack insisted on taking the wheel himself until we were through. Between the broken fangs of mountain on either side of the pass, uncountable millions of tons of water were rushing from the Bering Sea into a furious embrace with the Gulf of Alaska, generating the legendary riptides we had been warned about.

Cormack stayed up all night, handling the wheel as though it was made of eggshell, his head cocked slightly sideways, leaning forward the whole time, pressing all his senses down into the hull, feeling, groping, jiggling, and flicking the wheel like a man guiding an elephant along a high wire. Although the swells had subsided, there was a hum in the night air. Strange gurgling sounds came up all around the boat. "A lot of good men have gone down in this here pass," Cormack said. But mainly he didn't talk much.

Metcalfe managed to establish radio contact with a U.S. Coast Guard boat somewhere in the vicinity, called the *Balsam*. He requested permission to put in at Dutch Harbor, more or less around the corner from Unimak Pass on Unalaska Island. A reply came back to the effect that Dutch Harbor was a U.S. Navy base, therefore the request would have to be forwarded to the navy, and the navy would want to know all our names, nationalities, et cetera. In the meantime, we were to anchor and wait for clearance.

We came out of Unimak Pass about four in the morning, swung north around Akun and Akutan Islands, and dropped anchor three miles off Akutan. We had to wait twenty-four hours before word would come through from the navy via the Coast Guard cutter *Balsam*.

Saturday night, September 25, we broke out several bottles of rum and had a party. We had at least reached the Bering Sea. Considering the anxieties we had originally felt about the seaworthiness of the *Phyllis Cormack*, it was something to feel good about.

Emperor Hirohito had arrived in Anchorage that day; Nixon had arrived to greet him; and while we drank and sang in the galley, squawks of sound came bursting through the old wooden radio fixed to the galley ceiling. Crowds were cheering and chanting. Excited announcers babbled about what a "historic moment" it was, with the emperor of Japan stepping on North American soil for the first time.

Nothing was said about how, during the Second World War, thirty thousand Japanese had stormed and taken Attu Island, just two days sailing beyond Amchitka. Bohlen had been there on a repair-salvage boat to help clear the harbor of ships that the Japanese had filled with cement and sunk in a chain. He told us that the Japanese had refused to surrender and had to be wiped out to a man.

Moore opened a lemon-juice can upside down by mistake. Cummings screamed at him: "After ten days at sea haven't you fucking learned how to open a can?" Overboard went the can. We sang a couple of rounds of a newly invented song called "Bering Sea" to the tune of "Deep Blue Sea," several boisterous rounds of "I Left My Heart at Old Amchitka," and generally got ourselves plastered. Everybody loosened up. Fineberg got out his banjo, Darnell went to work with his harmonica, and Thurston and I started hammering away on some pots and pans. "We love you *Greenpeace*, oh yes we do. OH *GREENPEACE*, WE LOVE YOU." In the morning, suffering badly from a hangover and the runs, I staggered from my bunk only to find Metcalfe unconscious in the galley. In his red T-shirt, with his woolen ski cap still perched at a jaunty angle on his head, he was slumped in one of the canvas fold-out chairs we had brought along. The boat was rolling at a fifteen-degree angle. A rum bottle was clanking from one end of the galley floor to the other, like a glass rat. Everything was in motion, yet, somehow, Metcalfe remained perfectly balanced, head tilted on his shoulder, his silver-and-black goatee resting like a blue jay on his collarbone, his center of gravity just low enough so that he remained poised in space while the galley, the boat, and, indeed, the whole world went lurching and swinging around him.

The party had been good for us. Nobody had realized just how much tension and nervous energy we had stored up coming across the Gulf. Now we had blown it all out. The galley was a mess: cigarette butts, punched-open lemon-juice cans, dented pots and pans, crumbs, and the remains of a jigsaw puzzle scattered all over the floor. The puzzle had been donated to us by the United Church. We had drunkenly put it together only to discover it said: GOD BLESS YOU.

In the morning, a Sunday, Dorothy Metcalfe got through on the radio to let us know that the head office of the United Church in Toronto had authorized the ringing of bells across the country in a protest against Amchitka, and, more to the point, they would be saying prayers for the souls of us on the *Greenpeace*. Word also came through from the Coast Guard cutter *Balsam* advising us that the U.S. Navy had turned down our request to put in at Dutch Harbor for supplies and water. However, it was okay for us to go into the island of Akutan, where there was an Aleut Indian village. Dutch Harbor was military. Akutan wasn't. That seemed to be the gist of it.

"Huh," said Cormack, "I was in Dutch Harbor when they set that last bomb off at Am-*cheet*-ka back in 'sixty-nine, and nobody said anything then about it being a military base."

As the boat dove forward birds rose in immense sweeps, like parts of a single giant wing flapping and cruising over the water. I had never seen so many birds in my life. Their cries were all around, splintering on gusts of wind. Crab pots danced on the waves. Sea lions came up in packs of half a dozen, rolling and flopping along beside us. Seals, porpoises, jellyfish emerged. Patrick Moore was running from one end of the boat to the other with *Petersen's Field Guide to Birds of the Northwest* open in one hand, flipping madly through the pages, binoculars in the other hand, letting out yells like: "Wow—that one over there! With the bright-orange beak and the prominent white cheek markings on a black body, *that* one! That's a ... that's a ... crested puffin! Isn't that *beautiful?* Jesus Christ, what a heaven! Oh, I want to live here the rest of my life!" There were moments when the sky and the green carpet of the island were only flashes of color between the surging waves of feathers. The sounds of the birds enveloped the boat so completely that we had to shout to make ourselves heard.

Into Akutan Bay we chugged. Far down the neck of the bay was a tiny cluster, like a mushroom patch, of white wooden houses with green roofs. Already we were becoming aware of something strange but characteristic of the Aleutian Islands: there was grass that grew as high as your knees, and in a few places, as high as your thighs, but no trees at all. Apart from the grass, there were moss and fungus clinging to the rocks, and spruce trees that grew like plants in a Japanese garden, never more than a few inches high. Streams came sparkling down between gorges carved in the bedrock, but below the cliffs there was nothing that you could take a bearing on to judge the distance. It was impossible to figure how far you might be from any given object. A hill that seemed only a few thousand yards away might actually be five miles off. A meadow that seemed an hour's walk away might be reached in ten minutes. All the usual visual reference points were missing, and so the experience of arriving in Akutan Bay was a bit like landing on Mars. It was another world. The mind wobbled in confusion, not knowing how big a world or how small.

No sooner had Cormack let down the anchor than several of us shoved the skiff overboard and scrambled to be the first party to go ashore. Beyond the village rose lush green meadows topped by a darkly etched mountain. On the other side of the bay the hills were smothered in cloud banks that flowed like a foamy froth of white smoke, humping smoothly over rounded ridges, feathering in absolute silence down the slopes until they reached the water and then magically vanished as though into a huge hidden crack in the earth. It seemed that such a torrent of cloud must soon exhaust itself, but the surf cloud kept coming, rimming the mountains in a milky gauze, poised like a granddaddy tidal wave over the bay. *Very* awesome.

After a week on the gray Gulf, the colors that leapt out at us now seemed unbearably intense. Rainbows unfurled beyond the surf cloud. Cow parsnips were arrayed in whole galaxies at our feet. Moss, lupine, horsetails, blueberries, and daisies were scattered like flaming bits of nimbus. It was as though we had passed through a looking glass into a whole new universe.

A path leading up from the beach took us straight into the village graveyard, past wet ornate stones and mouldering wooden crosses to the doorstep of a magnificent Russian Orthodox Church with colored-glass windows. We knew that the Aleuts had been around for at least eight thousand years. Small island states had existed, wars were fought, empires had risen and fallen. It had all ended with the arrival of the Russians. The Aleuts had resisted fiercely but had been massacred. Dutch Harbor had once been known as the Bay of Women, because every male Aleut had been killed. Only about 3,500 Aleuts remained—scattered in seventeen communities throughout the islands. Some 1,000 Aleuts had once lived on Akutan alone. Now there were only 120. They were a race at the absolute bottom end of their history.

The village had looked enchanting as we were pulling into the bay, but now that we were ashore we could see the broken-down shacks along the beach, wrecked boats pulled up on the gravel. The green-roofed houses that had caught our eyes proved to be wartime prefabricated huts. The place was a scenic concentration camp, and the hills all around, which had once presumably crawled with people, were empty. There were not even any paths or trails that we could see, as though the people no longer bothered to venture beyond their huts.

An old Aleut man was sitting on a swing overlooking the bay. One hand ended in a mechanical contraption, complex enough for him to be holding a cigarette between plyerslike fingers.

"Hi," I said. "Beautiful weather you've got here."

"Been bad, been bad. Bad weather. First time the sun's shone since spring. Not a day of sun all summer."

"You mean today's the first time you've seen the sun since *spring*?"

"Bad. Bad weather. Yep."

"Well, I guess you wouldn't know about it, but that boat we're on, with the green sail, that's the *Greenpeace*. We're up here to, ah, protest against the, ah, bomb that's gonna be set off at Amchitka ... You know about that?"

"Yep. We know. Weather's been bad ever since they set off the last one."

"Right on. Well, anyway, the United Church of Canada said prayers for us today, so I guess all this good weather's a miracle, aye?"

He shook his head, looked away.

"Bad," he said. "Bad."

White Men did not seem to be too popular in the village. Most of the people we met as we passed along the boardwalk would only nod. No one seemed interested in talking. One of the younger men spat in our direction.

Once past the cluster of huts, we began to run madly up a hill, whooping and bounding. There were six of us—Bohlen, Moore, Simmons, Fineberg, Darnell, and myself.

The slopes were snagged with tundra and scrub willows and moss, a spring mattress of vegetation over which we went leaping, discovering that we hadn't quite got our land legs yet, so we stumbled drunkenly. The vegetation was like a trampoline. You could take seventeen-foot leaps and *bounce*. You could throw yourself around crazily and not get hurt. With the wind slapping our cheeks, we floundered up the hill following a stream until we were maybe six hundred feet above the village and seemingly out of anyone's sight. Immediately, we tore off our clothes and crashed into the stream. We lay there in above five inches of icy, sloshing water. Blinding ultrablue sky. Wind flattening the grass as though invisible helicopters were hovering. We stared out over the bay at the surf cloud and felt the crusts of sweat and salt being nibbled away from our itchy, sticky bodies.

"Oh, man," said Moore reverently. "That has gotta be the most beautiful planet in the universe."

Down at the boat later, we learned through Metcalfe that the very stream in which we were bathing was the one that served as the village's water supply. And at least one woman in the village had announced she wouldn't take another drink of water until those filthy hippies from the protest boat went away.

We were to remain in Akutan for six days, waiting for some word that might indicate the date for the setting-off of the bomb. There was only one telephone on the island—down at the grocery store. The ship's radio couldn't get our voices out because of the surrounding mountains, and the problem with the island's single radiotelephone hookup was that it was part of an inter-island party line. The main generator was on Unalaska Island at Dutch Harbor. Each island was allowed a ration of telephone time per day. With Metcalfe, Cummings, Fineberg, and I all wanting to file back dispatches, reports, press releases, and columns, there was an immediate logistical problem, for there were 120 Aleuts on Akutan who likewise wanted to use the phone. Priority went to Metcalfe, since his was the job of relaying press releases. The rest of us would have to wait until Friday morning, when a mail plane arrived for its once-a-week pickup.

Meanwhile, Bohlen and Simmons did what they could to place radiophone calls to the Don't Make a Wave Committee and to representatives of various environmental groups based in Washington, D.C., asking the same question over and over again: When is the bomb going to go off? Allowing decent weather, Cormack figured we were only four or five days from Amchitka. The problem was that we couldn't just go there and park for a month, not with the autumn equinox already past.

"The way fishermen work it up here," Cormack said, "is like this: Yuh wait for a break in the weather, then yuh rush out as far as yuh can, put down yer pots or nets, then hightail it back for shelter before one of them williwaws or a storm comes up. Wait again for another break in the weather, then make a run for it, grab yer gear, then git going back t' port as fast as yuh can. Weather's mean up here this time of year. So far we been lucky. That weather you fellas thought was so bad out on the Gulf, shit, that was fuck all. That was the flattest calm I've seen in forty years out there this time of year. But she'll start getting sloppy now."

"So we can't just get going now and wait it out?" asked Bohlen.

"For a whole *month*? Shit, no. Never last. Not unless we can run for shelter when the wind comes up."

"Well, we can't do that. It's a security zone around Amchitka. They won't let us in, that's for sure."

"Well then, you fellas are just gonna have to sit on yer hands and sweat it out. Nothing else yuh can do. Now we'll see how much of what y'd call *staying power* y'got ... *puh puh puh puh puh puh.*"

Frantically, Bohlen and Simmons and Metcalfe put through their calls, scratching like dogs for a single scrap of information. Reports were all contradictory. Some sources said the bomb would be going off in early October. Some insisted it had been postponed until November. Environmentalists in Washington were "confident" that the AEC was on the run. According to the White House, Nixon was reported to be "studying" the test. Six "leading scientists," including former presidential aides, had come out against the test because of "physical and political risks." Cannikin, they said, was "potentially the most destructive man-made underground explosion in history." Former Secretary of State Dean Rusk said he believed most of the American people were against it. According to an Alaskan senator, relations with Japan as well as Canada might be permanently damaged. One day newspapers would be carrying reports saying Nixon was "seriously considering" canceling the blast. The next day the headlines would read: N-TEST STILL ON. The *Saturday Review* had come out against it. *The New York Times* had come out against it. In Ottawa there was talk of an all-party Canadian delegation being sent to Washington to make a plea to Nixon. Under pressure from environmental groups, the United States Court of Appeals ordered a lower court to reconsider an earlier ruling allowing the test in the first place. While the ruling in itself did nothing to cancel the test, it had the effect of forcing the whole issue back into the courts. Like a seesaw, the battle swung back and forth.

The *Phyllis Cormack* by this time had become home. It was an uncomfortable, overcrowded home, but it was home nonetheless. Certain things had become fixtures in our lives: a photo poster of an atomic mushroom cloud pinned up in the galley opposite a Sierra Club poster with the motto, IN WILDNESS IS THE PRESERVATION OF THE WORLD. There was a blurred-out poster of Nixon, pinned to the wall with two green darts, and the quotation: "Let me make myself perfectly clear." In the bunk room there was a poster of a magnificent ancient sailing ship, ornate as a gingerbread house, titled the *Friendship Frigate*. And back in the galley, one more poster, which simply said, in large boldface type: MANGE D'LA MARDE. In theory, it was to be displayed should we be arrested.

Someone in the village learned we had a doctor on board. A couple of men came out in a boat to ask Thurston to examine a friend of theirs who had some kind of horrible illness. Back and forth he had been shuttled between hospitals, each time returning to the island with the stark diagnosis: chronic alcoholic. But Thurston wasn't told this, and thought it wouldn't hurt to improve our relations with the villagers. It was rainy the night he went with the two Aleuts into the village to diagnose their friend. The man was already drunk by the time they arrived at his house, complaining of chest pains, back pains, stomach pains, head pains. Catching on fast, Thurston figured he could get himself out of the whole mess by insisting on a urine test, knowing that in order to conduct such a test he would need a microscope. He hadn't brought one along, and guessed that no such device could be found anywhere on the island. But while the patient was staggering around in the rain in the front yard, trying to urinate into a beer bottle, his two friends suddenly remembered that the schoolteacher—the only white man living on the island—had a microscope. By this time they had plied Thurston with several glasses full of rye, so he was a bit wobbly himself. Off the four of them went, the chronic alcoholic merrily waving around his beer bottle half full of piss, the other two laughing and stumbling all over the place. Thurston said to himself: "Whoops, what am I *doing*?" But it was too late to escape. By the time they arrived at the teacher's door, it was well past midnight.

The teacher was an older man whose classroom was plastered with posters saying things like THIS IS *your* FLAG, BE *proud* OF IT. He had bragged earlier to Metcalfe that he hadn't read a newspaper in thirty years and had said that he thought the whole bunch of us should pack up and go home since the tests at Amchitka were none of our damn business. Now his doorbell was ringing in the night, and there were three hammered Aleuts on his doorstep, one of them waving around a beer bottle, which he explained was full of piss, and one of those hippies from the protest boat, claiming he was a doctor and wanting to borrow the school microscope to run a urinalysis test on this fellow here whom the teacher knew was the island's worst lush.

Grumbling and swearing, but somehow caught in the trap—since the crazy-looking foreigner was an actual doctor—he took them through the rain to the school, where Thurston hurriedly ran a urine test, mumbled some Latin incantations, and said he'd have to look some things up in one of his medical books back on the boat. Everybody crashed and stumbled back out into the rain, with the teacher swearing furiously after them.

"From now on," Thurston told us, "if anybody wants to know what I do, tell them I'm a *botanist*. Got it?"

Meanwhile, we realized that the young men in the village hated whites with a passion. It made no difference to them whether we were Americans or Canadians, for the bomb or against it. Walking past us on the wooden planks running through the village, they made a point of spitting on the boards ahead of us every time. There were several attractive young women who seemed much more friendly. When Cummings mentioned this—"Strictly as an observation, you understand"—the skipper pointed out that there was a one-thousand-dollar fine if you got caught with an Aleut woman on your boat, regardless of what she was doing or if she was doing anything at all.

Akutan was becoming depressing. The Aleuts were penned in like animals, forced to work on a crab cannery boat because there was nothing else to do. The fish had mostly vanished, and when the crab boat pulled out at the end of the season, there was nothing left for the adults to do except sit around and drink, which they did with a vengeance. After a few days, we started to avoid the village.

It was the Sierra Club directors—Bohlen, Darnell, and Simmons—who decided to organize a hike up the top of the 1,650-foot hill rising beyond the village. Moore, Fineberg, Keziere, Cummings, and I decided to join them.

At 8:00 A.M. on our fourth day at anchor, Birmingham shuttled us ashore in the skiff, and we started off up a slope that was already wrapped in fog. We were all wearing green rain jackets that had been provided by the committee. It looked as though we were in uniform. And, in a way, we were.

We started thinking about just who it was we represented. It was not just that we were "the unconventional symbol of all Canada's revulsion at Amchitka," we also represented one of the strangest alliances to have ever emerged on the West Coast. The unions were behind us. The churches were behind us. Students were behind us. And housewives, mothers, and politicians on every wavelength of the spectrum, from Trotskyites to the prime minister. Normally these factions fought like cats and dogs. Only the threat of an "environmental disaster" could unite them, and this gave rise to many political thoughts.

It seemed to Terry Simmons that the alliance we represented was a hint of a tremendous future power base, if any environmentalist could find some way to harness it. A transpolitical environmental alliance—that was what we wanted, and that was what seemed to have come into broad outline around the issue of Amchitka. As its vanguard, it was up to us to find some way to take this new coalition of energies and give it focus and direction. These were the thoughts, anyway, that streamed through our heads as we climbed.

There was also something superbly comical about it: here we were, eight green-clad amateur seamen, climbing a mountain in the fog, on our way to confront the deadliest fire of the age, like Hobbits bearing the Ring toward the volcano at Mordor. The alliance we represented was bizarre. At its extreme ends stood the Vancouver Liberation Front, which had led an invasion of Blaine, and no less unlikely a bedfellow than the Vancouver Real Estate Board, which had paid for television commercials showing an earthquake and a nuclear explosion, with the caption STOP THE TEST AT AMCHITKA. Later, when the story began to start moving in the major U.S. networks, these commercials would be incorporated as part of the news broadcasts, spreading the kind of near-panic that had enveloped the West Coast and right across North America.

It did seem, on Akutan Mountain, that we really were a small army, or at least an advance party—shock troops of the environmental movement— but also soldiers supported by real-estate sharks. In our own minds, we had a dozen multiple identities as a group. We were Captain Cormack's Lonely Hearts Club Band, the Warriors of the Rainbow, the Fellowship of the Piston Rings, the Akutan Branch of the Moody Blues Fan Club, Soldiers in the Army of the Landlord, and, finally, we had the inspired notion of calling ourselves the Greenhawks, an ecological version of the old comic book heroes the Blackhawks. *Hawkaaaa!* we cried, just like the Blackhawks used to when they leapt into action.

The Greenhawks leapt through the fog, clambering up to the tops of outcroppings of rock and throwing themselves into the air in apparent bids at suicide, only to land on a great natural trampoline of scrub willow and mush and blueberries, springing back up into the air, rolling over and down the slope, wrestling as none of us had wrestled since we were kids. *Hawkaa! Hawkaaaaaa!*

"Everything's getting unreal again," sang out Patrick Moore, for we had gotten into such a contact high that our minds had begun to interact wildly at dozens of different levels, and we were making instantaneous mental connections with each other. We seemed to understand one another perfectly. Bursts of ice like rain hit us, and the fog was all around, leaving us with the feeling that we were adrift on a patch of tilted prairie, with the rest of the whole earth wiped away. We sang: "*We love you* Greenpeace, *oh yes we do. Ohhhhhhhh* Greenpeace, *we love you.*" Floundering merrily, we reached the end of the grass and bunchberries and lupine. Gun-barrel-blue rock thrust above us. Jagged chunks broke off in our hands and went clattering down the hill, making no echo.

Now there were fewer cries of *Hawkaaa!* Harsh breathing. Ahead and above, barely visible, Bohlen and Darnell and Fineberg were climbing like ghostly spiders. I found myself lying at a twenty-degree angle, with my feet toward the top of the mountain, and the sensation of spinning, of being on a spinning planet, was so powerful that I thought I would drift away into space.

Then Moore was down on his knees beside me, digging gently through the moss into the wet dark earth with its complete civilizations of ants and caterpillars and worms. Moore, who had studied forest ecology; plant ecology; plant and animal genetics; plant, animal, and human biochemistry; plant physiology and morphology, was now delivering what seemed at the time to be the Ultimate Lecture on Ecology. Soon we were joined by the others, all lying about at crazy angles, pawing with a sense of wonder through the moss. Fantastic points were being made about how one life-form lives in relation to another, how this weird little creature couldn't possibly survive if this thing on that plant weren't around to turn this into that, and so on, ad infinitum.

To begin to understand the world, it seemed you could start anywhere, and the interwoven chains and bracelets of life would inevitably take you everywhere there was to go. There were no real divisions. Everything was One. Except for purposes of classification, there was no point where the plants and the insects ended and the animals began. "That means," said Moore—a grin spread from ear to ear, because he had never quite thought of it this way—"that *a flower is your brother!*"

Across the bay from the village, in the shadow beneath the rolling surf clouds, there was a cluster of abandoned buildings and large rusting tanks, which we thought to be water tanks. It was Birmingham who first took the skiff over to have a look, only to report back that the place had once been a whaling station. It had been closed down for probably two or three decades, perhaps even longer. "There's bones all over the place," he said.

We made several trips to the station to dig up massive ribs and vertebrae the size of toilet seats, hauling them back to the *Phyllis Cormack* as souvenirs. Once, Birmingham almost capsized the skiff by trying to carry a jagged chunk of skull that was so large it took four of us to hoist it onto the deck.

Told about the station, Cormack shrugged. There were similar abandoned whaling stations all up and down the coast, he said. "Used to be, when yuh came out on the Gulf, yuh could see them whales from horizon to horizon. They'd come up to the boat like big hush puppies. 'Course yuh don't see 'em anymore. They're *extinct*."

It was not quite true. Patrick Moore had reported sighting one whale during our crossing—one lonely whale off in the distance. But other than that, we had crossed the entire Gulf of Alaska without seeing any other sign of cetacean life.

A sense of futility crept into our minds as we stared at the wrecked buildings and the bones thrusting up out of the moss like pickets. The silence spoke louder than the bomb that would soon roar at Amchitka. Until then, we had been so intent on trying to prevent a holocaust, we had forgotten that for this other race of giant creatures the holocaust had already come.

Somehow, it no longer seemed quite so important to save ourselves and our children. The wind continued to stir the grass around the abandoned whaling station, as it would go on playing with the plants that would someday grow out of the ruins of our own cities and homes. It was as though we were being given a glimpse of our own future. Would mutated rats one day cart home our bones as souvenirs?

"Shit," said Thurston. "Why bother? Let them blow the fucking bomb off. Who wants to save the human race anyway?"

By Thursday afternoon, September 30, we were ready to take leave of Akutan. There was still no word about when the bomb would go off, but the most consistent reports placed the date somewhere between mid-October and the first few days in November, which meant we still had two weeks to a month to wait. After about a dozen strategy sessions in the galley we had decided to start on our way up along the chain, with the idea of making a scouting run around Amchitka, and then trying to put in at the island of Atka to resupply. Metcalfe and Simmons had gone ashore for the last time to get a message through to Customs and Immigration in Anchorage, advising them that we would be taking a "deep sea" route to Atka.

Now they were back, and Darnell was getting supper. Cormack was up in the radio room, flipping through the charts, a grizzle of white hair starting to grow like a mat on his face. "Seeing how all you *professors* and *reporters* got beards, thought I'd try one on fer size. 'Course it'll be a *man's* beard, not like that stringy thing on that Lyle Thurston fella's face."

Idly, Cormack looked up and saw a flash of foam and metal out in the ocean beyond the bay.

"Coast Guard cutter's comin'," he said matter-of-factly, just as he had said a week ago: "Test's been delayed."

Our long-awaited moment of confrontation with the American empire had arrived.

But it wasn't coming according to plan. Our strategy had hinged on our sailing a Canadian vessel in international water with the shield of centuries-old maritime law between us and the U.S. military. The confrontation was *not* supposed to occur one hundred yards offshore from Akutan Island, well within American territorial limits.

There was a fine, jittery mood on board. We all tried to maintain a casual surface, but I found myself going from the poop deck to my bunk at least half a dozen times to get my ballpoint and logbook, each time forgetting what it was I wanted. Metcalfe's hand was shaking with excitement as he fumbled through the boxes by his bunk for his film and tapes. Keziere muttered about what a rotten time of day it was, with the light fast fading. He tried one lens after another, darting in and out of the galley to check out his light readings. Fineberg arranged his papers neatly on his clipboard, flicking his pen nervously.

And then, suddenly, the engine started.

Birmingham was pulling up anchor, and the captain's huge bulk was surging from the engine room, yelling orders, and clambering swiftly into the Penthouse, spinning the wheel. The engine throbbed and clunked; water gurgled at the stern.

Immediately, Bohlen and Metcalfe were after Cormack, demanding to know what he was doing. But the old man ignored them completely, spinning the wheel, slamming gears, and yelling at Birmingham. It looked like we were making a run for it. Cummings, who was trying for the sixth time to light his pipe, looked at Thurston and asked, "Think maybe Cormack's flipped out?" Bohlen retreated from the Penthouse, shaking his head.

"Well, what's John up to?" Darnell demanded.

"Christ, I don't know. He's in one of those moods again."

"What next?" chuckled Keziere, holding his camera helplessly as we watched the Coast Guard ship slow down, then react with a blast of smoke as its skipper realized we were taking off.

"I'll bet they're as surprised as we are," said Metcalfe.

The Coast Guard cutter could easily do thirty knots. We could easily do nine. Not only that, but the cutter was coming in from the mouth of the bay, and there was no place for us to run or hide. Cormack's actions, again, totally baffled the rest of us. Then, as abruptly as we had started up, we stopped. The anchor went overboard.

We had only moved half a mile down the bay. It was Birmingham who confided later: "John just didn't want to be arrested in front of all those people in the village."

Bohlen got out the binoculars.

"Have they got a gun?" Darnell asked.

"Oh, yes, sure. Ah, I can see the name. *Confidence.*"

"What kind of a gun?"

"Dunno. It's covered up."

"No need to worry," said Fineberg sarcastically. "I advised the CIA in my last secret report that you guys would give up without a fight."

"I guess this part's real," said Moore.

"Another grand piano," said Thurston, looking serious for a moment, eyes roving over the surf clouds and the lovely cut of the wind across the water, breaking it up like an endlessly shattering mirror. He spent a moment absolutely still on the poop deck, absorbing the flavor of the experience, then drifted toward the galley. "Come on, might as well have some supper. It'll be a while before they get to us."

"We really are Hobbits," said Moore, chuckling quietly as we trooped into the galley and arranged ourselves around the table.

We broke out the last bottle of rum.

"We can't go to Amchitka now," said Metcalfe. "We won't have anything left to drink when the bomb goes off."

Darnell had the turkey out of the oven, Metcalfe was pouring the rum into the china mugs, the naked bulb in the ceiling made it all very bright and cozy in the white-enamel-walled galley, and we pretended to pay no attention to what was going on in the gathering dark. There was a lot of chuckling.

"When they come in, poking their Sten guns at us or whatever," said Metcalfe, "we should all burst out singing 'The Star-Spangled Banner' and demand political asylum from Canada."

It was a terrific supper. Halfway through, we spotted a motor launch bobbing across the water toward us. The *Confidence* had stopped about a half mile away. There were four figures standing in the launch as it approached. The launch cut across our bow, then puttered up beside the poop deck. At the last minute our collective cool broke down. Keziere and Bohlen and Metcalfe and Simmons rushed out on deck. Cormack remained up in the Penthouse, leaning with one hand on the wheel, the other shoved in his trouser pocket, with his fedora pushed down low on his forehead, trying to look nonchalant.

Birmingham crashed into the galley, pointing at the poster of Nixon. "Gotta take *that* down," he said, rushing to the wall. Moore and Darnell and Thurston doubled up laughing at his concern. "It's no laughing matter!" he raged, his normally pale skin turning bright pink. "No laughing matter at all. You damned young smart alecs can laugh all you want, but just you remember we're in *their* waters now, and we're in enough trouble as it is without going and getting them madder at us. How'd you like it if you went on an American ship and saw a picture of *our* prime minister with a couple of darts in his face?" He was shaking with fury, which triggered redoubled laughter. We laid back, gurgling, suddenly not giving a damn what was happening out on deck, while Birmingham scurried away with the offending poster.

Then Bohlen came bounding through the door, his eyebrows more than halfway up his forehead, yelling: *Whooopeee! Listen to this!* He was waving a cablegram from which he now read in a quivering excited voice.

DUE TO THE SITUATION WE ARE IN, THE CREW OF THE "CONFIDENCE" FEEL THAT WHAT YOU ARE DOING IS FOR THE GOOD OF ALL MANKIND. IF OUR HANDS WEREN'T TIED BY THESE MILITARY BONDS, WE WOULD BE IN THE SAME POSITION YOU ARE IN IF IT WAS AT ALL POSSIBLE. GOOD LUCK. WE ARE BEHIND YOU ONE HUNDRED PER CENT.

"*What's going on?*" screamed Thurston. "You mean they're on *our* side?"

We all jammed our way through the galley door to find out what was going on. It took a moment for our eyes to adjust to the gloom so that the launch could be made out, bobbing on the water, with three young seamen wearing wet suits standing in it. Metcalfe was jabbing his microphone at them, yelling, "Is it okay with you guys if we get this on the eleven-o'clock news?" Then he stepped back with his National Film Board camera, the other hand still holding out the mike of the tape recorder, while Bohlen stood beside the launch, the three frogmen in the background, and read the message out aloud again. Keziere was leaping about with his camera. Cummings was peeking over Bohlen's shoulder, frantically scribbling notes. Moore took one look at the three American sailors and chirped, "My God, they're freaks!"

Immediately we were exchanging passionate revolutionary handshakes, as though we had rendezvoused with our own troops along the edges of a lonely frontier. There had been talk in the counterculture for a decade about the Second American Revolution, and the blacks had for years now been walking about with carbines. One of these crewmen was black, and that cinched it in our Canadian minds—these three had to be part of the upheaval in the United States, and they really were on our side.

One of the sailors was jabbering excitedly:

"You know, except for a few of the officers, we're all against the war, man. There's a four-year waiting list to get into the Coast Guard, man, because if you make it in here you don't haveta go to 'Nam, dig? It's a drag, man, but you get to get back home alive. And, like, we're as down on that whole scene at Amchitka as you are, man. We got this petition together real fast and started getting guys to sign it. It was wild, man; guys were fighting to get to sign the thing. Everybody wanted to sign it, man."

For the crew of a U.S. Coast Guard ship to have signed a petition supporting a foreign vessel leading an attack on their own government was tantamount to treason. It laid them all open to the possibility of penalties, maybe even court-martial. It was a high, intense moment, enough to bring me to the edge of tears.

But in the meantime, Captain Cormack was up in the Penthouse having the book thrown at him by the commander of the *Confidence*. If it was nothing but contact high and good vibes at the stern, it was quite a different story on the upper deck. I had missed the arrival of the Coast Guard commander entirely and didn't realize in the euphoria of the first few minutes, that, in fact, the *Greenpeace* was being arrested.

The commander wore a wet suit, with a nylon jump suit on top, and an orange life jacket. His head was shaved completely bald. He had the air of an honest but world-weary highway patrolman who had just pulled over a speeding car. His name was Floyd Hunter, and he introduced himself as the officer in charge of the U.S. Coast Guard vessel *Confidence*, Seventeenth District, U.S. Coast Guard Command.

Having done his duty, he and the skipper climbed down into the galley, where he was introduced to Bohlen and Metcalfe, Bohlen identifying himself as the representative of the Don't Make a Wave Committee and the man whose instructions Cormack had been following. It was all very stiff and formal, very stylized, like a piece of medieval theater.

Commander Hunter explained politely, and with a trace of good humor, like a chess player who has just made an excellent move, that he was formally charging us with violation of the Tariff Act of 1932, according to which we should have notified customs of our arrival in Akutan within twenty-four hours. He explained that the act provides that vessels found in violation of Section 19 USC 1435 are liable for a fine of five thousand dollars and/or forfeiture of the vessel. In effect, we had crossed a border without reporting to customs. For all our elaborate planning, they had us on a technicality. Bohlen and Metcalfe tried to throw a few crafty legal questions at him, and Fineberg, assuming the role of our lawyer, demanded to know on whose authority this action had been ordered. But the questions bounced fruitlessly off the commander, who had all his answers ready. There was really no room for maneuver. We would now have to make "formal entry to customs" before leaving Akutan, and we would have to contact the district director of customs in Anchorage for instructions on further action to take if we were to avoid the fines or seizure of the vessel, which the commander was careful to refer to as "forfeiture."

When the commander stood up to leave, Cummings bolted out to the back deck to warn the rebellious young crewmen that the boss was coming. While the discussion had been going on in the galley between the chiefs, the rest of us had been scrambling around the boat, gathering up the poster of Nixon, the green darts, Canadian cigarettes, a small Canadian flag, peace pennants, and whatever books we could spare, to give as gifts to the crewmen who had signed the petition. The commander, of course, had not been advised what was happening behind his back. Quickly, the young Americans stuffed the gifts in their wet suits, and stood at attention when the commander appeared. They looked ridiculous with all the bulges in their outfits.

Once Commander Hunter was aboard, the launch headed back through the thickening darkness toward the steel-gray bulk of the *Confidence*. The three young crewmen flashed V signs at us whenever the commander wasn't watching them, until they disappeared around the stern of the cutter, which remained where it was as night fell, looking suspiciously as though it were blocking the mouth of the bay.

Our new position farther up the bay made it possible to get through on the radio. The eleven-o'clock news in Vancouver carried the story that the *Greenpeace* had been arrested, but that the entire crew of the U.S. Coast Guard cutter *Confidence*, in defiance of their officers, had signed a petition endorsing the position of the protesters. It made for excellent copy, because it looked as though a mutiny had taken place. But the fact remained that we no longer had freedom of action. Amchitka was slipping further and further away. The lights on the *Confidence* were ablaze, making her look like a hundred-eyed cat curled up just outside a mousehole, and inside the mousehole all we could do was squeak.

Furious arguments—the worst yet—broke out that night. Fineberg and Simmons argued that we had blown the whole trip through sloppiness. Metcalfe argued that it was the best thing that could have happened. At least now, he said, we would make it as a story in the U.S. media. And he was right: the mutiny of the crew of the *Confidence* was reported in American newspapers that had otherwise completely ignored our activities. But if the objective was simply to get stories in the U.S. press, this was an expensive way to do it. It looked now as though our chances of getting to Amchitka might be finished.

One by one, we drifted off to our bunks, depressed, confused, and angry. Up until this point, we had been generally impressed by our karma. On one of the first days out, Birmingham had emerged from the engine room to announce that "this boat runs on no known principles of engineering, it runs on shithouse luck." We had quickly translated that into a more modern term: *shithouse karma*. But now even our shithouse karma seemed to have failed.

In the morning, the *Confidence* was gone. As instructed, Metcalfe called the director of customs in Anchorage and was told that we would have to head back toward the Alaskan mainland to the nearest customs office at a place called Sand Point in the Shumagin Islands. There we could clear customs—maybe—and wiggle out of the trap into which we had stumbled. Either that or we would end up in Sand Point bogged down in yet more technicalities. But we really didn't have any choice. Accordingly, Cormack upped anchor in mid-morning and we headed out of the bay, back through Unimak Pass. For the first time, we were heading *away* from Amchitka. As we were heading down along the south coast of Unimak Island, a wintery wind whipped the waves and a few flakes of snow stung across the decks.

Bob Cummings emerged on deck, looking a bit like a figure in a Howard Pyle painting of the crew of the *Flying Dutchman*. He had run into a bad writer's block the night before while trying to put together a report on the arrival of the *Confidence*. From the point of view of the underground press, it was a great story. Coast Guard mutiny, solidarity in the assault on the American Death Machine, et cetera. But he had gotten paralyzed. He'd sat at his typewriter down in the engine room all night and hammered away at the keys, tearing sheet after sheet of paper out of the machine and crushing them in his fist, cursing and muttering. The deadline for the departure of the Akutan mail plane was 10:00 A.M. If he was to get his in-depth report back to Vancouver, it would have to be done by then or not at all. Whenever anybody had tried to hurry him, he had shouted back furiously: "Get off my back, man."

The Bering Sea was soon behind us. We had not got to know much about its mood except that it was getting mean. The ice packs were congealing. Winter was on the wind. The wheelhouse windows were fogging on the inside and covered on the outside with a wet, cold slush.

Coming across the slow mystical swells of the Gulf, we had gathered often in the little tramlike wheelhouse. It had been the center of daytime social activity. Now it was cold, like one of those old wooden streetcar waiting stations. Stiff gusts of wind blew through the door every time someone entered from the deck. The window-panes rattled. We had to wear gloves. We could see our breath. We had to tramp our feet up and down to keep them warm as we alternated between taking the wheel and going out into the sleet to wipe the slush off the glass.

At night, the radar was on all the time, but we still had to work at the windows to see the lighthouses and beacons. Lights moved blurrily in the puddles on the window.

Once we passed a Russian trawler. Voices came jabbering through the radio, her powerful generators overriding our tiny sideband completely. "Yep," said Cormack, "them Russian boats always has the latest stuff." Occasionally we heard Japanese boats calling to one another.

It was an odd lonely corner of the world. We were close to the American mainland, yet moving between peninsulas and little clusters of islands with names like Poperechnoi, Dolgoi, Ukolnoi, Iliasik, Chernabura, and Sanak. Between listening to the Russian and Japanese voices, with Cormack always on the alert for other Canadian boats he might know, the area did not feel the least American. It was only as we got closer to the Shumagin Islands that it became clearer we were approaching the good old U.S.A.

To the north, with great white scallops of wind-trimmed cloud around its peak, loomed Pavlof Volcano. The shore below was called Long Beach. After Cape Tolstoi came Seal Cape. After Belkofski Point came Bluff Point. North of Unga Spit was Lefthand Bay. Opposite Kaslokan Point lay Kitchen Anchorage. John Rock was only a few thousand yards from Olga Island. Going through Popof Strait we were just north of a bay called Saddler's Mistake.

Along the Alaska Peninsula, no one could quibble with the naming of the peaks, even as the Russian influence waned. After Pavlof Volcano came Mount Dana, then Hoodoo Mountain, Monolith Peak, Pyramid Peak, and Cathedral Peak, names that spoke of the scale of the place, with no signs of civilization beyond the flickering beacons. If the Gulf of Alaska had been a gray mercurial world after the Indian summer of the West Coast, and Akutan had been a whole other planet, we found, as we moved along Davidson Bank past Sandman Reefs and now to the Shumagins, that we had entered yet another dimension, and the season was early winter.

The guys who were awake tended to stay in their sleeping bags, wearing sweaters and sometimes scarves, and mainly gathering in the galley to be near the big iron-topped stove. The four-hour watches that had seemed long enough coming up the Inside Passage, and that had stretched out into stony trips on the Gulf, now became eternities of waiting, stamping your feet, for a streetcar that never came.

We were all starting to get tired. Conversations moved more slowly. We were no less nervous than before—Cannikin still loomed in our minds—but a kind of numbness was setting in.

One afternoon, Metcalfe and I were on watch. He was sagging his hefty bulk against the wheel, looking half-asleep, like a cat melting into another world.

"The question is, Bob," he said, speaking from far away, "is this an Ahab trip or an Ishmael trip? You know, to this day, I still don't know which one it is for me."

Metcalfe had been in the newspaper business twenty-five years longer than me. We had both gone through the same hoops in the peculiar small world of western Canadian journalism, from the Winnipeg police beat to Vancouver City Hall. We had both dropped out of high school on the prairies and had sweated through the same late-night despairs of booze and hamburgers and stained copy paper.

We shared a feeling of brotherhood. We saw it as a media war. We had studied Marshall McLuhan. Metcalfe had a street-fighter's understanding of public relations: "It doesn't matter what they say about you, as long as they spell the name right." I had pretensions of being a media theorist in my own right. Not long before Irving Stowe had first contacted me about coming on the *Greenpeace*, I had finished writing a book that suggested that a radically new consciousness had evolved in the postwar period and that this consciousness had taken as its task the goal of creating "ecological awareness" in the mass mind. I had predicted the emergence of "Green Panthers," and so had not been able to refuse an offer to join what we now only half jokingly referred to as the Greenhawks.

For Metcalfe, the voyage of the *Greenpeace* was a campaign like others he had run in the past when he had been paid by politicians to apply his knowledge of the day-to-day mechanics of journalism to get them elected. Image was everything. He knew exactly how to grab headlines, how to drop catchy phrases that would be reprinted, how to play on the reflexes of bored editors. While I had far less practical skill in this department, I had my theory of mind bombs—powerful new images delivered via the media—changing mass consciousness. I saw the *Greenpeace* as an icon, a symbol, as a kind of wobbling control tower from which we might affect the attitudes of millions of people toward Amchitka specifically, and their environment as a whole in general. The screeching static-choked radio on the *Phyllis Cormack* and the lone radiophone at Akutan had taught me some of the limits to McLuhan's concept of a "global village," but nothing had yet happened to convince me that the boat was not, after all, a mind bomb sailing across an electronic sea into the minds of the masses. Madison Avenue and Hitler had changed the face of the world through application of the tactic of image projection, and the environmental movement could hardly attempt to do less.

For now, we were merely cold and tired, and the abstractions of mind bombs and mass media seemed remote. It was enough to hang on to the end of one's watch, and then hustle below to climb into a warm sleeping bag until dinner.

Into the Shumagin Islands we sailed on October 2, the day we had originally expected the bomb to go off at Amchitka. Instead, we were a thousand miles away, coming around the tip of an island called Unga, not Amchitka, moving up through Popof Strait to the cannery town of Sand Point, with the test probably still a month away.

There was little excitement as Cormack angled the boat in toward a dock, looking for a place to berth among the dozens of fishing boats. Sand Point did not look like much, even though it was the largest community we had seen since Vancouver. It did not seem that we had been on the boat a mere seventeen days. It seemed more like months. There were moments when the boat seemed like a prison, and I could feel my own determination crumbling at the edges. I could see the resolve fading like old linen on the faces of the others. Maybe Sand Point would prove to be a good place to get our energy back while waiting another whole month before finding out what was going to happen at Amchitka. But the prospects weren't good.

Cormack jostled the boat up against the wharf at low tide. The pilings reached up over our heads, and we found ourselves staring between them into murky cavelike glooms. The water was milky and full of orange chips, with hundreds of gulls *yowk*ing and crashing up and down on the other side of the wharf to get at the bits and pieces of waste being discharged from the cannery. There was a wooden sign nailed to one of the pilings beside the ladder leading to the top of the dock. The bottom half of the sign had been broken off, so it said simply:

PLEASE DO NOT

It was a very negative greeting. The air was cold and dank, and there was a peculiar unpleasant odor. Did we have to tie up right next to a stinking cannery? It seemed there was no other place to go. Up on the wharf we found ourselves confronted by a long white warehouse with the name WAKEFIELD'S printed on the wall. And underneath, in smaller print: ALASKA KING CRAB. A forklift was trundling back and forth between huge wooden boxes and the entrance to the cannery. With each trip, the forklift carried a metal container. Over its edge we could see prickly armored crab legs and claws waving frantically. Then, with a pulpy eggshell-breaking sound, the container was flipped on its side and about one hundred Alaska King Crabs came clicking down in a heap into a chute that took them into the cannery.

"Oh oh, this is gonna be a bummer," warned Doc Thurston.

Five of us gathered at the edge of one of the wooden bins and looked down. At first there was a rush like vertigo, and I had the sensation of being on a ledge looking into the land of childhood nightmares peopled with these clacking mandible-waving creatures. Backward and forward over each other they plowed, heaving desperately upward, falling back with a splash. There were probably a thousand of them crushed in each bin, eye stalks snapped, limbs twisted at horrible angles, pincers groping around blindly. Their purplish shells were cracked like chinaware. There was a constant rattle and *clickety-click-click* like distant typewriter keys. Small bulbs of broken-off eyes floated in the water. I reeled away from the bin expecting to be sick. Moore and Thurston and Keziere and Cummings looked no less shocked.

"That's fucking awful, man," was all Keziere said.

Thurston got that hopeless defeated look on his face again. "Let the goddamn bomb go, man. The sooner this planet is rid of us the better."

The record for an Alaska king crab was sixty-one inches across. That was a crab, spread out, that was larger than most tables. But that had been ten years before. Since then, the boats had been going out in hordes. With the whales gone, the herring close to oblivion, the shrimp populations diminishing, and the halibut getting so scarce that boats were having to push into uncharted waters to find them, the fishermen had turned their attention to the crabs.

At first, it was a bonanza. The giant creatures had been larger than German shepherds. But each year, as the plunder continued, the crabs dropped in size. Each year, the "minimum allowable size" had been lowered. Now, in the autumn of 1971, a decade after the Alaska king crab industry had become "commercially viable," it was on its last legs. The crabs in the bins into which we were staring were only about an average of seven inches across the shell. The workers at the cannery admitted that they were now being taken at six and even five inches. From sixty-one-inch giants down to a pathetic fifteen inches in just ten years. It did not take much imagination to see that the Wakefield cannery would soon be abandoned like the fish canneries we had seen along the coast, like the whaling station at Akutan, and these fish boats, too, would be washed up on some gravel beach, and the oceans would be empty, dead.

"Christ," said Moore despairingly, "are they all blind? Can't anybody see what's happening?"

Sand Point was indeed going to be a bummer. Redneck-looking characters were shambling along the wharf. When we went down to the general store, a guy wearing a cowboy hat and mirror sunglasses came out, looked our way, and spat on the planks. He got into his truck, started the engine, and then the truck leapt toward us. For a second, we were frozen in shock, not believing what was happening. Almost too late, we realized he had no intention of slowing down. We lurched out of the way. The truck barreled past, missing Thurston and Moore by inches.

"You motherfucker!" howled Thurston. "Come back here and I'll kick your sunglasses down your fucking throat!"

"Cool it," said Fineberg, looking nervously around. There were about a dozen equally redneck-looking characters hanging around, staring at us icily. We picked up the distinct impression they would have been pleased if the truck had run us over.

Back at the boat, Bohlen called us together in the galley. "Look, I think we'd better recognize the fact that we're not in friendly territory. Most of these people don't know anything about us except that we're a bunch of political radicals, and from what I can tell, this is a pretty Right-wing place. I think we'd better not make any assumptions about how welcome we are here, especially you four with long hair. I think it would be wise if none of us went out alone. Let's stick together until we get out of here."

The customs officer, a retired fisherman named Lyle Hansen, came down to the boat to let us know that the business of our customs violation was being dealt with by authorities in Anchorage and that he would tell us as soon as possible what the decision was. In the meantime, we would have to put up a thousand-dollar bond, and there was a distinct possibility that the nine Canadians on board might yet be fined a thousand dollars each for going ashore at Akutan. Bohlen put through an emergency call to the Don't Make a Wave Committee to find out how much money we had left. Irving Stowe reported that he was expecting some donations from groups in the U.S., but no money had come through yet. After posting the bond, we were just about broke.

"The hell with it," said Thurston that night, when we were all feeling just about as low as we thought we could feel. "There must be a pub in this burg somewhere. I don't know about the rest of you, but I sure could use a drink."

"I wouldn't advise it," warned Fineberg. "I've been in a lot of these little coastal towns. They can be pretty rough. This isn't Haight Ashbury, you know. And it ain't the Summer of Love, my friends."

Thurston said, "Fuck it. I'm thirsty."

It was he who led the way through town, followed by Keziere, Cummings, Moore, Darnell, and myself. Within less than ten minutes we had sniffed out a bar called the Sand Point Tavern. It *looked* bad, crowded with short-haired suspicious-looking hard rocks, a couple of actual peroxide blondes, guys with cowboy hats and buckskin jackets, scar-faced Aleut Indians in lumberjack shirts, beefy meat-hook-handed whites off the boats, laborers from the cannery, a few miners, and a couple of men in flight jackets who undoubtedly worked for the airline company that had a contract to take stuff out to Amchitka. The waiter had a scar that cut across his face from forehead to jaw, winding crookedly over his broken nose. A jukebox blared raunchy Western music, but no one was dancing. At exactly the moment when the six of us came through the swinging door in our Greenhawk rain jackets, the music whanged to a halt, leaving only an electronic buzz in the air, the squeak of chairs, and the odd clink of a glass, as everyone in the joint turned around to stare at us.

There was nothing for us to do but keep on coming in, allowing Thurston, who seemed fearless, to remain in front. Staying close together, we clumped in our rain boots across the floor and noisily drew up chairs near the bar, far enough from the door so that it wouldn't appear that we were afraid to stay. A moment of paranoia settled over us, while we wondered whether we would get served, trying hard to look and sound casual despite all the looks we were getting.

Then the music was blaring again, everybody had gone back to their drinks, glasses were clinking, and the scar-faced barman was trotting over to take our orders. We ended up staying at the Sand Point Tavern for five hours, getting to know a lot of the locals, ignoring others and being ignored in turn. We blew about five dollars on the jukebox. Thurston got up and danced with a fifty-year-old woman who later joined us at our table and confided that she was the town whore. When a drunk who seemed to be almost seven feet tall got up and lumbered toward us, snarling and swearing, saying "you fucking pinkos," she got up in front of him and screamed that he'd better "fuck right off, Charlie, these're my friends." Grudgingly, he backed down into his seat. That was the end of the big confrontation. We had an ally. We ordered drinks. And then, surprisingly, someone from another table ordered a round for us. The next thing we knew, fishermen were crowding around our table, talking about the "goddamn bomb" and how it was "no damn good for the fishing."

"Did'ya know the halibut catch per man per boat is down fourteen percent this year, and even the shrimp are starting to avoid the banks around the Sanak Islands where they've *always* been runnin' this time a'year? It's that goddamn bomb, fer sure. Hope you boys can do something about it."

We seemed to become involved in a dozen overlapping conversations at once. Over everything roared the music from the pearly fluorescent baleen teeth of the jukebox. Thurston leapt about with our new lady friend like a madman. Finally, about three in the morning we stumbled back past the house trailers and prefabricated houses and the corrugated sheet-metal Quonset huts, past actual normal-sized spruce trees that were flapping in the wind. We skidded and banged and half crashed down onto the deck of the *Phyllis Cormack*.

Bohlen and Cormack and Birmingham were sitting in the galley, looking exactly like three righteously outraged fathers. Cormack had his big hands wrapped around a china mug. Birmingham was pretending to read a Western novel. Bohlen sat there in his long underwear, breaking up toothpicks and cutting up little pieces of paper, gluing them all together into tiny geodesic domes.

"Well, where the hell *you* dummies been?" Cormack snapped. "Thought y'mighta been lynched by now. It'd serve yuh bloody well right, goddamn loonies."

Bohlen, not wanting to admit to any paternalistic feelings, said: "Well, I was just going to hit the sack."

Birmingham wanted to know all about it. "Hot dang! Never been to one of those places, you understand, all them hot-to-trot gals and all that, go-go girls and all."

In the morning there were a few hangovers and a generally numb mood about the boat. We pecked at our typewriters. Metcalfe worked longhand as usual on his Monday morning CBC commentary. Others wrote letters home. Bohlen built more toothpick-and-paper geodesic domes. The customs officer dropped by to chat. Several fishermen dropped down to give us gifts of halibut and crab. We sipped a few beers. Darnell cooked up a tremendous supper. The galley windows were steamed. Soon it was dark again outside. Boat engines were starting up around us. Small waves slapped at the hull. The cannery was closed, but we could still hear the restless hopeless *click-click-click* of the king crabs trying to escape from their bins up on the wharf. A thin rain puddled the deck. The wind was cold. And still there was no word about when the bomb might go off.

It was easy to see that we were getting unnerved by all the horror stories we were being told at the Sand Point Tavern, such as the one about Old Jock, a part-Aleut fisherman whose ship got caught in a bad storm up near Unimak Pass.

Expecting the boat to break up, Old Jock had pulled five or six layers of clothes over himself, wrapped towels around his head, and tied life jackets around his limbs. Then he leapt overboard. Another fishing boat found him three days later, still barely alive, but as they peeled away the layers of clothing, whole colonies of sand fleas began to emerge. They had gotten to Old Jock early. There was hardly anything left of his stomach, where the sand fleas had conducted an open-pit mining operation, and his legs and arms had been nibbled almost to the bone. There was nothing that could be done for him except to let him die as quickly as possible on the deck, while the sand fleas seethed out from among the blood-stained rags.

Our dreams were filled with legions of sand fleas, and huge hopping king crabs bent on revenge. It was not just the storms and the williwaws that worried us now—let alone the bomb itself, which seemed more and more remote—but the thought of the creatures under the sea who would get to us *after* we sunk.

On Wednesday, October 6, word came through from Vancouver that an anti-Amchitka demonstration in front of the U.S. consulate general's office had attracted some ten thousand high-school students. From all reports, it had been an uprising. Principals and school boards throughout the city had refused to allow time off to attend the demonstration, yet from all over Vancouver and surrounding municipalities, and from towns and villages as far as ninety miles away, the kids converged, skipping classes in droves.

As their numbers swelled, they blocked traffic for miles around. They filled up the street in front of the consulate general's office, overflowing into nearby parking lots. For an hour they were treated to one of Irving Stowe's impassioned speeches.

On board the *Phyllis Cormack*, we hung around the radio room the whole afternoon, jamming together around the single side-band, hanging on every word. All our wives and girlfriends and kids were at the demonstration, of course. Several of us broke down and wept. I found myself out at the bow, crying as I had not cried for years, with joy, the kind of crying people did during the war when news came through of a great victory. We all felt the kind of awe and love that mass movements inspire in those who take part in them, a feeling of unaloneness.

Yet these emotional binges were beginning to take their toll, for it would only be a matter of hours before one was back in one's foul-smelling little bunk, feeling claustrophobic, staring out the porthole at the wharf pilings covered with barnacles, listening to the guys in the galley arguing or babbling, whichever it was.

By this time we had been living closely for so long that each of us seemed to know more or less precisely what the other was going to say. It was maddening to have to go any longer through the ritual of pretending to listen or, worse, of having anything to say yourself. Everything that was said seemed transparently obvious. Communication had become an agony of shuffling along a treadmill. The voices were all too familiar. The responses were all too familiar. The jokes and attitudes were all too familiar. Everybody had a usual routine, and by now we knew each other's usual routines so well that a certain amount of contempt had set in. Each man seemed hopelessly locked into the trap of being himself and not able to escape or get free. No real growth seemed possible, no evolution. The stench from the crab cannery had gotten into our pores and seemed, somehow, to have gotten into our minds.

On the faces of the others I could see the same wild look as I saw on my own face when I dared to look in the mirror—which was not often, for there were lines all over my forehead like road maps, and hollows had begun to develop under my cheekbones. My cleanshaven chin had vanished into a gritty, ugly stubble. Cummings struck me by then as looking a bit like Count Dracula. Thurston had allowed himself to grow crankier and crankier. He shouted at Cormack, shouted at Birmingham, shouted at Cummings. But they were all shouting themselves. Cormack *always* shouted; you hardly ever heard him merely talk. But at least he had been that way all along. It was the rest of us who were getting more and more like *him*. There were times you thought you were casually conversing, when suddenly you'd realize from the expression of the man in front of you that you were *shouting* again.

We were getting cabin fever. With absolutely everything going wrong that could possibly go wrong, the energies we had mobilized to carry us to Amchitka were finding themselves with no outlet other than each other. The test seemed every day to be slightly more delayed than the day before. New lawsuits were being launched. Rulings were overturned in subcourts. More politicians were speaking out. Demonstrations were building in pitch and size.

The worst part of it was that we couldn't just get our confrontation with Cannikin over with. We could only hope to hang on to our position until the issue was resolved one way or the other, and we would either have our face-to-face encounter with the bomb, or, for the first time in history, such a test would be canceled.

It all seemed so important, yet the issues that captured my emotions now were petty things, like Cummings's habit of using *my* typewriter, Fineberg's habit of droning on and on about some fine point of law or other, Simmons's insistence on talking furtively, Cormack's inability to stop shouting. Personalities had begun to wink in and out of phase. One moment Birmingham would be playing the part of a dumb old gaffer with only half his marbles, the next he would be expounding brilliantly about hydrodynamics. Bohlen alternated between moods of elfish humor and a kind of bruised despair. He seemed to have aged several years. Moore, who I had generally thought of as a cast-iron radical, now metamorphosed into a flower child. Thurston's laughter had gotten so weird it jabbed through the air like an electric shock.

As the days passed, the frustration levels mounted. Pages torn from notebooks began to appear taped on the galley walls with aphorisms scribbled on them:

FROM TERROR ONE ESCAPES SCREAMING
BUT FEAR HAS A STRANGE SEDUCTION

REALITY: A FEDERAL OR PROVINCIAL RESPONSIBILITY?

MASTURBATION IS THE PREROGATIVE OF THE
ESTABLISHMENT PRESS

A CRAB IS YOUR CAPTAIN

The arguments around the galley table were beginning to get jagged at the edges, like broken beer bottles. The old days of the hassles over the false rumor of Fineberg being a CIA agent seemed far behind and almost quaint. The dickering had always had a gamelike quality to it, and in the back of my own mind there was always an observer taking notes.

But things were changing. The observer was being driven out from hiding. It was happening to us all. We were being flushed into the open.

Now we would find out who we were. The remarks we made about each other had a razor quality to them. One night Thurston announced loudly: "When I get back home, I'm going to set up the Lyle Thurston Fund to send John C. Cormack to charm school." Cormack hesitated only a few seconds before grabbing the doctor in a bear hug that was so powerful we thought we could hear Thurston's bones cracking, and for a moment it looked like we might have to get a crowbar to rescue him from the skipper's grip.

The meetings while coming up the Inside Passage had had a quality to them that suggested the boys had just finished work and were down at the pub talking things over. Out on the Gulf, the meetings were less chummy, more formal. By Akutan, the meetings were definitely heading in the direction of encounter sessions. And now, in Sand Point, they had taken on the tone of psychodramas.

Against the apocalyptic background of nuclear bombs being detonated and earthquakes and tidal waves being unleashed, it seemed bizarre that the most trivial issues triggered our emotions most easily. The issue of personal incomes, for instance, was enough to spark furious outbursts. Bohlen stood to lose his job if he wasn't back within six weeks. Keziere and Moore might lose their positions at university. Birmingham was beginning to express a desire to be paid to cover his bills. Metcalfe said his free-lancing business was going to hell. Thurston was paying out one hundred dollars a day just to have his practice maintained while he was away.

A new division opened in the ranks. While some stood to lose their jobs, others among us were in fact being paid to be there. Cummings and myself were both on assignment and so could afford to stay out forever. Neither Simmons nor Fineberg had anything else to do at the time, although Fineberg had turned down a $10,000 post-doctoral fellowship to come on the voyage. Bill Darnell was still getting his pittance from the Company of Young Canadians.

The hope of achieving "consensus" became less than a hallucination. It became a fairy tale. Without quite being able to pinpoint the moment when it happened—or how far back its roots went—the moment came when we were no longer the Greenhawks. Suddenly we were twelve men looking at "reality" from twelve different angles, with vastly different backgrounds and different minds. Fights were starting to erupt daily. I'd wake up to the sound of Simmons screaming wildly at Metcalfe in the bunk room, then have to listen to Birmingham snarling at Darnell,

Thurston bitterly complaining about Cummings, and the skipper grumbling about us all.

The crunch was coming. As Metcalfe said in his broadcast from Sand Point,

> The crunch is coming and they know it. They know that soon, very soon, they'll have to decide whether they should try to wait out bureaucratic delays of the Amchitka test and take the risk of being ploughed under by the sands of time, or whether to call their own shot, which would mean realizing that Canadian opinion against the bomb is at its peak now and that they can move more usefully by going home to help that opinion there than stay and watch it decline. ... So now the *Greenpeace* must contemplate its nettle. Whether they grasp it will depend to some extent on whether their voyage was planned as a practical protest to help raise public opinion in Canada or whether it was a hero trip for the gratification of a few egos.

Bohlen was tending to side heavily with Metcalfe. "To go from this point, from Sand Point," he said, "to Amchitka and back in the later part of October and early November is bordering upon a lunatic expedition. You may recall that when we left there was a vast segment of public opinion that thought we were loonies. And we proved that we weren't loonies by being able to do to this date what we said we were going to do, and it is essentially, in my opinion, because we proved that we are not loonies that we have this mass base of conservative public opinion behind us. It is also my opinion that we stand a large chance of *becoming* loonies. ..."

Metcalfe could hardly be blamed—in my case at least—for raising the issue of "ego gratification." Certainly, of all the people on board, I was the one who was getting the most consistent exposure in Vancouver for my part in the voyage. An editor had already advised me by phone that my readership had quadrupled since the beginning of the trip. Wasn't I going to look a royal fool if I had to start explaining in print why we suddenly copped out? There was a definite "loss of face" problem. In some cultures, the whole lot of us would have been expected to throw ourselves on our ballpoints.

I happened to be the most public figure. But the rest were squirming too, from Cormack on down. It had been pleasant to be heroes back in Vancouver before we left. Few, if any, of us had ever been heroes before. Now we had to wonder how it would feel to be famous craven cowards.

As the decisions wracked us all, our laughter seemed to be pain. People seemed to be moaning and crying in front of each other, but we remained hypnotized into believing that these agonized sounds were really signals indicating merriment or joy. Mentally, we had arrived in yet another zone. It was a season without weather, color, or shape, without aim or purpose. It was the useless searching of a compass needle on a world with no magnetic poles. More bits of graffiti went up on the galley walls:

COWARDLINESS IS NEXT TO GODLINESS

TEETH WILL EXPAND IN DIRECT PROPORTION
TO THE SPACE ALLOWED THEM

WHAT MAN CAN HALLUCINATE, MAN CAN DREAM

A DEATH WISH WOULDN'T BE CONSCIOUS, YOU KNOW

Late one night, Thurston and I crept over to the wooden bins beside the cannery with the object in mind of liberating at least one Alaska king crab. The crabs were clacking and clicking in their usual hopeless manner, and we were very much aware of their huge pincers, which could easily take off a man's finger. It took several tense minutes before I managed to get hold of one by his back legs, paralyzing the claws, and hoist him out of the bin before any of the rest could get a grip on me. We made a mad, nervous dash across the wharf under the floodlights, with the crab thrashing powerfully in my hands, like a huge armored spider. At the edge of the wharf, I simply let go. Down the crab whirled through the air for at least twenty feet before hitting the inky water with a sickening smack. Immediately, it disintegrated, limbs snapping off and shell splitting open, leaving a soupy puddle as the bits and pieces sank into the darkness.

"So much for Crab Liberation," muttered Thurston.

Then it was Tuesday, October 12, and we held our final meeting. It was conducted without the tape recorder. We had gotten down to the level of physical jockeying, letting our tempers run as they would. Simmons had already been shaking his fist under Metcalfe's nose. Bohlen had reemerged as the tough navy vet, ready to fight on the decks with his bare hands. Cormack's wrath bloomed like great wings, driving the rest of us back against the walls. Birmingham's massive workingman's muscles had flexed with the tension of not strangling any of the "damned young smart alecs." I had studied Butokan karate for two years, rising to purple belt, and was starting to feel so angry that I *did* want to give somebody a front-thrust kick right in the face. Even normally tranquil Darnell was kneading his knuckles, his face rippling with emotion. At one point, Bohlen, Birmingham, and I were on our feet at once, screaming at each other and smashing at the table with our fists.

In the midst of it, Thurston got called away in town because somebody was having appendicitis. He missed the final maneuver, wherein the "kamikaze" rebels tried to trap Cormack into insisting that the boat could still get us to Amchitka. Our theory had been that in a crunch, Cormack would never admit that there was anything he or his boat couldn't do. He was a man who couldn't say the word *defeat* aloud. If we could get the argument down to a question of whether the boat was capable or not, and then put it to Cormack to tell us the answer, we were certain his pride would prevent him from admitting that it couldn't be done. Ergo, there could be no rational argument for not continuing the trip. Sure enough, when he was brought in to answer the question, Can we still get to Amchitka? his reply was exactly what we had guessed. "We can make it," he said angrily, as though the suggestion that there was the slightest doubt was an insult in itself. "There's not a damned thing wrong with this boat. She's the finest sea boat on the whole West Coast. She'll take you anywhere! Ain't no kind of weather this boat can't handle!" I thought for a moment we'd won.

Then Cormack added, in the most desperate, pleading tone of voice we had ever heard him use: *"But you'd be crazy to try it!"*

The look of honest terror on Cormack's face was all the rest of the crew needed to confirm their worst fears about what lay ahead. Here was the man who *knew*.

Only four of us voted to continue, an unlikely alliance of two Canadians, Darnell and myself, and two Americans, Fineberg and Simmons, none of whom were under any serious personal pressures to go home.

We all sagged in our seats, more astonished than anything else that the decision had finally been forced. Bohlen looked pained all over. "Well," he said, "I guess that's it."

On the way out of the galley, I smashed the top part of the door off its hinge, then went straight to the battenclaim and cried with anger and frustration for three full minutes. It had really happened. We were defeated. We who had had the paranoid grandiosity to think we could stop a megamachine were truly and thoroughly beaten.

"We're all disappointed," said Metcalfe later, "in ourselves and each other." At midnight—scarcely an hour after the vote to turn around—we were down at the tavern, drinking bitterly and wildly, when I suddenly realized it was October 13, my birthday. I was now thirty years old. Hours ago I had been twenty-nine, on my way up to Amchitka to save the world, and now I was thirty, an old man, heading home in defeat. Thurston laughed so hard he fell out of his chair. Metcalfe used the coincidence to send back a press release about my "birthday party," deftly burying the news that we had decided to turn around. Sure enough, the radio and television coverage back home the next day was about the joke of a *Greenpeace* crew member turning thirty during a protest against the establishment. The media still thought very much in terms of the "youth revolution" of which we seemed so visible a part, and it was much easier to report a silly cheerful story like that than to carry yet another dreary Amchitka protest report. Using all the skills he had applied to build us into a big news event in the first place, Metcalfe, the PR man, now set to work putting up as many smoke screens as possible to cover our retreat.

Out past the Sand Point grocery store, there was a small gill-netter beached on the gravel, a splintered bone of a boat with its windows and portholes all smashed. Inside was a rusting old engine, which the wind played like an instrument. In green, now peeling, someone had sloshed a crude peace symbol on the hull. Our final green peace symbol. The boat was tilted over as though frozen in the moment when it was taking a wave over the bow. It seemed a perfect expression of our mood and situation.

It was to this old sarcophagus of a boat that Keziere and Thurston and I retreated the next day, waiting for the pills that the doctor had prescribed to calm me down to take effect. When they did, everything became simultaneously dreamlike and yet more real than ever. The winter light pressed down through ragged holes in the roof, like a shroud settling over our once-great ambition.

"The trick," said Thurston, "is to systematize your delusions." It seemed at the time like the most profound piece of advice I'd ever heard. It suggested glimpses through the veil of Hindu maya. "Reality is a crutch, man," the doctor added.

We had walked Fineberg to the airline company office, where he used the last of his Alaska checks to buy a ticket to Anchorage. "I can do more at home," he said. "I believe I got scapegoated for raising the kinds of questions that needed to be raised. I got kicked in the teeth. Since we all came together with a common purpose, that's a pretty heavy trip to work through." Both Terry Simmons and I had gotten up in the morning, badly hung over, with the same intention—to fly away. But we got talked out of it. Keziere suggested the best thing to do was "see the thing through." Now, hours later, under the influence of the doctor's medicine, it was possible to see that nothing really mattered anyway.

Ultimately, everything was a joke.

The Greenhawks. Mind bombs. Guerrilla musical comedy. The Fellowship of the Piston Rings. The Warriors of the Rainbow. The *Greenpeace* Church.

Absurd. All of it.

"How far?" Keziere shouted when we got back to the *Phyllis Cormack*.

Hoarsely, we responded:

"ALL THE WAY *HOME!*"

During the long blank retreat from Sand Point toward Kodiak, our next stop, the boat was like a museum late in the afternoon. We took our turns at the wheel like zombies. All the energy had spent itself, had blown through the rooms of the boat like a wind leaving a ruin. The weather was rough all the way. There was a nutrient-tank atmosphere, a restless, punchy, twitching mood wherein we shot barbs back and forth at each other, pouring out our mutual scorn. An incredible tangle of psychological wrestling matches were going on at every conceivable level. We saw through each other like wraths. We knew each other's every number, every act, every routine, as though we had acquired blazing X-ray-vision eyes.

There were quite a few personal duels taking place. They took strange forms:

One afternoon, Metcalfe climbed up into the wheelhouse, book in hand, not saying very much. He was still immaculately dressed, with his goatee trimmed, still in possession of all his worldly cryptic wisdom. If there was any one person on board who personified the image of the upper classes, it was Metcalfe, theater critic and bon vivant, connoisseur and literati. Metcalfe's sophistication had stood in such contrast to

Captain Cormack's overwhelming working-class style that it seemed at times difficult to believe that they really were on the same boat together. Cormack had not been the only one to detect a trace of superiority in Metcalfe's attitude toward the skipper. Now, as Metcalfe was lounging in the corner of the Penthouse, reading quietly and waiting to take his turn at the wheel, the old skipper ducked into his cabin, emerging a few moments later with a box full of assorted chocolates in his hand, part of his secret stash of candies and bars and bags of jellybeans.

"Gentlemen, help yerself," said Cormack.

Bohlen, Darnell, and I were there at the time. Everyone helped themselves, except Metcalfe.

"No thanks, John," he said, scarcely bothering to look up from his book.

"No, no, Mr. Metcalfe, sir, I insist."

"No, really, John."

"Now, Mr. Metcalfe, surely yuh ain't *above* havin' a chocolate?"

Metcalfe glanced up for the first time, nodded, and took one.

I think he understood perfectly what was happening, but was just too tired to fight it. A moment later, the skipper was offering another round. The rest of us turned it down and Cormack just shrugged. But when it came to Metcalfe, he pushed the box forward. Something was going on, all right. Cormack had somehow got Metcalfe in a corner where he couldn't say no without its being a put-down. And no sooner had Metcalfe been forced to take a second chocolate than he was having a third pushed upon him—same tactic—and a fourth.

Finally, he looked up from his book, face gone pale, as though for the first time he was getting a taste of seasickness in his mouth. He lurched.

"I don't think I can take this watch," he said quickly. "Excuse me." With that, he staggered down the wooden ladder into the bunk room—and stayed there three days in his bunk, staring out the porthole, not eating, barely getting up to go to the shitter.

The moment he collapsed out of the wheelhouse, Cormack burst out in a joyous stream of laughter.

"By God, did yuh see *that*? Did yuh *see* that? That Metcalfe there, did yuh *see* the look on his face?"

There had been no problems with Customs and Immigration upon leaving Sand Point, once they understood we were heading away from Amchitka. The next port of call would be Kodiak, simply because a message had been passed through to us from Vancouver, explaining that the mayor had invited us to stop in as a gesture of opposition to the

Amchitka test. Even if we were no longer on our way to park next to the bomb, we were still considered the "cutting edge"—somewhat blunted, but still visible—of anti-Amchitka opinion.

I was amazed at how little criticism was coming at us from back home over the decision to turn around. Rather than attacking us for our failure to carry through, the media had simply dropped us as a story. In terms of print, the only coverage we were getting now was in my columns and Cummings's reports to the underground press. Metcalfe was keeping us just barely alive on radio, but otherwise the focus of the Amchitka protest had shifted to the legal maneuvers taking place in the U.S. and to huge petitions that were being gathered throughout Canada. There were rumors of further border blockades in the works, and the politicians had not stopped talking about the issue. Metcalfe was milking the theme that if we had failed at all, it was only in one respect. While the crew of the *Greenpeace* "never dreamed they could stop the Amchitka test by themselves, they did believe that Mr. Trudeau [the Canadian prime minister] would eventually come through with his own personal protest. It seemed to them quite natural for Mr. Trudeau to speak for Canada on this matter. ..." That was the game plan for covering our retreat. Blame the prime minister. It wasn't our failure, it was *his*.

As we approached the dock at Kodiak, a motorboat came out to greet us and guide us back to a large crowd that was waiting, holding up signs like THANK YOU GREENPEACE. People were waving and cheering. Thank us for *what*? We had become so intensely involved in our own little drama that—except for Metcalfe—we had lost touch with the general image that was still being projected beyond the confines of the boat, in the strange other world inhabited by people who read newspapers, watch television, listen to radios.

So far as the people in Kodiak were aware, we were still the protest ship that was trying to stop the test at Amchitka, and whether we were en route or in retreat, it made little difference. They had at least heard about us, and that made us celebrities. Beyond the fact that we were an acknowledged symbol of opposition to the test, it was up to them to project whatever fantasy or expectation they wanted onto us. Since they were against the bomb themselves—Kodiak had been hardest hit by the last great earthquake—they were basically cheering for their own side, and we were simply the "home team." The home team can do no wrong.

There in the crowd at the dock was one familiar face: Rod Marining, the Yippie who had sworn at dockside back in Vancouver that, one way or another, he would be on board. He grinned up at us through the ragged dog hair of his beard, shaking his head, tape recorder slung over his shoulder, headphones around his neck, wearing an old leather jacket.

"Boy!" he yelled, "I don't *know* you guys! You look *weird*! You look like spooks, man, you really do!"

All along, it had been one of the finer thoughts of the voyage that Marining would suddenly show up riding on back of a whale or zipping past us on his own boat. Through the two years of preparations that had gone into the trip, Marining had managed to take enough time off from his usual Yippie tasks—liberating land about to be grabbed by real-estate sharks, constructing artificial twelve-foot joints of marijuana to be handed out at smoke-ins, dodging the Vancouver police—to attend every meeting of the Don't Make a Wave Committee. He had done all the joe-boy jobs, like painting the sail and cleaning out the hold. At one point he had tried to finance the purchase of his own little boat to travel to Amchitka. His plans had included levitating the island; casting voodoo spells on the bomb; broadcasting appeals for help to Namor, the mythical comicbook prince of Atlantis; calling in Aquaman; and getting on with all the mad Yippie media tricks he could imagine. Now, with Fineberg gone, Marining would finally get his wish to be on board. Even if we were heading in the wrong direction, here at least was one vision that was not completely obliterated.

Marining had flown to the coastal port of Prince Rupert after we left Vancouver, betting that maybe somebody would be seasick enough by the time we got there that he'd be able to replace him, but we didn't stop at Prince Rupert. When that failed, the Yippie had gone back to Vancouver, taken part in the border blockades, then hitchhiked to Seattle, caught a plane to Anchorage, and had hitchhiked from Anchorage to Kodiak, arriving just in time to stir up publicity and generally get a welcoming committee in motion. One of Marining's radio interviews—where he was identified as a *Greenpeace* "organizer"—had led to the mayor of Kodiak getting excited enough to extend a welcome to us. It had taken a certain amount of guts to hitchhike across Alaska in 1971, where only a year before a hippie had been beaten up so badly he lost the sight of both eyes.

Whatever the mayor of Kodiak had had in mind when he invited us in to dock, it quickly became evident we were not what he expected. There were twelve mostly scruffy-looking characters, almost half of whom could be mistaken for weirdos. At the dock, a contingent of young Coast Guard types had gathered, remembering that the crew of the cutter *Confidence* had defied their officers in order to sign a message of support. They were waiting in the crowd, with the gift of a three-foot-long fur-covered dope-smoking pipe with a surgical mask at the end, which they openly presented to us. This caused concern among the *Greenpeace* brass, who were worried about turning off the mayor, but a fair amount of delight among the psychedelic kamikaze nuts on board. From beginning to end—fortunately only three days—the visit to Kodiak was a study in schizophrenia. We were feted with a banquet in our honor, paid for by the city council, and ended up at a long table, with the likes of Rod Marining sitting opposite the police chief while toasts were proposed. The chief and several of the councilmen nervously munched their food, stealthily observing that not *all* of the crew of the *Greenpeace* wore white lab coats, nor were they exactly a model of Quaker-like passive resistance. Nevertheless, the mayor, police chief, and councillors had taken the daring step of committing themselves to a policy of endorsing opposition to Amchitka, and they were stuck with the allies they had acquired. With good grace, they still went through with their intention of presenting us with an Alaskan flag to fly from the rigging, made several speeches in support of our action, and were thankful when Bohlen made a rational response that did not immediately call for the smashing of the state. We thrust all our scientists forward—Bohlen, Moore, and Simmons—while the rest of us retreated from the banquet as fast as possible, tracking down the Coast Guard freaks who had presented us with the pipe at the dock.

The crew of the *Confidence*, we learned, had indeed all been slapped with fines, and a couple of junior officers who had signed the message of support had been demoted. But all the other Coast Guard "volunteers"— who were doing nothing more than trying to avoid being drafted and sent to Vietnam—had refused to take over the positions of the demoted junior officers, with the result that they were reinstated. As for the fines—one hundred dollars each—a group called the Alaskan Mothers' Campaign Against Cannikin had raised the money to pay, and so the crew had escaped the consequences of the mutiny without a scratch.

There were several public meetings while we were there that were attended by Alaskan legislators and even a woman senator. Again and again, Bohlen got up at these meetings and started tirelessly into his "alphabet" of crimes against the environment, starting with "A" for the arms race, "B" for bombs in general, "C" for Cannikin specifically, "D" for doomsday, "E" for ecology, and on and on and on. Across town, several of us wound up at the home of a Gestalt therapist whose wife was busy throwing the *I Ching* for us, getting fantastic readings that could be interpreted to mean that somehow or other everything we were doing was exactly correct. The host, whose name was Tony Stikel, had written a Gestaltian truth, which he had tacked up on the wall, and which seemed appropriate to our state of mind at the time:

> Whatever has been, has been
> Whatever must be, shall be
> Whatever can be, may be
> Whatever I was, I was
> Whatever I must be, I will be
> Whatever I can be, I may be
> But whatever I am—I am

A couple of days later, we were back on the Gulf of Alaska, swinging two hundred miles out from land toward Juneau, where the governor of the state was reported to be waiting to greet us as a gesture of solidarity with the anti-Amchitka forces—potentially a politically important move, since it would put more pressure on Richard Nixon to cancel the tests.

Gray pastes of clouds, which had been on the horizon ahead the day we left Kodiak, moving swiftly away from us, were now backing up and bearing down on the boat like dark fleets of bruises, their fringes rimmed with throbbing flares.

"See them clouds up ahead?" Cormack had said when they were still running away over the horizon. "That's the storm that's just passed. It's moving away maybe seventy, maybe eighty miles an hour, and we're just poking along at nine knots behind 'er. Wal, long as she keeps moving *that* way we're okay. Nutthin' coming up behind us, far as we know. Only thing we have to worry about is if them clouds up ahead turns around and starts headin' fer us. We're pretty far out from the beach, maybe two hundred miles to Cape St. Elias ... over *that* way."

"Getting a little rough, ain't she, skipper?" Rod Marining asked, tape recorder dangling around the chest of his sword-fighter's blouse, woolen cap perched like a green pumpkin on top of his curly lion's mane of hair, a mane that had completely hidden away his ears and the back of his neck and most of his face. He loved to taunt Cormack—gently, of course, quivering nicely on the fine line between friendly jibes and insults. Marining bugged the skipper something awful. When he had first come aboard and sat down in the galley with his headphones in place, Cormack had come in, taken one look at him, stopped, and snorted:

"What's *that*? A *spaceman*?"

After a single night of Marining sharing the radio room with him in Fineberg's place, Cormack had ordered him out.

"Goddamn gooney bird! Sleeps there on the floor instead of the bunk. A man can't goddamn move without tripping all over that goddamn gooney bird lying there like a big stupid bug with them goddamn space things on his head."

If some of the rest of us had been a bit hard for Cormack to take, we were solid citizens compared to the Yippie. His one startlingly dark brown eye and one seashell-glittering blue eye did not help him in the least, because, talking to him, you would inevitably come up against the feeling that—depending on which eye you were noticing—you were dealing with two completely different people. Marining's style was to defend himself against people like Cormack by simply staring back at them intensely as long as it took for them to get nervous and give up.

"Getting a little rough, ain't she, skipper?" he asked again. Cormack ignored him, and kept his own eyes fixed on the sky. He was jittery—no doubt about it. He kept rubbing his chin and muttering to himself. I felt tempted to ask him: "Is *this* nothing, John?" But the old boy didn't look in much of a mood to put up with wisecracks.

Down in the galley, Moore was saying: "John's really uptight. We must be in for it. Oh, Jesus." Metcalfe had stirred from his bunk and was staring out through the open galley door at the lime-green water moving like floods from broken dams across the poop deck. Thurston muttered, "That figures. If there was ever going to be any time when we'd be killed for sure, it *would* happen on the way home. We won't even get to be heroes."

Now that we were returning, we had all been in such a rush to finish the trip off we had pressured Cormack mercilessly to speed things up. Maybe by that time he was tired and fed up, too. Whatever the case, he decided to take a little chance.

By rights, now that the Gulf was being swept by the first of the winter storms, we should have avoided going out into the deep. We should have followed along the shore, taking advantage of the islands that fishermen normally used for shelter. But, in our rush to get home, we had told Cormack over and over again that our karma was terrific, there was nothing to worry about. We were charmed. Hadn't the weather on the way across been a sign? For Cormack, terms like *karma*, once translated, were familiar. It was just a variation on the great theme of luck—and, yeh, our luck had been pretty good, weatherwise. Maybe we *were* charmed.

So now, counting on that luck, we were two hundred miles south of Cape St. Elias, in three thousand fathoms of water, midway between Kodiak Island and Juneau, no shelter at all, and the storm that had swept the Gulf ahead was backing up and rolling over us. We were starting to take green waves over the bow, and the water was beginning to "smoke"— meaning that the wind was turning the tops of the waves into a mist.

Up in the wheelhouse we had to hang on to the wheel with both hands. So much water had splashed in through the port doorway—which kept banging open—that even with boots on we would be slipping and sliding around. Sometimes as you clung with all your strength to the wheel, you found your feet planted briefly on the port wall—the boat was going over *that* far. Nobody except Cormack could hold the wheel for much more than half an hour at a time. There was nothing languid or slow motion about the movement of the *Phyllis Cormack* now. We came down the sides of the waves like a one-hundred-ton surfboard. The bow was smashing the upcoming wall like a battering ram, and with each crash the whole boat shuddered.

"We're starting to torpedo," Bohlen announced. Cormack himself wasn't joking any longer. We were finally in trouble. The wind was coming across the hunching shoulders of the waves in sandblast bursts up to eighty miles an hour. One moment the windows in front of the wheel were translucent ectoplasms of water, the next, a gray nothingness as we tunneled through a whole mountain of sea. After a while, Cormack took the wheel himself and refused to give it up—slapping his great belly against the spokes to hold the wheel steady while his tree roots of arms went grappling, spinning the wheel this way, and that, and *more* that way. Grunt, heave, *wham*!

"Gotta watch fer the freak ones," he said between grunts. "They're the ones come up like a pyramid—like that one over there—all green. ... One of them comes up under the bow a certain way and she can flip yuh right over. Hup! Uh!" The impact of the waves was like massive boot kicks against the hull.

"It's a cat-and-mouse game now," said Cormack as he scanned for freak ones, banging the port door open to stick his head out into the whistling keening wind to check to see if there's been any change in direction yet. First, he tried ploughing straight into the waves, but we were starting to torpedo too often and too deep. Then he hauled her over to the left and tried taking them at an angle, but we nearly got caught in a couple of those narrow green pyramids. One threw itself toward the sky like an explosion just off the bow, rattling the whole boat. We could feel a sudden amazing change in the way she was moving. For just a few seconds, we seemed to be skating on ice, an out-of-control kind of skating, then we were tumbling back into normal water. Outside, it looked like we had arrived on yet another planet—Jupiter, probably. The wind was a baying moaning thing. The boat moved like a chess piece on a table in the process of being overturned.

Bohlen was breathing like a man in an icy needle shower, going "ah, ah, ah" from the ecstasy of it. Terry Simmons had dug in like a bulldog, and even though he kept slipping and falling, he grimly took his turns at the wheel during the breaks when Cormack went below to check the engine and pump the bilges. Thurston played Beethoven full blast in the galley. Darnell and Keziere retreated mainly to their bunks. Birmingham hurried back and forth between the engine room and the wheelhouse, seeming quite cheerful now that we were into the thick of a real sea situation. While the politicking and dickering had infuriated him, he now seemed calm—to the point of an almost catlike relaxation. Patrick Moore buried his nose as much as possible in *The Lord of the Rings*.

We all kept thinking of that damned two-by-four wedged between the old Atlas engine and the hull, which seemed to be all that stopped the machine from toppling over. The bilge-tank pump broke down and the engine room started to flood. Cormack was down there in a flash, cursing and working angrily to fix it. The engine spluttered and coughed, making exactly the kinds of noises it had uttered before conking out completely more than a month ago coming across the Gulf. It hadn't mattered much then, since we just bobbed around in the water like an apple until it was fixed, but *now*. We held our breath until our chests ached. Darnell wanted to know if everyone had said their morning prayer to the Engine God.

Then Cormack was back at the wheel, and the engine was chuttering and clunking away as usual.

"How do we watch for the freak ones when it gets dark, John?" I wanted to know. He shrugged.

"It's a checker game now." He talked mainly to himself, rubbing his chin, "Now how the hell do yuh win?" He was as agile as a cat. A gray-and-green tuque was jammed on his head, and he wore a torn purple-and-gray plaid shirt, the sleeves completely shredded, and big, open, leather boots; white whiskers prickled out all over his jowls; he concentrated like a grand-piano player at a concert, steak-sized hands wrapped around the spokes, poised like a surfer, legs bent and braced. The years had sloughed away. The floppy old man was gone, and here was a muscular barrel-chested Hell's Angel of a captain brawling his way across the screaming barroom sea.

Near the end of the day, October 20, after thirty-five hours of the boat chopping away like an axe bit at the waves, everyone was battered and bruised from being thrown against walls and cupboards and ladders. There was nothing left of the contempt we had felt for the *Phyllis Cormack* when we first saw her tied up in the Fraser River, looking old and broken and beat. Now she was our goddess. She moved with a beauty and grace that made us want to reach out and touch her cedar walls as you would touch a holy object.

That night the ocean smashed at us with a power that made our teeth rattle. Cormack retreated to his bunk, where he lay fitfully while Bohlen, Simmons, and Moore fought the wheel. At fifteen-minute and half-hour intervals, the skipper would shout out fresh instructions:

"Feels like she's running stronger this way now. Take her over a little more to the left. Keep 'er there steady, 'less yuh can't keep the wheel up, then let 'er fall a bit the way she wants to go, and try to ease to the right fer a while. Sometime yuh kin kind of zigzag yer way through."

By morning there were only five of us still able to get out of our bunks. Cormack gave the order to swing her around. Bohlen and I tackled the wheel together, using all of our combined strength to hang on as the boat bucked and threw herself over so far that the water was running directly below the port door. If the door were to pop open now we would plunge through into the ocean. Long incredible seconds followed where we were being shaken like wet cloth dolls by the convulsion. The old boat bounced like a stand-up punching bag weighted at the bottom, and almost went over—*almost*. Once we had steadied the wheel again, we were running with the sea. An entirely new kind of motion had taken us over.

A following sea, unlike a sea that is tackling you head on, is one that lifts you like a leaf and lets you waft this way and that, always rising lazily, then slowly settling you down like an infant into the bathtub. Ahead of us, a path appeared, like an abandoned trail across rolling foothills. All we had to do was keep the bow pointed into the trail, while the wind swept by laying long streamers across the waves. The "trail" was the absence of streamers directly in front of us, a kind of wind wake in reverse, and we were driving back into it. It was a gentle angel-cradling motion, a feather movement, a love act.

"Gotta watch this kind of thing," said Cormack, spoiling the mood because he looked so less tense than he had when we were ploughing head on.

"What's the danger now, John?"

"Wal, yuh git a following sea like this, and y'don't haveta *work* so hard, if you know what I mean, but yuh kin git caught nappin'. Them waves're still out there, even if yuh can't feel 'em so much on account of we're not joggin' into 'em anymore. But the stern ain't built like the bow y'know, and yuh don't have't'have many of 'em break over the stern 'fore she starts t'break up."

So we handled the wheel like a great butterfly, holding her oh so gently, just trying to let the bow flow the way she wanted into the wind wake, making only the tiniest adjustments when the stern swung out too much. We could feel the tons and tons of water pulled by planetary forces, opening and closing like a vast dark lens at our backs.

For a whole day we drifted through the angel dance of the following sea. Rod Marining chortled over the idea that maybe fate had decided to take a hand in things and was driving us, like it or not, back toward Amchitka.

But in time, the storm swept past us. "Hard around!" barked Cormack. We churned wildly in the troughs again, then the *Phyllis Cormack* straightened like a wing and started whacking her way back through the surges of sea. "Got 'er licked now," said the skipper.

Into Juneau we came chugging, late in October, to glaciers on the dark mountains, moonlight on ice, and the lights of the city emerging like campfires beneath the arctic cliff faces. The governor of Alaska would not be coming down to greet us after all. He sent a message saying: "I'm glad you made it back into these waters safely. I admire your courage in carrying out your convictions and I think you have proven your point." But he had gotten word from Kodiak that we were a "motley crew," and had had second thoughts about being seen in public with us. Instead, there was to be a welcoming committee of ordinary citizens, and a town meeting was planned. Cormack had shaved and decked himself out in his fedora and Sunday woolen jacket, with gray slacks and polished black shoes. Icy rain was pouring down.

The problem was where to tie up. We had not been given specific instructions. Cormack finally picked out a cannery wharf at random. We pulled in; Darnell leapt up onto the dock and neatly fastened the line. Moore got another one out from the stern, and we had a moment of feeling quite pleased with ourselves. Some of our previous landings had been near-disasters, since few of us knew anything about ropes. Finally we were getting the hang of it.

Three men wandered over, looking at us quizzically. One of them asked: "What kinda fish ya bringing in? We're closed today." Darnell had started explaining that there was supposed to be a welcoming committee, when a couple of young men appeared: "Hey, *Greenpeace*, you're at the wrong dock! Go back up the bay and under the bridge and turn right at the first pier." Darnell started to untie the rope, but by then the boat had drifted out too far for him to get back.

"I'll go with *those* people," Darnell yelled, throwing loose the rope and galloping away. But the well-wishers had vanished. Darnell came running madly back toward the boat. Cormack tried to get in reverse to slip back toward the dock. The boat crunched against the pilings, then bounced out again, leaving Darnell dangling from a ladder, reaching out for my hand but not quite making it. The boat wallowed around. Darnell got back up on the dock, backed up twenty paces while Cormack shouted obscenities at him, then came charging back toward us while everybody yelled: "No! No! No! Don't try it!" Off the end of the wharf he came flying ... flying ... looking as though he were going to fall short and splash into the water. But instead, he crashed with an awful *thunk* on the poop deck. Cormack slammed into forward gear, and away we steamed.

Coming under the bridge we discovered it was high tide. The masthead scraped the underbelly. Cormack failed to spot the pier where a small rain-soaked crowd of people was waving madly. We sailed on past, then banked slowly around, taking fifteen minutes to angle back toward them. Cormack finally nosed in beside a crab boat. By this time there were only a dozen completely soggy people left to welcome us. Three hours earlier—when we were supposed to arrive—the crowd had numbered several hundred, but most of them had gone home.

Off we went to another meeting of concerned citizens and Alaskan Mothers Against Cannikin. Bohlen went once more through his patient speech about the ABCs of environmental ruin. There were two hundred or three hundred people present, many of them fantastically beautiful young women who seemed to want to rub their bodies up against the crew of the protest ship.

"Oh, boy," said Cummings, "*Greenpeace* groupies."

The last thing I remember was several of us arriving at a dance pavilion called DREAMLAND. The name was emblazoned in immense neon lights, and the ceiling was aglow with artificial stars. Except for the plastic palm trees, it was precisely like being inside an interstellar spaceship. After a few drinks it was easy to accept as reality the impression that the women were angels and that we were dancing in a huge neon pinball waiting room just outside the gates of heaven. The music completely engulfed us. A light show came on. DREAMLAND! Oh, it was, it was lovely.

Then it was late morning again and one by one we crawled back from DREAMLAND into the grim crystal reality of the *Phyllis Cormack* still chugging past glaciers and frosted rock. It seemed we would never get home.

We were halfway between Juneau and Ketchikan, Alaska, when the whales appeared.

The temperature that day was below freezing, but the sky was cloudless. The sunlight glinted on every ice crystal from the tops of the mountains down to the pebbles on the beaches. Everything had an aura of exaggerated realism, as though creation had come close to overdoing itself in richness.

Five of us were standing around with our hands in our pockets near the bow, scarves around our necks and caps pulled down over our ears. Most of our clothing by this time was almost as grease-covered as Cormack's. Everything we wore was rumpled. Our faces were mostly covered with ratty beards. Now that the trip was almost over, several of us were coming down with colds. We hacked and coughed weakly, like pensioners. We were still stiff from the battering during the storm. We moved as little as possible.

The five out at the bow that afternoon were the ones who happened to have independently all listened to Roger Payne's famous recorded album, *Songs of the Humpback Whale*. The sounds had truly haunted. They gave you the same feeling you'd have if there was a ghost on the other side of a darkened room. It had been impossible for any of us to totally shake off the impression that we had, in fact, been listening to the voices of huge *entities*. Several of us had noted that our wives, when they listened to the record, had cried, and so had many female friends. At least two members of the crew had taken LSD while listening on headphones to those majesterial tones, coming as though from across some astral plain or through a crack into another dimension.

Now, up ahead, a great gray-black shape was moving down the channel toward us.

We were electrified.

"No!" hissed Patrick Moore. "There's more! Three ... four! They're *sperms* ... No! *Finbacks*! No ... I dunno. Wow! Whales!"

Someone thought to ask Cormack: "How close will they come, John?"

"Huh. They're not dummies, ya know. They go ta school. Hell, they won't come any closer than a couple of thousand feet."

On the bow, Keziere said: "Well, let's call them in."

Immediately we began to fill our heads with the most loving thoughts and images we could imagine, consciously attempting to "beam" the collective feeling outward toward the still-distant dark shapes. The *Phyllis Cormack* was strangely quiet, with half the loonies up at the bow, their eyes shut, and the skipper shaking his head on the upper deck. We glided steadily forward. The whales continued to advance. It was not until they were pulling almost abeam that one changed course toward us, its back coming up out of the water like a sleek giant snake. It was easy to see how ancient mariners had gotten the impression of great serpents in the sea. As the whale drew nearer, our excitement intensified. It was impossible to avoid the feeling that this one whale, at any rate, was responding.

"It's our vibes," said Thurston. "We've got good vibes."

The whale came within roughly two hundred feet, enough to leave Cormack impressed and the rest of us ecstatic. Then it dived under the boat, came up another one hundred feet off the starboard stern, dived again, and resurfaced far down the channel alongside the other whales, which had chosen to ignore us.

"Wal, they're not *all* extinct," muttered the skipper.

Down along the Alaska panhandle we went to the town of Ketchikan. Another anti-Amchitka demonstration. More public meetings. More speeches. Then on to Prince Rupert, touching at last back on Canadian soil. Another wild round of meetings and speeches—the whole affair had turned into something more resembling a seagoing election campaign than a protest. There were interviews with the press and radio and television. The crew was by this time punchy and exhausted. At each new port, another wave of people surged aboard, their energy crashing against our heads.

In Prince Rupert, there was ice on the dock and snow on the beaches.

It was here that Cormack got even with me for a crime I hadn't even been aware of having committed. Apparently I had gone out on the upper deck one night to take a leak, not knowing that the skipper was standing in the darkness below. He hadn't said anything at the time. Now, early in the morning, I was climbing onto the dock on my way to town to do a radio interview. As I picked my way carefully over the slippery planks, Cormack appeared on the bridge and called me. I stopped beside the boat, directly opposite the hole through which the bilge was pumped.

"What is it, John?"

Cormack hit the bilge button. A huge stream of brackish water abruptly splurted out of the side of the boat, covering me from head to toe. I was left standing there, gasping and steaming, while Cormack howled with laughter. My interview had to be delayed while I changed clothes and washed all the goop out of my hair.

It was not until we arrived at Prince Rupert that we realized our decision to turn around had not slowed the momentum of the anti-Amchitka movement in the least. Canadian newspapers were full of headlines about the test. Environmentalists in the U.S. were launching bigger legal attacks than ever. In eastern Canada, authors Pierre Berton and Charles Templeton had organized a petition with 188,000 signatures on it—the world's longest telegram—and had delivered it to the White House. Demonstrations had been mounting in intensity right across the country. U.S. consulate's offices were under siege from Vancouver to Fredericton. The border had been blockaded again, and there had been threats of bombings of American-owned companies. If our goal had been to arouse Canadians against the bomb, we had succeeded. At each B.C. port we visited, the level of excitement increased. Rather than the struggle having come to an end, it seemed instead to have only begun.

When we reached Alert Bay, we were met by forty Kwakiutl Indians and taken to their longhouse on a hill overlooking the bay. The world's tallest totem pole still lay in the grass outside the building, not yet finished. The earlier promise to carve our names on it was quietly shelved. Instead, a ceremony had been arranged. Inside the longhouse, a cedarwood fire was flaring. Great Kwakiutl totems held a massive roof on their heads. Women danced in blood-red beaded robes. Old men with hands like roots pulled from wet clay were beating on a huge wooden drum. The Greenhawks were ordered to stand in a row.

We felt like pygmies beneath the overhanging Thunderbird beaks of the totems, their wings fanned outright as in the moment of the leap into the sky. We did not know quite what to do. Each of us stood in a different posture. Bohlen had his head bowed. Simmons stood as though waiting his turn at a water fountain. Moore looked like a kid standing respectfully before his grandfather. Cummings, his hair slicked back like a 1950s greaser, somehow managed to look very innocent. Metcalfe had his feet planted apart and his arms across his chest. Marining slouched like a shaggy prophet. I had my hands crossed and my hair combed like an altar boy. Doc Thurston looked like a wise old mandarin who understood exactly what was happening. Darnell had his arms at his side, a bit like a boy scout. Birmingham was standing almost at attention, but with an air of expecting a lecture from a teacher. Old Cormack hung back until the end, then shyly came forward, wearing the grumpiest expression possible, while Keziere took pictures of the historic occasion, then joined the lineup with the expression on his face of a lab technician who cannot quite believe what he has just found under the microscope.

The Kwakiutl ritual was normally reserved only for weddings, funerals, and the election of chiefs. We were asked to remove our hats. We were anointed with water and eagle down feathers and made into brothers of the Kwakiutl people.

Vividly, I recalled the prophecy of the old Cree grandmother, Eyes of Fire, who had said a time would come when the Indians would teach the White Man how to have reverence for the earth, that we would all go forth as Warriors of the Rainbow ... Until a decade before, the Kwakiutl religion had been outlawed by the Canadian government. It had had to go underground, so few white men had witnessed these rituals, let alone taken part in them. Now her prophecy began to come true.

Three dances were performed. The first urged men to let go of their egos, and that had a very special meaning for us. It was as though the Kwakiutl were able to read our minds or had somehow succeeded in understanding perfectly the experiences we had just been through. The second dance symbolized a voyage at sea. The third was a dance of peace, and we were invited to join in. The brilliant beaded robes were slipped over our shoulders. The drums pounded. We danced— awkwardly at first—through the smoke around the fire, while the vacant eyes of the totems watched. The eagle down feathers that had been sprinkled on our heads were allowed to drift through the air, touching everyone present.

Afterward, we wandered about town, shaking hands with people, being taken into various homes for drinks. Thurston had decided to wear the Alaskan flag given to us by the police chief in Kodiak around his shoulders, and I donned an American flag. We strode about like comic-book caped crusaders. It quickly became evident that while it was one thing to get one's land legs back, it was something else to regain one's "land mind." Our language was still so in-group that others we met could offer only blank stares to comments that set us off laughing wildly. "Yer acting like a bunch of loonies," Cormack warned.

We were on the last leg of the retreat to Vancouver, heading down Johnstone Strait, south of Alert Bay, our forty-second day on the *Greenpeace*, when word came through that Nixon had finally made himself perfectly clear. It had been announced from the White House that Cannikin would be detonated on November 4.

It was now October 27. If we had stayed in Sand Point, as Fineberg, Simmons, Darnell, and myself had wanted, we would now be in a position to run for Amchitka. Granted, the weather would probably be as bad as we had experienced it between Kodiak and Juneau, but at least the option would still exist. As it was, we were two thousand miles from Amchitka, without a hope of getting back up there in time.

The mood on the boat snapped immediately back to the same bitter level it had reached at its lowest point, when the fight was progressing over whether or not to turn around. There were few words exchanged as we crowded around the radio room, listening to Metcalfe's wife reading out the new report. But the tension was palpable. There was a new layer of numbness laid down on top of all the other layers of numbness. It was one last, hard smack in the face.

But now, a final grand piano was about to come flying through the window. Having told us the date of the test, Dorothy Metcalfe was now reading a statement advising us that the Don't Make a Wave Committee was readying a converted minesweeper, the *Edgewater Fortune*, which would leave Vancouver within a day to attempt to reach Amchitka by November 4. Some four hundred volunteers from all over western Canada had already called up to apply to be on the boat. Of course, preference would be given to any crew members on the *Greenpeace* who wished to transfer to what would be called the *Greenpeace Too*.

"*What?*" screamed Thurston. "A *minesweeper*?"

There was a moment of stunned surprise. Then, oddly enough, snorts of disgust filled the radio room. Our immediate reaction was anger, even loathing. We were still so disappointed with ourselves that the idea of anyone else setting out to undertake such a trip provoked nothing but a sense of dismay. All we really wanted at this stage was to jump off the boat and escape from one another. Now—and here was the real source of the dismay—we would each have to go through the turmoil of deciding whether or not to transfer to the new vessel.

"Argh," said Keziere. "I can't stand it."

There was a moment when Bohlen and Metcalfe and Darnell and I stood facing each other in the galley, the tension of the old ego conflicts rising like a sunburst. The unspoken message was: *If you go, I'll have to go.* It broke the moment we all grabbed hands and Bohlen shouted: *"How far?"*

And we shouted back: "ALL THE WAY HOME!"

The announcement of the launching of *Greenpeace Too* had not been quite so spontaneous as most of us thought. As far back as Sand Point, Jim Bohlen had been making arrangements by phone with Vancouver for "second-strike capability" with a fast ship. He had only confided in Metcalfe. The rest of us were left in the dark. Behind the scenes at home, Irving Stowe had been frantically searching for a bigger, faster ship, and he had finally tracked down Hank Johansen, owner and skipper of the *Edgewater Fortune*. Two years of searching in the first place had produced no other captain than John Cormack, but since the launching of the *Greenpeace*, the public at large had been bombarded with images of a courageous crew on its way to stop the bomb. It was a new ball game. Hundreds of people wanted to be on board. Money was pouring in. Television stations were clamoring to send camera crews along. Everybody wanted a piece of the action. Only four of our crew opted to take up bunks on the *Greenpeace Too*. Terry Simmons said simply: "I intended to go to Amchitka all along. I still intend to go." Rod Marining never hesitated—the Yippie's whole voyage so far had been in the wrong direction, and now he could get back on course. Birmingham received a special appeal from Vancouver because the new ship badly needed an extra engineer. "Well, shucks," he said, "if I'm needed, I'm needed. That danged hippie kid of mine would never forgive me if I didn't go."

The fourth person to decide to transfer was Bob Cummings.

The Kwakiutl had spoken of men "letting go of their egos." But it was easier said than done. Cummings's decision to transfer stung me. Maybe I was giving up too easily. For the first time, I got on the radio and begged my wife to tell me what to do. But she had been standing out in the rain every night in front of the U.S. consulate general's office, and her commitment to stopping the bomb went back all the way to the Aldermaston marches in England. She had resisted fierce lobbying from some of the other women to try to talk me into agreeing to turn around, back when the main struggle was happening at Sand Point, and she would not betray her effort or mine by making the decision for me now. "Whatever you decide to do," she said, "I'm with you."

But the thought of climbing on a boat with dozens of fresh people and having to go through the debates and decisions all over again was more than enough to offset any lingering feeling of competitiveness I might have had with Cummings. Let go, let go ... .

A day later, the *Edgewater Fortune* was coming up on us fast in the gray dawn.

It was as though the Coast Guard cutter *Confidence* had come back to haunt us. Here was a military ship bearing down on us in wet, cold gloom. If we seemed to have arrived in the midst of a new movie, its theme was more warlike than peaceful.

"*Look* at that big mother," whispered Keziere.

"Oh, lovely, lovely," said Bohlen.

"Isn't that nice," grinned Moore.

"Is *that* on *our* side?" asked Cummings.

"This is certainly something, ain't it now?" said Birmingham.

We were standing out on deck, cheering—the cheers were amazingly hoarse—and there were dozens of strangers on the minesweeper cheering back as the huge ship glided effortlessly toward us, looming larger and larger until new men were looking down at us as though from the roof of a building. The ship was at least 130 feet long, white painted with black trim, with a vast array of antennae lifting from its superstructure. It was a tremendous powerhouse compared to the funky old *Phyllis Cormack*. And its crew was a crowd—with a few familiar faces from early Don't Make a Wave Committee meetings, such as red-haired Chris Bergthorson, who had designed the Greenpeace symbol, and young Paul Watson, the muscular seaman who had been rejected from the original crew because of his support of the Red Power movement, the Black Panthers, and the North Vietnamese. Like Rod Marining, he had finally found a way to get on a boat going to Amchitka.

There were more cameras on board the *Edgewater Fortune* than you could count at a glance. As the two boats closed together in the middle of the Strait of Georgia, a thrill ran through us all, a feeling that the war veterans—Bohlen, Metcalfe, and Birmingham—had probably known, but which was new to the rest of us. It was expressed perfectly in a ceremony when Bohlen hauled down the tattered *Greenpeace* flag and handed it over to the *Greenpeace Too*'s Captain Johansen. Several of his crewmen rushed to run it up on their own masthead. Everyone cheered. The adrenaline was pumping. Revolutionary handshakes and embraces. It was as though Canada had finally declared war on the United States and had been joined by rebel Americans fighting the tyrant who had taken over their country. It was breathtaking how close to the edge of such a reality we seemed to be.

We were moved and shaken at that moment by the same great force of patriotism that has always driven men, but in this case—perhaps for the first time—our loyalty lay not with a country, a religion, or a language, but with the planet itself. The ranks of the Warriors of the Rainbow had swelled. And now we had a warship. The Kwakiutl would be pleased.

Climbing on board that immense supercraft—which is how the minesweeper seemed after six weeks on the little *Phyllis Cormack*—was a bit like visiting a spaceship from a vastly more technologically advanced world. Our old halibut seiner could have been picked up and stowed in the *Edgewater Fortune*'s hold without slowing the larger craft by so much as a knot. But there was little time for anything more than hurried handshakes, hurried interviews, and much wild laughter. Everybody had a sense of urgency. This new ship had only a week in which to make it twenty-four hundred miles across the gale-haunted Gulf to Amchitka. We threw them our remaining posters, whatever foul-weather gear we could find, and, of course, our Geiger counter. There were twenty-eight men—again, no women—on the *Fortune*. Rod Marining climbed away, headphones slung around his neck, carrying his little white suitcase containing tapes and batteries and games like Risk and chess. And, then, before we quite realized what was happening, the *Edgewater Fortune* was sloughing away in the drizzle, and we were heading the other way.

Coming into Vancouver at 7:30 P.M., October 30, car horns were honking from the bridge.

It was low tide. There was some doubt whether we would be able to make it through the final channel into False Creek. We all stood out on deck and tried to levitate the boat. Slowly, slowly, Cormack pushed through the channel, one eye cocked on the depth sounder, saying:

"Got about one fathom clear on the bottom. ... Nope, now she's down to nutthin', we're touchin' bottom." Levitate! LEVITATE! Magically, the boat swilled her way through loose mud. "First goddamn time I ever got through *there* when the tide's right out," commented the skipper, rubbing his chin again.

And suddenly we were bearing down on a dock jammed with several hundred people, including all our relatives and friends. The boat seemed to be whirling forward, and going through all our minds was the single thought: *Are we going to crash into the dock and kill everyone?* Snow was falling. Moore threw a line, as the dock with all those people on it waving and yelling came streaming by like a jammed streetcar waiting station, and the boat was moving *awfully damned fast*. Cormack was roaring at someone on the dock to *"git that goddamn rope around something!"* The rope caught and whanged and started to split. With a torturous sideways heave, the *Phyllis Cormack* crunched heavily to a stop, less than a dozen threads of string having made the difference between stopping and sailing on past the dock to crash into a concrete sea wall. HOME!

With the launching of *Greenpeace Too*, the Amchitka protest had taken a quantum leap. Some explosive mix of images had begun to clash in the media. Now the apocalypse had form. The largest crowds in the histories of Toronto, Vancouver, Winnipeg, and Calgary turned out to stalk the streets in protest. Hardhats marched with hippies beneath the offices of American consulates. The Rainbow Bridge at Niagara Falls was plugged with thousands of fist-waving students. Down a street in Toronto marched a long phalanx of middle-aged men in stockbrokers' overcoats and bowler hats, like a whole trainload of little windup toys, bearing a sign that said: STOP AMCHITKA. A group of Canadians, including former prime minister Lester Pearson and Chief Dan George took out a full-page ad in *The Washington Post*—an open letter to the American people—urging cancellation of the test. Thousands of pulp workers downed tools in British Columbia. Fishermen swung their boats out into the Strait of Georgia, tooting their horns in protest. Carpenters and plumbers were on the march. Members of Parliament, mayors, senators, aldermen, clergy—all were together in the attack. Briefly and magnificently, the transpolitical environmental alliance came into view.

Amchitka, which had begun as a West Coast story grown to the level of a Canadian story, had now established itself as an international story. Demonstrations grew larger in Tokyo. British newspapers carried articles on the "threat to Canada," which had come close to making a shambles of traditional cozy American–Canadian relations.

The *Greenpeace Too* had mounted a stage far larger than any from which the crew of the *Greenpeace* had played. The new ship's progress was the subject of nightly national TV broadcasts and daily front-page reports.

But the North Pacific was still a tunnel of raving sea witches and "freak ones," moans out of the void, smoking waves, sleet and axe blows of green. The weather by this time was far worse than the storm that had caught the *Phyllis Cormack*. And Captain Johansen, who had never taken his ship out into the open sea before, was suddenly learning a lot about both the ocean and his vessel.

He had only recently set himself up with a partner in the charter-boat business. A veteran, he had been affected by the anti-Amchitka mood running among even his most Right-wing colleagues. Johansen was forty-one. He had never been involved in any kind of protest before. And while the exposure that his ship was going to get could hardly do his business any harm, there had been more in his mind than a straightforward charter agreement. Like any man involved with boats, he had a taste for adventure. He was a tall, strongly built man, not much of a talker. He was a member of the New Westminster Chamber of Commerce.

The *Edgewater Fortune* had been one of several "Bay"-class minesweepers built in the early fifties by the Royal Canadian Navy, then decommissioned hardly more than a decade later for the simple reason that Canada could see no more mines to sweep. She was 154 feet long and the equivalent of about four stories high. To avoid being blown up as they probed ahead of the fleets for mines, the minesweepers had been built with wooden hulls and aluminum superstructures. They were beautiful ships to look at. They had fierce rapier-like lines. They could easily do twenty-two knots. They were as maneuverable as well-bred racehorses. Unfortunately, they were narrow-beamed and tended to roll dramatically in any kind of weather. From the flying bridge, it seemed as though you were standing on the roof of a small apartment block, yet the vessel drew only eight feet of water and she had a flat, planing hull. From the moment the *Fortune* cleared the head of Vancouver Island and began pounding out into the open water, there was no one on board who could be certain for even a minute that the whole ship wasn't about to topple like a domino.

Captain Johansen had picked his crew the morning of the day the boat was to depart. Some fifty volunteers had arrived with their duffel bags, and he had let them all work for several hours before picking out those who seemed to have been working hardest and most knowingly. But as it was, when he left Vancouver—apart from a six-man CBC-television camera crew, a *Time* magazine photographer, a reporter from the Vancouver *Sun*, and two radio broadcasters—his actual "crew" was small. Out of twenty-eight men, only nine could reasonably be called experienced seamen, unless you wanted to include Marining, Simmons, and Cummings from the *Phyllis Cormack*. A few others had been out sailing before, but the experience of a minsweeper was something new to everybody. Even the captain himself had so far only taken his boat along the Inside Passage and had not been quite prepared for the ferocity of the elements that confronted him now.

By the time the morning of the first day in the open sea dawned, the *Edgewater Fortune*'s huge galley was empty except for three bodies curled up in sleeping bags in the corners and under the furnishings. Rod Marining's first impression as he crawled out of his stateroom was that everyone had abandoned ship. After the *Phyllis Cormack*, with its single, cramped little bunk room, this new vessel seemed huge and cave-like and strangely uninhabited. The fact that he had a whole stateroom, which he only had to share with one other man, was itself enough to give him the feeling that he had entered a brand-new dimension. The other crewmen were either up in the wheelhouse, down in the engine room, or, mostly, clinging to their bunks. There were some men he didn't see for days at a time. Marining had been assigned to the job of dishwasher. But from the time the *Edgewater Fortune* cleared Vancouver Island, the cooks were all knocked out with seasickness, no meals were being prepared, and so there were no dishes to wash. The Yippie was free to prowl the ship at will. And while on the *Phyllis Cormack* he had been the junior crewman, now he was a veteran of the first trip, with a certain amount of status as a result. At least he had been to sea before and was still functioning despite the pounding. That put him in a minority immediately.

Also, something peculiar had happened to his mind.

The only word he could think of that described what was happening to him was *telepathy*. The few lonely wanderers that he found staggering and sometimes crawling through the caverns of the ship seemed to him to be broadcasting their thoughts without having to say a word. He would hear words in his head and *know* that they were coming from the middle of the mind of the man in front of him. While much of it was jumbled, he could hear it as clearly as though it were coming over his own headphones. He found by experimenting that he could leave the headphones on, with music playing in his tape recorder, and still be able to make out what the people around him were thinking. He didn't bother trying to explain it to anyone at the time, because he sensed that they wouldn't understand. So, on top of all their other woes, the others had to deal with the apparition of a weird, shaggy Yippie with two different-colored eyes and a pair of headphones covering his ears, nodding and staring at them intently as if he understood, even though they *knew* he couldn't hear a word they were saying. Marining also spent a lot of time out on the deck and up on the flying bridge, watching in awe as waves that would have swallowed the *Phyllis Cormack* came sluicing like opened dam gates over the decks of the *Edgewater Fortune*, as the great swordlike bow wagged at the sky, and as the ship's bottom smacked like an ironing board against the water, sending up huge explosions of foam.

He felt completely unafraid. While most of the rest of the crew were curled up in panic-stricken fetal balls in their bunks, Marining calmly hopped along the decks, music and other people's thoughts playing through his head, listening to the pulsations of an even deeper force that to him was so obviously at work there could be nothing in the world to worry about.

Few of the others were doing as well. A young radio broadcaster from Prince George, hundreds of miles inland from the coast, had only been seen once since the departure from Vancouver. A few people had spotted him crawling along the corridor on his way to the head. Afterward, some of the others couldn't remember having seen him at all. Others remembered that he had, indeed, come aboard and that he had said something about his wife having just given birth to a baby, and about how he had quit his job and had driven all the way into Vancouver to "do something" to preserve his infant's future from the horrors of Amchitka. The ship's doctor, Joseph Stipec, said the man would "live," but that was about all. He was the worst seasickness case Stipec had ever seen.

The CBC camera crew was tougher. Its chief was a bulldog-like older journalist, Doug Collins, whose fame lay in the fact that he had escaped from German POW camps no less than seven times during the war and had written a book about it. Collins and Metcalfe were contemporaries, but while the one had moved for a time into the dubious realm of public relations, the other had tenaciously stuck with the old values of objectivity and straight reportage. Collins would cover whatever happened—would go to hell and back to get a story—but would not take sides. Amchitka was just another assignment. Whether the bomb went off or was stopped was not his concern. His bearing was still military. He trusted no one.

But even with Collins's leadership, the CBC men were having problems. They had been given a stateroom up near the bow—the worst possible place to be, because it was there that the pounding was most vicious and the sensation of flying in all directions through the air most pronounced. Rod Marining caught glimpses of them in their bunks; bottles of Scotch in their fists, dealing with the crisis in the classical objective-journalist fashion—by getting plastered as fast as possible. Marining's greatest joy of the voyage came the second day out when he stepped into the camera crew's quarters only to notice that their bodies were being left for a few seconds *floating in the air* as the bow dropped out from under them on its downward plunge. Marining could actually pass his arm under their bodies before they settled back into the depths of their bunks.

A few of the crew held their own even through the first hellish two days of battling the waves. Among them was Paul Watson, twenty-one, the merchant seaman who had wanted so badly to go on the *Phyllis Cormack*. He had served on Norwegian freighters around the world, and while a minesweeper was as new to him as everyone else, he at least did not have to worry about seasickness. The doctor appeared only briefly the first day out, gray-faced and apologetic—astonished that his profession had somehow failed to make him immune. Also, his supply of Dramamine had disappeared, because he was too sick to notice who had taken it.

Dr. Stipec had thought himself exceptionally well qualified for the voyage, partly because he was a ham operator and had brought a transmitter onto the ship to talk to any sympathetic ham who was listening. There were many. Once he managed to get in contact with an operator in Central Siberia, who said: "Best of luck." Stipec listed himself as a conservative, and remarked: "I can't be one of those guys who does abortions, but I think the profession has to get itself involved." He had had to make roughly five hundred phone calls to arrange care for his patients while he was gone. He was forty years old.

Johansen had chosen two men as his main assistants. One was Bill Smith, thirty-five, who listed his occupation as .beachcomber, which meant he had a license to pick up logs that were washed up on beaches and return them for a fee to logging companies. Until eighteen months before, he had been the manager of a company in Britain that sold microwave systems. "I'm retired now," he said. "I'm doing what I want to do." He had moved to Canada and purchased a forty-six-foot workboat, which was now his home. "I wouldn't carry a placard in a demonstration or anything like that. I don't know how to do that. But I do know about the sea, and since this bomb could hurt the sea, I'm against the bomb."

The man who did the most work to keep the boat running, however, was Will Jones, a white-haired mild-seeming man with wire-rim glasses who had not even heard about the voyage until an hour before the ship sailed. "Hot damn," he had said as he watched the television coverage, then got his sons to help him stuff his duffel bag, promise to phone their mother in San Francisco and to call the computer center at Simon Fraser University to tell them that Mr. Jones would not be in to work for a while. He made it to the dock just as the boat was ready to leave. Johansen had already picked his crew and was not willing to take another person—until Jones outlined his qualifications. He was a retired U.S. Navy lieutenant commander and a trained navigator. The *Edgewater Fortune* was a bit smaller than what he was used to, but, yes, he could navigate it to Amchitka. His last navigation job had been on the U.S. battleship *Iowa*. Once out of the navy, he had worked as the head of IBM's Pacific region, earning eighty-five thousand dollars a year. The previous summer he had quit, and like Jim Bohlen and Irving Stowe, had fled with his draft-age sons to Canada. Jones was also a Quaker. "I left the States," he said, "because I'm disgusted at the arrogance of the country. I chose Canada mainly, I think, because I doubt that this country will ever get big enough to push anyone around."

The most silent man on board—apart from the mostly forgotten seasick broadcaster from Prince George—was Jim Hunt, the machinist. A shy, tense person with a crew cut, he confided to reporter Doug Sagi that the reason he had decided to join the voyage in the first place was because his eight-year-old daughter Wendy had told him she didn't like the world. "She said she wanted to get off the world," said Hunt. "What the hell can you say to that?"

In addition to being a member of the New Westminster Chamber of Commerce, Captain Johansen was also a Rotarian, a Shriner, and senior ward in his Masonic lodge. Normally he chartered his boat for fishing and office parties at a rate of $550 a day. He had agreed to do the Amchitka trip for $12,000, about a third of the regular charter fee for such a trip. "I figured it would pay expenses, provided nothing went wrong and we could get a crew to sail without pay."

But now, two days out in the North Pacific, with wind armies holding the sea, many things were going wrong. Even though the radar dish perched atop the mast was some six stories above water level, it had been smashed by the fury of the waves. That impressed—or depressed— Johansen considerably. So did the news that the metal railings along the flying bridge had been bent, as though some superman had taken them in his hands. But the worst news came from the engine room. Johansen scrambled up and down the ladders from his quarters back of the wheelhouse to the engine room every hour, his eyes almost bugging out with horror. The V-12 General Motors diesels ran at 750 revolutions per minute. Each of the two immense machines delivered some twelve hundred horsepower. Running at sixteen knots, they burned roughly one hundred gallons of fuel per hour. That meant twenty-four hundred gallons per day. The ship carried twelve thousand imperial gallons, but at the rate the fuel was being spent, it did not take Captain Johansen long to figure out that this trip was soon going to be costing eight thousand to twelve thousand dollars worth of fuel alone, to say nothing of the damage that was already starting to occur. No one in the chamber of commerce needed to have the message spelled out for him: one hundred gallons of fuel per hour was a hell of a lot of fuel.

Worse, the boat was hardly moving. As the wind rose to a steady scream and the waves piled up higher and higher, the vast bulk of the *Edgewater Fortune* slowed to a crawl. From the wheelhouse, it was possible at times to detect bubbles breaking the surface ahead of the boat, which meant that even though the two great diesel engines were pounding along at what should have been several knots, the vessel was at times actually being driven slightly backward. Some three hundred miles out to sea, with the fuel being gobbled voraciously, the minesweeper was only able to hold her position. Amchitka was not coming any closer.

Like Cormack before him, the time came when Johansen had to say: "Hard about!"

There was not much of an argument. Chris Bergthorson tried to argue in favor of carrying on. Marining's telepathy told him that the gods were still keeping an eye on things, and so there was no serious problem—why not keep going? But the boat belonged to Johansen, and he could see his investment fragmenting like a wet cookie around him. Also, there was the problem of the seasick broadcaster whose condition was edging in the direction of terminal. There was one other crewman who had been curled up on the floor in the galley for three days. Both men needed to be put ashore.

It was not until the boat was getting close to the northern port of Prince Rupert that the majority of the crew began to emerge from their hiding places, demanding to know why the boat had turned around. The media contingent was generally discontented because the assignment, after all, had been to go to Amchitka, not Prince Rupert.

Now all the subtle and not-so-subtle macho arguments came out into the open. Who was chicken around here? Why had we turned around? The crew had so far fed themselves on the knowledge that they were the ones who were still carrying on when the original *Greenpeace* had quit. They had a bigger boat. Faster. Tougher. By implication, *they* were bigger, faster, tougher. But now the big, fast, tough boys were turning around too, and after having only gone a few hundred miles out to sea. Now that the vicious pounding of the waves had reversed itself, and they were experiencing nothing more than the vast angel dance of the following sea, Johansen's crew felt very guts-ball again and wanted to get on with it. For the most part, the skipper hid out in his cabin. It was below, in the galley, that the arguments began to flare.

For the veterans from the *Greenpeace*, the arguments had a familiar ring. It was a bit as though they were back at Sand Point, with the kamikazes pitted against the "older, mature" men, except that now there was neither a Bohlen, a Metcalfe, nor a Cormack to make the decisions. Moreover, there was already a major new fragmentation process at work. The volunteers on the *Greenpeace Too* had almost to a man been turned on to the anti-Amchitka cause by the massive waves of publicity that had been pounding the coast ever since the departure of Cormack's old halibut seiner. They were a second generation—one that had grown up under the media spell cast from the boat by Metcalfe. While the original crew had not really believed its own propaganda, and had, in fact, understood perfectly that we were engaged in a propaganda war, the bulk of the second crew was composed of people who had accepted our propaganda as truth. It was not that we had ever lied—that's one thing you must never do in modern propaganda—but we had painted a rather extravagant picture of the multiple dooms that would be unleashed if Cannikin ever went off: tidal waves, earthquakes, radioactive death clouds, decimated fisheries, deformed babies. We never said that's what *would* happen, but that it *could* happen. Among ourselves, we had always understood that Amchitka was not going to bring an end to the world, but we had had nothing more than images to hurl against the AEC, so we threw the heaviest, most horrifying images we could. By the time those images passed through the mass communications system, they assumed tremendous proportions. Children all over Canada were having nightmares about bombs. Many of their parents were half convinced that the end was nigh. And the men who had leapt aboard the *Greenpeace Too* came for the most part from the ranks of those whose minds had been affected by our little exercise in propaganda.

They had also been following the adventures of the noble little *Greenpeace* on television, radio, and in newspapers for over a month. It had looked very heroic. Thanks to the Unity Rule, none of the bitter internal wrangling had spilled over into the open. From the shore, we had looked like twelve brave men out fighting the elements and the bomb itself. So when Marining, Cummings, Simmons, and Birmingham had climbed on board the *Greenpeace Too*, it was as though they had just stepped out of a myth. Their new crew mates stood somewhat in awe of them.

It was not that the "original" *Greenpeace* crew felt superior. The problem was more that their individual communications systems had all but broken down. Marining was so strange to most of the new crew that they could hardly understand him at all. And Cummings had the definite feeling that "my brain was about six feet back from where my face was."

"There was a quick rivalry between the crews of the first *Greenpeace* and *Greenpeace Too*," he explained later. "Without meaning to, we were laying it on people that our trip was where the action had been. The *Too* crew got defensive. It was as if we'd been saying: 'Ah, yes, but you weren't there for the Second Coming.'"

There was another problem.

Everyone had been openly expressing their doubts and fears on the *Greenpeace* for so long that it seemed natural to continue to express oneself as honestly as ever. The new crew was not ready for this. They were mainly either "believers" or objective journalists. In front of the objective journalists, it was not wise to discuss one's views about propaganda, lest they get the idea that they were somehow being manipulated. And face-to-face with the volunteers, it quickly became evident they didn't want their illusions shattered, not now that they had put their lives on the line. Cummings found himself having to keep all his doubts to himself. That made them so much harder to handle. The gallows humor that had developed on board the *Greenpeace* was not particularly welcome here. The crew was suffering the usual paranoid grandiosity about their ability to stop the bomb and save the world and did not want to hear somebody like Cummings wondering aloud whether it was all worth it. It was bad enough that the "heroes" from the original *Greenpeace* seemed like near-basket cases. Worse, the full storms of winter were upon them, and the physical chances of reaching their goal had never been slimmer. And, except for one, there was no man on board who did not experience momentary rushes of terror and near-panic in the crashing of the waves.

For his part, Rod Marining remained utterly convinced that the weather was no danger to them at all. He had had a vivid recurring dream of a huge chamber with a Greek-style marble table in the middle. Seven figures were gathered around the table, discussing a situation in which he, Marining, was intimately involved. The effect of the dream was to tell him there was "something else" at work here and it was nothing so simple as the weather. He was still into his "religious space."

None of the others gave any sign of having such certainty. But they were all sure of one thing—that this was no time to quit. Johansen found himself repeatedly at the butt end of a microphone, or with a camera swiveling to bear close-up on his face, while tough old Doug Collins shot pointed questions at him. Was he giving up? What was the problem? So long as the crew wanted to continue, and so long as Chris Bergthorson could state that the Don't Make a Wave Committee still had funds to cover the costs of the fuel, Johansen really had no way out, short of admitting that he himself couldn't take it anymore. The boys back at the Masonic lodge would hardly buy that.

So at Prince Rupert, there were only two things to do. First, the seasickness victims had to be put ashore. The broadcaster from Prince George was taken off on a stretcher, and the chief engineer who suffered a nervous breakdown had to be led away, kicking and screaming wildly. The ship's tanks had to be topped up and interviews given—even though there were more reporters on the minesweeper than in the whole town. Then they were ready for sea again.

But as they tried to clear Dixon Entrance, they found that the weather was worse than ever. In no time at all, they were virtually motionless in the water while the waves rolled over the foredecks, and the ship went back to its ironing-board pounding. The fuel gauges registered the usual one-hundred-gallon-an-hour loss.

There was no practical choice but to come about again and head north into the labyrinth of islands and fjords down which the *Phyllis Cormack* had retreated a week before on its way back from Juneau. The channels

119

of the Inside Passage were still peaceful, and the *Edgewater Fortune* could make good time without wasting her fuel. The route was longer, but it was far better than taking a lot of useless punishment in the open Gulf of Alaska. With each day, of course, the date of the expected firing of the bomb at Amchitka drew closer, and the crew was gripped by the frustration of believing that they might already be too late.

Of the men on board, only Marining and Terry Simmons fully understood how flexible that date was. It depended entirely on weather conditions and political conditions in the United States. There were reports that the Supreme Court was going to be asked to rule on whether the test should go ahead or not. Environmental lawyers in Washington, D.C., were still confident that, so long as public pressure continued, the test might yet be canceled. "Public pressure" meant press coverage, and press coverage was hitched to the onward momentum of the ship. But the crew of the *Edgewater Fortune* was not nearly so well versed in the details of the lobbying taking place in America as their predecessors had been. Their doubt that there was any hope of reaching Amchitka in time sapped their determination, left them with a feeling of futility, and made it that much harder to keep pressing forward with all their strength.

Arriving in Juneau, the ship once again had to top up with fuel. Hank Johansen was out on the stern shouting orders back to the bridge as the great craft attempted to back into a narrow docking space. The rest of the crew was running around with ropes or else jumping about with their TV cameras to get everything on video. Photographers were perched in every corner they could find.

Rod Marining lingered near the wheelhouse, watching the whole scene with the feeling that he had arrived in the midst of a new kind of super-high-fidelity wraparound movie theater with eight-track sound. It was not just that he felt amazingly detached—it was more as if he had just landed in some wholly new perceptual continuum. What a movie! This huge gray-and-white warship attempting a landing in the early-winter Alaska gloom, and all these weird, tense people scrambling about, shouting at each other ... What did it all mean? What planet was this? Were all these cameramen making some kind of a movie? Or was it already a movie about people making a movie ... about reality ...? Too many dimensions ...

Johansen was yelling:

"All right! Stop!"

But the ship kept backing up, closer and closer to a whole row of barnacled telephone-pole pilings.

"*All right! Stop!*" Johansen yelled again, much louder.

Gracefully, with only the sound of steady throbbing engines, the *Edgewater Fortune* drifted back, back, back.

"ALL RIGHT! CUT THE ENGINES!"

The next sound was a tremendously resonant *Ping!* A piling was bending in slow motion around the vessel's stern, which was still performing its grand backward swing.

And another *Ping!*

And another.

There was much debate later about the exact number of pilings the ship had taken out. The damaged dock belonged to the father of a girl Rod Marining had met the night the *Phyllis Cormack* stayed in Juneau and her crew had all gone crazy at the DREAMLAND dance pavilion. The young lady herself had been passionately against the tests at Amchitka, and presumably equally passionately *for* Marining, although he was not certain that her father shared her politics. Once the *Edgewater Fortune* had finally docked, Marining ran to the nearest phone booth to call her. But it only took three hours to refuel, and in that time he never managed to contact her.

Then they were back out at sea, heading on a great circle route for Amchitka, the last leg of the journey. They were into the region where the *Phyllis Cormack* had had her roughest hours, but now the Gulf of Alaska was calm. The skies were ragged and bleak, and storms were known to be stirring in the Bering Sea, but for the moment the *Edgewater Fortune* had exactly the kind of weather she needed. Her knifelike bow slit through hissing water, the V-12 diesels pulsed, and a broad lime-colored wake fell back straight and wide as a city street behind them to the horizon. The tattered *Greenpeace* flag snapped in the wind. There was an air of grimness aboard. For, calm as the weather might be, everyone's nerves were stretched tight.

The bomb had been scheduled to go off on November 4, a Thursday. But by nightfall of that day, Cannikin had still not awakened. Instead, the Supreme Court of the United States was in session, hearing special last-minute pleas from environmentalists. Harry Bridges, president of the International Longshoremen's and Warehousemen's Union had telegrammed Nixon warning him of boycotts of American ships in Canadian ports. Thirty-four senators had demanded that the tests be canceled. More than one hundred Canadian members of Parliament had sent telegrams to the White House. Church bells tolled in dozens of cities. In the final hours before the bomb was to have gone off, some 5,000 protestors sealed the border at Windsor, Ontario; 3,000 at Sarnia, 1,500 at Niagara Falls; and 150 at Cornwall. U.S. riot police had to be stationed at all border crossing bridges. Some 8,000 more protestors gathered outside the U.S. consulate's office in Toronto.

By far, the most significant moves were being made in Washington, D.C. Russell Train, Nixon's chief advisor on environmental policy, came out against the test, just as the seven judges of the Supreme Court were sitting down. Train's objections were based on the fear that "this underground explosion could serve as the first domino in a row of dominoes leading to a major earthquake." The governor of Alaska deployed a destroyer and a helicopter carrier in the direction of Amchitka "in the event of any emergency." He also asked the secretary of defense to have rescue forces standing by in the event that native villages in the Aleutian Islands had to be evacuated. A Royal Canadian Air Force plane was sent winging out from Ottawa to attempt to reach the site to monitor for any radioactive leakage. Three Nobel Prize winners signed a statement urging all American citizens to join with them "in insisting the president postpone the test."

At the moment when the bomb was supposed to have gone off, November 4, the *Edgewater Fortune* was fourteen hundred miles from her objective. But she was narrowing the gap rapidly, and the weather was holding. While Chief Justice Warren Burger was promising reporters in Washington that the court would come up with an "early decision," the crew of the *Fortune* were nervously pacing the decks, hoping for exactly the opposite. If the test could be delayed just three more days, they might yet come skidding into the test zone on time.

A jet had passed far overhead bearing the chairman of the AEC, James R. Schlesinger. In a desperate effort to counteract the bad press the test was getting, Schlesinger had announced that he would demonstrate his confidence in the safety of the blast by taking his own children with him to the island, where they would remain through zero hour. The effect of his demonstration of confidence was somewhat dissipated when it was learned that the Schlesinger family would in fact be holed up in a concrete bunker mounted on steel springs huddled behind a mountain range on the far side of the island. "It's a nice place for the kids to have a picnic," Schlesinger was quoted as saying.

On Friday, while the Supreme Court was still in session, a storm hit the island, delaying the removal of several hundred workers who were to have been evacuated.

As darkness fell that night, the *Edgewater Fortune*'s throbbing engines had brought her to within nine hundred miles of the site. And while furious winds were lashing Amchitka, the *Fortune* was moving through a charmed zone of near-perfect sea conditions as though the forces that had kept her pinned down so helplessly off the West Coast had changed their minds and were letting her have a clear shot at her target. The men on the boat had wild nightmares that night. They had been nine days at sea, rushing almost full tilt every moment of the way and now they seemed within a hairbreadth of their goal. Their bodies and minds ached from the tension of unconsciously straining forward, even in their sleep.

By dawn, on Saturday, November 6, the minesweeper was within seven hundred miles of the island. Dark swollen clouds were racing overhead, although a patch of clear sky lay ahead in the west. Dr. Joe Stipec was at his ham receiver, while voices crackled through the wheelhouse with a steady stream of messages.

The news was all bad. By a vote of four to three, the Supreme Court had refused a request by environmental lawyers for a full hearing, which would have had the effect of postponing the test indefinitely. Within minutes of the announcement of the decision, an "arming party" left the Amchitka camp by car to drive twenty-three miles to the test site to make the final connections in the firing circuits of the warhead. Winds close to one hundred miles an hour were from the north, leaving only a brief opening in the predawn light for the planes to arrive to remove the unnecessary workers. The arming party itself had to drive because helicopters couldn't get off the ground. Six U.S. Air Force and two U.S. Navy aircraft converged on the site to take photographs. Battleships and carriers took up their positions in the test zone.

Visibility around the island was ten miles under a partial cloud cover hovering at fifteen hundred feet.

Scientists at the seismological observatory at Palmer, Alaska, were standing in front of rows of instruments, with hookups to both the AEC's new center in Anchorage and to Amchitka and Adak, alert for the slightest sign of a developing earthquake or tidal wave. After all the petitions and hearings, blockades and demonstrations, vigils, marches, headlines, newscasts, speeches, prayers, and threats, a hush suddenly descended.

What was to come would be the largest underground nuclear test ever set off by the United States, and it had attracted more attention than any blast in history except for the two that went off over Hiroshima and Nagasaki. It would be 240 times as powerful as either of those.

Captain Johansen had ordered the ventilation intakes and hatches sealed and had made the arrangements for crew members to go out on deck in oilskins with the Geiger counter.

At 9:30 A.M., the last workers on the island hustled, bent over against the wind, through the waning three-hour daylight period into their jeeps and roared away across the moonlike barrens of Amchitka, through the mountain range, and leapt into the spring-mounted concrete bunker.

At 11:00 A.M., Bering Sea time, James R. Schlesinger gave the "go," triggering a nuclear phenomenon whose force was measured at five megatons, flaring to the temperature of the sun, instantaneously creating a cavern the size of four football fields in diameter out of what had until a microsecond before been solid Aleutian rock, unleashing a shock wave that registered 7.2 on the Richter scale, meaning that an energy pulse had been generated through the earth's core that was equal to the pulse that leveled San Francisco so many years before. The ocean around Amchitka churned like a milkshake, rocks slipped down fifty-foot cliffs, whole sections of earth fell into the sea, and a crater almost half a mile wide was sucked out of the center of the island. The bunker containing Schlesinger, his family, and his technicians bucked like a cable car on its steel springs. Rare peregrine falcons and bald eagles sitting on rock ledges had their legs driven up through their bodies. Some one thousand sea otters died in the Bering Sea, their ears split by the shock wave, their wet brown bodies left to wash up on the shore for weeks afterward.

Four minutes after the blast, AEC Public Affairs Director Henry G. Vermillion announced to the world media: "Everything has gone exactly as we expected."

On the *Edgewater Fortune*, seven men were standing, braced, in the wheelhouse. The others were gathered around the radio hookup in the galley, except for a few who had retreated to their bunks. The mood in the wheelhouse was completely without humor, even though nobody really expected anything to happen. The swells were moving with the steady rhythm of deep-sleep snoring. Bob Cummings had an overwhelming sense of "lethargic nothingness." Rod Marining's telepathy told him that, in fact, a weird break in the pattern of the swells had occurred, like a twitch, as though the planet itself had been stung. The Vancouver *Sun* reporter, Jim McCandlish stated: "No one on board saw or heard a thing."

It was not just that the voyage had ended without a bang, there had not even been a discernible ripple. The mood down in the galley turned more sour, more rancid, than even the worst mood that had ever been experienced on the *Phyllis Cormack*. Somebody smashed a china plate against the wall. Like a slow-motion psychological body blow, the realization set in that they had *failed*. Right up until the last few hours of the night before, they had clung to the hope that the Supreme Court ruling might at least delay the bomb a few days more, and that had been all they needed to get there. "Getting there" had meant—somehow, however illogically—*winning*. No one had known whether the ship would be immediately seized the moment it entered the zone, or whether it would be ignored and left to sit, if it liked, in the path of any radioactive leakage that might occur, or whether the whole military circus would have ground to a halt, exactly as though a monkey wrench had been thrown into the dead center of the gears. But until that morning when the word had come through that the Supreme Court had let the beast loose through its fingers, the fantasy had remained as vividly alive as ever that the bomb might really and truly be stopped.

For the media contingent on the boat, it was relatively easy. They were *pissed off*. The great Amchitka holocaust story had ended with headlines that trumpeted:

AMCHITKA N-BLAST

'SAFE, SUCCESSFUL'

For Captain Johansen—as with all the rest—there was a certain deeply felt joy that one was still alive, despite all the dreams and moments of doubt and fear and paranoia. But there was also the overriding consciousness that the arguments put forward by one's own side—that the world was taking a giant step closer to Armageddon—had been booted aside. Everything continued. The AEC had been proven right. No earthquakes. No tidal waves. No leakage. Only dead otters and birds. And, after all, who cared about *them*? Marining argued that the boat should carry on, even though it was too late to park in front of the bomb, in order to "bear witness" to the deaths of the otters. But the CBC camera crew, the newspaper and radio reporters all wanted to go immediately home. The story was over.

Time magazine was to write, under the headline, AMCHITKA'S FLAWED SUCCESS:

Seldom, if ever, had so many Canadians felt so deep a sense of resentment and anger over a single U.S. action. For once the cries of protest were not confined to the radical Left, but came from a broad spectrum of Canadian society distressed by the environmental risks and the other possible hazards of the test.

Yet ... the feeling of depression and/or embarrassment on the *Edgewater Fortune* was not quite so gut deep as it might have been. If the vessel's forward charge had seemed Don Quixote-like to many, there were many others who perceived that even if this one particular test had not been halted there was plenty of evidence that suggested too much of a stink had been created for the AEC to dare to try to use Amchitka as a test site again. Within days of the test, "U.S. government spokesmen" were being quoted as saying that "they were pleased with the results of the Amchitka explosion and plan no further tests on this bleak Aleutian Island between the Bering Sea and the Pacific Ocean." There was nothing official yet, of course—it would hardly have done for the AEC to admit too quickly that the site was finished, not when they could bask in the glory of everything having gone "as expected." But the fact remained that Amchitka had been billed back in 1969 as a location where a series of *seven* tests were to be conducted, each one bigger than the last. After just three—Cannikin— the plans were sharply adjusted. Four months of silence were to follow the November 6 detonation before the AEC finally announced that the Amchitka site was being abandoned "for political and other reasons." Amchitka Island would henceforth be turned back into what it had been before: a bird-and-game sanctuary.

On the way home, none of the crew of the *Edgewater Fortune* knew for certain that while a battle had been lost, the war had been won. They talked a lot about what to do about the next test, presuming there would be one. If the talk was unenthusiastic, it had to be, because, deep inside, most of the men were convinced that the issue was finished. Yet however much one felt it, it would be madness to announce, in the days or weeks or months immediately after the blast: "We won." The aura of failure hung heavily over the boat, over the whole West Coast environmental movement.

For those who had rallied around the cause in more distant places than British Columbia, the issue was quickly forgotten. They had not, after all, put years of effort into it. On the West Coast, it would take longer for the memory to fade, if for no other reason than the fact that there had been a much greater investment of time, energy, and passion. In the worst hours of disappointment after the "safe" firing of Cannikin, the men of the *Edgewater Fortune* were the bearers of the deepest sense of defeat, but they were also, in those hours, the custodians of the tremendous vision that had been at the root of everybody's frantic activity for more than two years—the notion that maybe, after all, everybody's worst nightmares had been wrong, and that the nuclear holocaust could really be avoided. Amchitka had merely been an expression of the broader dread. Now that there was an intuitive sense that Amchitka had in fact been beaten, there started to be discussions about what to do next.

It was not until late February 1972 that the AEC officially announced that the Amchitka test site had been abandoned "for political and other reasons." By that time we were already launched on another program, one that was to turn *Greenpeace* into an international catchphrase for environmental activism in a way that the assault on Amchitka had never quite managed. Only in the final weeks before the test had much anxiety been stirred in the U.S.; until then, it had been mainly a Canadian issue. With Amchitka now turned back into a wildlife sanctuary, our attention was already focused on a South Pacific atoll, some seven hundred miles from Tahiti. The place was called Mururoa, and it was being used by the French government as the site for atmospheric nuclear tests. Our plan was to send a boat to park inside a forbidden zone the French declared every year around the atoll during the tests.

There was, of course, a huge difference between Mururoa and Amchitka. The French bomb was detonated out in the open, and anyone in the direct path of its fallout was doomed.

In the months following the safe return of the *Phyllis Cormack* and the *Edgewater Fortune*, there were dozens of meetings. With the addition of the crewmen from the minesweeper, we had a much larger group than we had at the beginning. It also meant there were more quarrels, more internal power struggles. It was Jim Bohlen, more than anyone else, who set the course of the events that were to come. Acting on a suggestion I had made to him back on the boat, he arranged things so that the original three-man Don't Make a Wave Committee dissolved itself, changed its name to the Greenpeace Foundation—intended to be an all-purpose ecological "strike force" rather than a one-issue ad hoc committee—and appointed Ben Metcalfe as chairman.

Thanks to Irving Stowe's meticulous bookkeeping, the foundation started life with nine thousand dollars in the bank. The original idea to go to Mururoa, as far as anyone can trace it, was the brainchild of Joe Breton, a crewman on the *Edgewater Fortune*. But from there, Metcalfe took up the cause with a vengeance, despite heavy opposition from other "veterans" who either felt the voyage to be impossible or who resented Metcalfe's style, which was impulsive, individualistic, and, at times, heavy-handed. Whatever the criticisms at the time, it was Metcalfe's bullish determination, more than any other factor, that launched what was to become *Greenpeace III*. The key to the campaign lay, however, in finding a boat and a skipper, and since it was unlikely that French authorities in Polynesia would allow any antinuclear protest boat to set out from any of their territories, it meant that the voyage would have to be initiated from New Zealand, some thirty-five hundred miles away from Mururoa. How does a person sitting in Vancouver get such a trip in motion on the other side of the world? Metcalfe's solution was the essence of simplicity. He placed ads in the New Zealand newspapers, announcing that the voyage would take place and asking for volunteers.

Within a matter of days, a telephone call came through from New Zealand. A thirty-nine-year-old former Vancouverite, David McTaggart, just happened to have a thirty-eight-foot ketch that was capable—maybe—of making the trip. And he just happened to be in New Zealand. Further, he just happened to be at a critical juncture in his own life that left him free of commitments and searching for something to do with himself.

The final coincidence was that he was a Canadian and his vessel, *Vega*, was Canadian registered. He was, moreover, a relatively famous Canadian, having been a world-champion badminton singles champion in the late 1950s, before moving to the U.S. and working his way up to millionaire status in the construction industry. He'd built a lodge in California's Bear Valley ski area and managed to put the place on the map with a controversial SKI BEAR poster that featured his third wife schussing the slopes in the nude. Later he moved to Colorado, where he took over a ten-thousand-acre Aspen Wildcat ski area. But in 1969, back in Bear Valley, a propane explosion destroyed his lodge, badly maimed an employee, and the resulting litigation wiped out his assets. He fled to the South Pacific with what was left of his fortune, bought his ketch, and had been sailing among the islands and atolls since then.

Although the voyage was Greenpeace initiated and mostly Greenpeace financed, with help from the New Zealand Campaign for Nuclear Disarmament, and even though Metcalfe flew down to join the boat as a crewman, the success of the undertaking rested entirely on the shoulders of the former badminton champion. In every essential respect, once it was in motion, it became McTaggart's trip. Greenpeace's contribution lay in Metcalfe's ability to whip up media interest and keep the publicity ball rolling while McTaggart fought the seas and the French navy.

Vega—renamed *Greenpeace III*—set out from New Zealand on April 27, 1972, with five men on board. Two were Englishmen: Nigel Ingram and Roger Haddleton. Two were Canadians: McTaggart and Metcalfe. And one Australian: Grant Davidson. Even though the boat moved swiftly in southwesterly winds, radio trouble developed and personality clashes quickly came into focus. The two Englishmen couldn't get along and neither could the two Canadians. Sensing that there would be serious trouble ahead, McTaggart changed course and headed for Rarotonga, where both Metcalfe and Haddleton left the boat, with Metcalfe leading a group of Vancouver Greenpeacers, including Patrick Moore, Rod Marining, and Doc Thurston to France to organize demonstrations against *la Bombe*. Metcalfe's stay in France proved short-lived. He and his wife Dorothy were immediately arrested and escorted out of the country. They had more luck in Rome, where they succeeded in getting the pope to bless the Greenpeace flag, triggering a debate between bishops and generals in France itself. The others sought refuge in Notre Dame but were dragged away, and Rod Marining found himself being taken by French plainclothesmen to an empty parking lot and punched repeatedly in the stomach. "Are you a Red?" the cops demanded. "No," gasped Marining, "I'm a Green!" He was given a warning to get out of Paris.

Back in the South Pacific, McTaggart carried on with his reduced crew, bucking headwinds most of the way to the test zone at Mururoa. It was not until the beginning of June that they crossed the line of the one-hundred-thousand-square-mile cordon that France had declared around the atoll, despite the fact that the International Law of the Sea does not recognize any nation's right to declare such a zone.

Buffeted by fierce winter storms, *Vega* tacked back and forth for weeks, while food and water supplies ran low. Then, on June 19, the crew saw a huge helium-filled balloon rising over the horizon, the very balloon that would carry the bomb aloft. When that failed to scare them away, the French navy sent a small fleet of ships, including a six-hundred-foot cruiser, minesweepers, and tugboats, out to play a brutal game of high-seas "chicken" with the tiny ketch, charging at it from every angle, even closing in on both sides, very nearly crushing *Vega* between the hulls of the warships. Eventually driven back by winds, the protestors were unable to maintain their position downwind from the bomb, which the French fired during *Vega*'s sixtieth day at sea. The fallout missed the ketch by a margin of no more than fifteen miles.

Hearing over the radio that one bomb had gone off, McTaggart struggled to get back in position, and was, in fact, heading into another radioactive cloud when the French navy sent out a minesweeper, which came up on *Vega*'s stern and rammed her, leaving her paralyzed in the water.

Later the French towed the crippled ketch into Mururoa Lagoon, made minimal repairs—just enough to keep her afloat—then towed McTaggart and his crew to sea, leaving them to limp back to Rarotonga while the rest of the bomb tests were carried out.

The following year, McTaggart set out again. By then, opposition to the French nuclear tests—triggered by the sensational press coverage that had surrounded the protest, most of it orchestrated by Ben Metcalfe—had created a whole new mood in Australia and New Zealand. Conservative governments in both countries had been swept out of office and replaced by socialists who pledged action against France. The New Zealand government went so far as to send a navy ship into the test zone, duplicating the action of *Greenpeace III*. By summer 1973, there were so many protest boats en route to the test site that journalists began to lose count. Back in Vancouver, McTaggart's disagreements with Metcalfe had reached a point where, if we were to launch the *Vega* again, the Greenpeace Foundation would have to undergo a leader change. Metcalfe agreed to step aside, and Rod Marining and I assumed command, working closely with McTaggart to raise the thirteen thousand dollars that would be needed to repair his boat and get him back out to sea. This time, he took only three other people: Nigel Ingram and two New Zealand women—Ann-Marie Horne and Mary Lornie.

By the time they arrived in the test zone at Mururoa, the rest of the flotilla of "peace boats" had either departed or been seized by the French navy, leaving only *Vega* standing between *la Bombe* and its detonation. The balloon was already aloft. And this time, the French were angry. On August 15, they sent out three warships and launched a high-speed rubber inflatable boat filled with commandos armed with truncheons. McTaggart and Ingram were grabbed and savagely beaten. One Frenchman jammed the butt of his truncheon into McTaggart's right eye, damaging it so badly that it was permanently injured. *Vega* was taken in tow with the girls aboard and brought once again into the lagoon. McTaggart was flown to Tahiti, where he was held in hospital under armed guard.

Fortunately, Ann-Marie Horne had managed to get photographs of the attack and smuggled the film in her vagina past unsuspecting French guards. Later, we released the photos to the world's wire services, embarrassing French generals and politicians who had insisted up to the last minute that McTaggart had not been beaten. In the volatile atmosphere of international outrage over the tests, this evidence of French military savagery came as a kind of final blow. Not three months later, in early November 1973, France announced that, after one more year of testing in the atmosphere over Mururoa, all future tests would be conducted underground.

La Bombe had been driven into the darkness.

Twice, we had scored incredible victories against military superpowers, and there was now a great deal of excited thrashing about taking place in Vancouver as we pondered our next move, little realizing that the course had—in its own mysterious way—already been set, and it had been set as far back as the day the *Phyllis Cormack* nosed into Akutan Bay in the Aleutians, near the ruins of the abandoned whaling station. Although none of us yet quite realized how deeply we had been affected by the sight of that potent symbol, it was permanently fixed in our consciousness, deepening our sense of the real scale of the damage that had already been done to the planet. Inadvertently, we had borne witness to the evidence of a menace that was finally greater than even nuclear bombs: the pervasive day-to-day industrialized destruction of every major link in the chain of life, beginning with the largest creatures in the world.

Even though our attention had been focused for two full summers on Mururoa, we had been joined by a man who was to shift Greenpeace onto a whole new course, opening the gates on actions that were to take us to a new and deeper level of struggle, not just to protect our own kind, but to protect other forms of life as well. ...

His name was Paul Spong, a New Zealander in his early thirties who had earned his degree in physiological psychology at UCLA, where he had met his wife, Linda. Moving to Vancouver with their son, Yashi, Spong had landed a job studying a captive *Orcinus orca*—better known as a killer whale—at the Vancouver Public Aquarium. While he was considered something of a renegade in local academic circles, he was a bit of a legend among eco-freaks on the West Coast, having been fired from the aquarium for daring to state in public that the whale he had been studying—named Skana—"wanted to be free."

Having been kicked out on the street for that radical statement, he had moved to Hanson Island—not far from the Kwakiutl Indian village of Alert Bay—where he had set up a whale-watching station, swimming among the great toothed whales, playing music to them, and coming eventually to the conclusion that their intelligence, while of a different order than ours, was far greater than anyone was willing to believe. Spong wanted Greenpeace to use such clout as it had to join him in his campaign to end the mass killing of whales. It was a notion that was still ahead of its time, even among Greenpeacers. But while McTaggart's voyage to Mururoa had been absorbing our time, energy, and money, Paul Spong had been quietly moving about in the background in Vancouver, planting the seeds of what would eventually transform us from an antinuclear protest group to a true environmental movement.

2

The Great Whale Conspiracy

*I believe change can happen
at the speed of thought.*

—PAUL SPONG

By November 1973, Greenpeace could claim to have been the catalyst that had not only brought American nuclear testing in the Aleutian Islands to an end but had set in motion the chain of events that was now forcing French atomic weaponsmakers out of the skies over the South Pacific—a track record that no other environmental organization on the planet could match. Yet, technically, Greenpeace did not even exist. The only piece of paper we possessed that gave us any kind of legitimacy was a notice dating from the autumn of 1971 authorizing the changing of the name of the Don't Make a Wave Committee to the Greenpeace Foundation. Attached to that were a dozen pages of standard bylaws and procedures as set down under the British Columbia Societies Act, including instructions for holding annual general meetings, auditing, establishment of a board of directors, rights of members, and so on. Instead of a board, we had two "interim chairmen," Rod Marining and myself. We had not held any annual general meetings. The "books," such as they were, amounted to an untidy pile of scribbled account numbers, bills, bank statements, debits. And we had no members. It was a situation that would bring a glow of sheer joy to the cheeks of an anarchist.

Irving Stowe made a passionate speech about how cigarette smoking was a form of pollution that could not be tolerated in a true environmental group and moved that it be banned during our meetings. McTaggart let it be known he intended to carry on with his plans to bring his case against France to the courts, even if it meant going to France alone to set it all in motion. Spong wanted to stage a "Greenpeace Christmas Whale Show" to raise money to travel to Japan to make the case for a ten-year moratorium on whaling, but there were fears that the show would lose money, so it was proposed that Spong post a bond guaranteeing to pay Greenpeace back for any losses. Nobody wanted to back up McTaggart's plan to go to the courts.

The whole attitude seemed tightfisted, unimaginative, and petty. After all the internal hassles during the summer of 1973—the fights over whether to support McTaggart in the first place or not, the fights over whether to cut him off because he did not seem to be doing his job, then the fawning over him that had taken place upon his triumphant return, now the refusal to support his case, the refusal to support Spong's trip, the low-level nagging about such inconsequential issues as smoking or no smoking, the stubborn insistence on going back to Mururoa when, so far as I could see, the issue was a dead duck—the feeling came back that the creative energy of this group was again exhausted. How long had it been since I hadn't awakened to the thought of Greenpeace and gone to bed haunted by the thought of Greenpeace?

After several of these uninspiring meetings, I decided on two things. First, I would resign as interim chairman. Second, Spong and I would set up the Stop Ahab Committee and begin exploring ways and means of getting an expedition together to save the whales despite Greenpeace. If the rest of the group wanted to run off and tilt uselessly at the windmills of Mururoa, God bless 'em. I did not see that there was any way to force France to end her atmospheric test program any faster than she was already doing so, and now that the commitment to quit had been made, the diplomatic heat was off. Short of someone getting killed in the test zone the next summer, I did not see how we could draw any more attention to the unfortunate atoll than we had already done. And as Spong pointed out repeatedly, the whale populations were being whittled down every year before our eyes. The time to act was now. It was galling that hardly anyone, even in this particular gang, could see the direction in which we *had* to move. We were now faced with the choice of a new chairman. Rod Marining was much loved by everyone, but was still considered a bit too far-out to be chairman himself. Most of the others were either too new to the situation or too dogmatic one way or the other. A new polarization process had developed, with the antinuclear element pitted against the pro-whale element. Basically that meant Irving Stowe versus Paul Spong. The choice of either would have split the group.

Our man was clearly Hamish Bruce, the long-haired lawyer in whose house most of our meetings had taken place that summer. While Bruce had a habit of dressing most of the time in Salvation Army garb, outlandish even by West Coast hippie standards, and while his blond hair fell halfway down his back, he was nevertheless capable of putting on his work clothes—meaning a suit and tie, or appearing in court in his black robes—and he could be articulate in the straight sense when he chose to be. He was ambivalent about Mururoa because he had applied for a visa to move to New Zealand but had been turned down because his wife, Kimiko, was Japanese-Canadian, and the good New Zealanders had a racist immigration policy that forbade "Asiatics." After being rejected on those grounds, Bruce was as much inclined to advocate low-yield atomic bombing raids on New Zealand as he was to try to save the place from radioactive fallout. Bruce had to think about the chairmanship for a long time. It was a dilemma that called for a very high order of morality and compassion. In the end, he agreed, with the provision that when it was all over he would be allowed to issue a nasty statement about the racist pigs down in New Zealand.

On the whale front, he wanted to organize a "call-in" for whales along Vancouver's English Bay. The call-in would involve hundreds of people going down to the beach and using telepathy to attract the whales from all over the world and providing them with a sanctuary. Our candidate's mind was an amazing blend of legalistic practicality—he was a tough, brilliant, radical Left-wing lawyer during his working hours—and unfettered mysticism in private. From time to time he had tapped into the "collective unconscious." He was a devotee of the *I Ching*, ancient Indian prophecies, Jungian concepts of synchronicity, and a believer in magic, a true hard-core West Coast freak. There was one thing about Hamish Bruce on which we all agreed: he had *soul*. He also had a small fan club of his own, consisting of younger lawyers in the city who admired the nerve it took to appear in court defending dope dealers with hair hanging halfway down their backs. The police themselves had an affectionate regard for him, and so did quite a few of the judges.

The vote for him as chairman was unanimous.

We had now successfully passed through our fourth change in leadership: from the triumvirate of Jim Bohlen, Paul Cote, and Irving Stowe to Ben Metcalfe to Rod Marining and myself to Hamish Bruce. It was either a sign of healthy vitality or instability—hard to know which. At any rate, under Chairman Hamish the faction that wanted to "keep up the pressure" against the French at Mururoa would get their way, and Spong and I would be allowed to carry on with our own antiwhaling committee.

The harsh reality for David McTaggart was that he was broke, in debt, his boat was still in the hands of the French, and he did not even have a place to stay. It was not until December that negotiations between the Canadian Department of External Affairs and their French counterpart resulted in a decision on the part of the French to release *Vega* to McTaggart in Tahiti. The Canadian government had been getting so many letters complaining about the incident at Mururoa that External Affairs had had to hire two full-time employees just to keep up with the mail. Prodded by blistering editorials, Canada offered to pay the twelve-thousand-dollar cost of shipping *Vega* by freighter back from Tahiti. McTaggart's fare down to Tahiti to pick her up was also covered by the government. By December 12, he was standing on a dock in Papeete, flanked by a French secret service man, watching as a 372-foot French navy vessel slipped into harbor with little *Vega* mounted on its deck. The French were far from happy. By then, the announcement that atmospheric testing would be suspended had been made, and the officers charged with the task of handing *Vega* over to her owner could barely restrain their fury at the man who had done more than anyone else to force them to change their plans. He had already been declared an "undesirable alien" in all French territories and was only allowed into Tahiti on a special permit. As soon as he had made the arrangements for his boat to be transferred to a freighter heading for Vancouver, he was ushered back to the airport and told never to set foot in French Polynesia again.

Back in B.C., he turned his attention to pressuring the Canadian government to support his case against France, but the best he could get out of Ottawa was a promise to espouse the case *if* he himself managed to "exhaust all local legal remedies," which meant he would have to go to France, find a court in which he could lay a charge against the French navy, and carry the burden of all the legal expenses on his own. The Greenpeace Foundation was still not interested in a court case, so his only course of action was to take *Vega* up to Buccaneer Bay and set to work putting her in good enough shape to sell. A trickle of donations arrived from time to time in the mailbox, and his book, *Outrage: The Ordeal of Greenpeace III*, had been issued by a small West Coast publisher, but neither the book nor the donations amounted to anything like the money he would need to pursue the case. Ann-Marie Horne flew up from New Zealand to join him. They moved into a small cabin at Buccaneer Bay, worked on the boat, wrote letters to lawyers, politicians, editors, and supporters and laboriously began the task of assembling the evidence against the French navy.

At the end of December, Paul Spong staged the "Greenpeace Christmas Whale Show," and while it did show a profit, it was not enough to get him to Japan. He had to contact Canadian folksinger Gordon Lightfoot, who contributed five thousand dollars more to allow Spong to take the show across the Pacific.

He was not ready to go until February. During that time, we worked out several of the details for the expedition we planned to launch the following year to save the whales. Apart from the idea of a boat going out to confront the whaling fleets somewhere in the ocean and a favorable *I Ching* reading, we had lacked any specific plans. It was to the specifics of the protest that we now turned our attention, and it was McTaggart's experience at Mururoa that finally gave us the key element we had not been able to dream up on our own. Looking over the photos of the boarding of *Vega*, we realized that the French would have had a difficult time seizing the ship without the high-speed rubber inflatables that had carried the commandos so swiftly and effortlessly across the water. Called "Zodiacs," these inflatables—a French invention—were apparently capable of speeds up to thirty knots on a flat sea, they were extremely maneuverable and were stable in the water—even if they flipped they would still float—and Spong recalled that undersea explorer Jacques Cousteau had used identical craft to get film footage of live whales. They could be launched easily from a larger ship. They were perfect. All we would have to do was get close to a whaling ship, drop a Zodiac into the water, and then race ahead of the harpoon, positioning ourselves between the gunner and the whales, making a clear shot impossible. Shortly before he left, Spong decided he needed to stir up some publicity to get things moving before he arrived in Japan. To that end, we decided to announce the launching of the traveling "Greenpeace Whale Show" from poolside at the Vancouver Public Aquarium, the place from which Spong had been fired for saying that Skana, the captive orca, wanted to be set free. There was little chance the aquarium officials would agree to let him near Skana unless enough newsmen could be rounded up to turn the thing into a media event. With *Time* magazine and several local newspapers and television stations present, the aquarium might be reluctant to refuse to allow the "whale man" to play his flute at the side of the pool. It was, after all, a harmless-enough gesture. As a columnist, I would of course attend. It would also provide an opportunity for Spong to introduce me to Skana. I had no direct contact with whales, had not had "the experience," and was still prone to question Spong at length about whether or not they really were as intelligent as he claimed. "Fear is the greatest barrier between whales and man," he had said. "What you have to do is trust them completely. And if you trust them, they'll trust you. Don't worry about them making any slips; they seem to have perfect physical control."

The official purpose of the visit to the aquarium was to hold a Greenpeace press conference, at which Spong would play music to the whales and announce the whale show was going to Japan as a first step in a major campaign aimed at bringing about a ten-year moratorium on whaling. He would travel to two dozen Japanese cities, showing slides and film, playing whale recordings, and lecturing on the need to preserve the whales. If that failed, the Greenpeace Foundation pledged itself to put an expedition to sea to tackle both the Japanese and Russian whaling fleets. It was enough of a news story to attract a half-dozen reporters. Spong promised not to say anything about Skana's desire for liberation, and the aquarium in turn agreed to let us use the whale pool as a backdrop to our announcement.

It was 11:00 A.M. on a Tuesday morning, February 26, 1974, when Spong climbed down on the metal platform a few inches above water level at the pool. Skana and a second, smaller whale named Hyak rushed over to him immediately, nuzzling up against him like sleek immense dogs. It was very much as though they did remember him from years before, when he was supposed to be studying them. The exchange of greetings between Spong and the whales was almost embarrassing to behold. Cameras clicked and whirred, and Spong became so ecstatic I thought for a moment he was going to cry. He was on his knees, patting them, pressing his face against them, making incomprehensible sounds, and backing off every few minutes to play a few notes on his flute. The whales seemed no less ecstatic. Some of the reporters looked at each other questioningly, and I knew they were wondering if the man wasn't out of his mind, fawning so passionately over a couple of whales, *killer* whales at that. Then Spong signaled for me to climb down on the platform beside him.

I realized my body was trembling. Not an unpleasant trembling, but a sure sign of tension. "The whales might test you," Spong had warned. "They test by checking out your level of fear. If there's no fear, they'll love you." What he hadn't said was what they would do if they discovered a large dose of fear. It's not all that easy to completely trust a creature probably one thousand times as powerful as yourself, equipped with a formidable array of pointed teeth. Yet the moment I got down on the platform and put my hands out to Skana, she moved lightly—like a woman—against me, allowing me to stroke the underside of her jaw. The fear fell away within seconds, to be replaced by a sense of marvel and excitement. Before I quite realized what had happened, I was rubbing my forehead against hers, stroking her, feeling nothing but sensuousness. Despite her colossal weight, she balanced perfectly and effortlessly in the water, and I had the feeling she could have taken a razor blade in her teeth and shaved my chin and throat without leaving a scratch. It was very much as though the whale was extracting feelings of happiness from deep at the bottom of my mind. I became oblivious to the horrified stares of my journalistic colleagues, the glares of the aquarium staff who had not been warned that this was what Spong and I were going to do. After several minutes of this kind of delicious contact, I thought there was no fear left in me at all. I trusted her like I would trust my own mother.

Then she dived backward and came up with her mouth open, the round rubbery bumpers of her lips closing around my jaw and behind my left ear. Astonished, I realized that half my head was now *inside her mouth*, and I was able to sense the power in her, power enough to nip my head off like an eggshell. Yet her touch was still lighter than the lightest kiss. My only reaction was to blow gently against her tongue and around the pink contours of that cavern into which I was now looking, seeing the water beyond through rows of teeth. She backed down into the water again and I stayed frozen on the spot with my head held out as though waiting for a guillotine to descend. She shrank into the water, then came up like a projectile, jaws agape, and this time she took almost my whole head into her mouth and held me for several seconds. I could feel the teeth making the slightest indentation against the back of my neck. It was not possible for me to move a fraction of an inch. I was like a crystal goblet in a vice. I experienced a flash of utter aloneness, knowing that if she chose to chomp, there was nothing anyone on earth could do to save me. I was completely at her mercy. Fear exploded in my chest, yet the feeling of trustful happiness continued in my head. As though satisfied, she let go and sank away—ever so gently—with a handful of my hair snagged around two huge teeth. Oddly enough, there was no sensation of pain. The hair had come away from my skull as though I had been administered a local anesthetic.

But after that, my nerves were so jangled I had to back away, contenting myself with blowing air on her skin—amazingly warm flesh, not cold or clammy at all, warm and firm like a great breast—providing a sensation she obviously enjoyed. She writhed lovingly and offered her jaws several more times, but the loss of that shank of hair had badly weakened my confidence in Spong's assertion that she had complete physical control. When I finally climbed back off the platform, she followed me around the pool, making high-pitched whistling sounds, very agitated, and I was absolutely certain she was saying: "Come back here, boy, we haven't finished." But I was finished for the time being.

The aquarium staff scolded me. Spong said: "She was telling you to get a haircut." The reporters got their pictures and story. And we all trooped outside.

I was left in a daze. It was all irrational—completely irrational—but I was convinced that Skana had taught me something that no therapist had ever been able to do. I had to wander off by myself into the park, feeling as though my insides had been gutted. That whale had found out a lot more about me than I had found out about her. She had manipulated me, and not just physically, but psychologically, as though she had reached into my brain and flipped a switch. Fear! She had shown me exactly where my courage ended and my fear began, and it was not just the fear of her teeth or size or strength. It was my fear of everything, my inability to step out from a job whose excitement had worn off, to remove myself from a marriage that had proved unworkable, my fear of other men, my fear of the sky falling, my fear of failure, all my fears wrapped into one. It was as though she had drawn a great chalk line across my psyche, then told me precisely how far I could go and where the point was that I could go no further. I could see *myself* more clearly than ever before. Irrational! Irrational! No animal could illuminate the secret corners of my soul like that! Desperately, I told myself that I was overreacting, that maybe I was cracking up simply in response to damn near having my head bit off, but no matter how I phrased it to myself, it remained that I found myself standing among the trees alone, crying like a baby with relief because at last, *at last*, I could see exactly who I was, what my limitations were. It was a gift. I also knew that, apart from Paul Spong, there was no one to whom I could talk about it without running the risk of being locked up. Certainly, I was going to have to be very careful what I wrote in a family newspaper.

Yet I could not have felt more shaken up if a flying saucer had landed and an alien had stepped out and zapped me between the frontal lobes. I had been through marathon t-group therapy sessions and emotionally exhausting workshops with the great Gestalt therapist Fritz Perls, but neither experience had been so far out of the framework of my understanding that it left me as shaken as I was now.

Could it be that there were serene super beings in the sea who had mastered nature by becoming one with the tides and the temperatures long before man had even learned to scramble for the shelter of the caves, but who had not forseen the coming of small vicious monsters from the land whose only response to the natural world was to hack at it, smash it, cut it down, blow its heart away? Had the whales enjoyed a Golden Age lasting millions of years, before their domain was finally invaded by a dangerous parasite whose advance could not be checked by any adaptative process short of growing limbs and fashioning weapons? What, indeed, could a nation of armless Buddhas do against the equivalent of carnivorous Nazis equipped with seagoing tanks and Krupp cannons?

While this vision was bursting upon my mind, Greenpeace itself was busy experiencing its annual nervous breakdown. We were trying to come to grips with a real-life biological and evolutionary apocalypse. There could scarcely be any greater issues in the twentieth century, or any prior century, than the extermination of species after species and the construction of nuclear weapons. And however clumsy, ineffective, and fragile Greenpeace might be, it was the only tool to which any of us had access that promised to give us a grip—again, however feeble—on the overall situation. It *did* appear we had had an impact at Amchitka and Mururoa. That meant it was at least within the realm of possibility that we might score similar successes in other environmental areas. The feeling could not entirely be dismissed as paranoid grandiosity. So we all clung to the tool rather desperately and strained with all our vigor to apply it properly and effectively.

In our minds, all of us had broken through to what was, in fact, a relatively new level of consciousness: call it "planetary consciousness." Even with a group such as Greenpeace there were lesser and greater forms of planetary consciousness. Many of the people in the antinuclear faction could quite easily see that plutonium was a threat to the whole earth, and to nonhumans as well as humans, but their moral and spiritual revulsion was still essentially centered on their concern about the impact on people, not whales or sea lions or lizards. Irving Stowe, for instance, was a planetary patriot, but he was still an agent of the human race, looking out for the human race's long-term interests, especially his own great-great-grandchildren. Paul Spong, while no less a planetary patriot himself, could see in the destruction of the whales a greater evil than the mere destruction of a given subspecies of humans, of whatever race, religion, or nationality. In his view, the human race and all its minorities already had legions of champions and saviors. It was time for some energy to be channeled into saving the whales and the porpoises, too.

In an interview with me just before he left for Japan, Spong set out his case this way:

> Who knows what the importance of the large whales is in the life-systems of the ocean? They're big. They consume vast quantities of plankton. Sperm whales have these gigantic battles with enormous squids and octopuses that live in the ocean depths. Now I don't know what the importance of eating plankton is particularly, and I don't know what the importance of killing giant squid is in the oceans, but say you completely removed the whales and their particular contribution to the balance of life in the oceans? You know, it's really hard to predict what would happen. Maybe the oceans get taken over by giant squid. ... Let's start to think of the whales as our ocean neighbors, the highest-evolved life-forms that exist in the oceans of our planet. We don't know too much about them at this point. But what we do know suggests that they are really highly evolved creatures with an intelligence that is probably different from ours but obviously of some high order. There is even a question here that has not really been asked, and it's this: Can we learn anything from these whales? And I think the tentative answer is this: The whales we have been watching and studying do not kill one another. They may hassle but they do not kill one another. They exist in communities that have long-term stability, and the communities are comprised of families that have long-term stability. Whale families, so far as we know, stay together for life. And not only that, they very effectively mobilize group energy to satisfy individual needs. Everyone in the whale family, the whale community, the whale society, has all his needs provided for life because they work together. It seems to me that in time we might be able to learn an awful lot from the whales. But if this generation allows the whales to be wiped out like the dinosaurs, future generations will never have the opportunity

to make the discoveries that are possible. It will be too late. So we have to do something now. That's all there is to it.

Most of the people who now came to the Greenpeace meetings, and therefore controlled what the organization did, had been attracted by the opportunity to "do something" about the atomic plague. Whales were a chic topic of conversation, and whale freaks like Spong had a lot of sex appeal in the counterculture, but ... should the pathetic resources of one of the world's most effective antinuclear coalitions be drained off into a relative side issue like whaling? Weren't there already nine hundred species on the endangered list, the whales being just one? And weren't they all doubly threatened by the proliferation of nuclear energy and weaponry? The general feeling of the group was that Greenpeace itself could only afford to have one obsession at a time. Every meeting now turned into a battle for the hearts and minds of the Greenpeacers.

The battle took specific turns. Since the power to do anything is closely hitched to money in the bank, the main power struggles emerged less as Platonic debates than nasty little fights over who got to sign the checks and who got to authorize what money got spent where.

The resources of the Greenpeace Foundation came down to one thing: a name. A brand name, if you like. The word *Greenpeace* had a ring to it— it conjured images of Eden; it said ecology and antiwar in two syllables; it fit easily into even a one-column headline; it had a track record; and it was identified with radical and militant action, albeit of a Gandhian order: flower children, civil rights, Ban the Bomb, conservation, Save the Redwoods, Wooden Ships, Woodstock, et cetera. (We had learned in Alaska that king crabs only became a marketable commodity after their name was changed from spider crabs.) In other words, our marketable, zippy, neat, easy-to-remember name, with the good image it conjured, was all we had. In terms of physical assets, there was not so much as a single typewriter. What money that came to us fell into the category of unsolicited donations. Considering we had no advertising program, no mail-outs, no leaflets, not even a newsletter, and our address changed every time the leadership changed, the miracle was that any money came in at all.

We were probably one of the hardest organizations in town to get hold of—despite the fact that we were also, at an international level, the most famous organization in town. There were millions of people around the world who had never heard of the Province of British Columbia but who had definitely heard of Greenpeace. Somewhere out there, foreign admirals and commanders fumed with rage whenever they heard mention of us. The Japanese whaling industry was already summoning its chiefs into boardrooms to begin planning a counteroffensive against the attack they had heard we were preparing to launch against them—yet at home, in Hamish Bruce's living room, we could not even get agreement over the placing of a single long-distance phone call, or, when on occasion we did agree, it was usually doubtful that we could afford it.

Spong was absolutely convinced that another trip to Mururoa was a waste of time. But while he was an often-brilliant orator and a good showman, he was not particularly effective in an intimate group setting, at least not one of a political nature where the art of bringing people over to your own side is paramount. Spong's style was both pushy and apparently devious—it was as though he couldn't quite trust the others he was dealing with to render an intelligent decision, and once he had made up his own mind he was intolerant of contradictory points of view.

In opposing Spong's operations, Irving Stowe would reach for the rod of righteous wrath, speeches that probably would have worked beautifully had he been addressing a mob of at least sixty thousand or one hundred thousand people, but that only had the effect of exhausting and depressing a single roomful of individuals who had already heard it all. Finally, two separate bank accounts were set up—one for whales and one for another boat to Mururoa.

Spong took off with his wife and kid for Japan, carrying at least half a ton of slide projectors, dissolve units, tape recorders, screens, and cameras.

It was hard slogging in Japan. When Spong presented the first showing of his films and slides in Tokyo, it was to discover that half of his audience consisted of people working for the whaling companies. Their main purpose was to try to find some way to discredit him. Among other charges they threw at him was an accusation that he was an agent working for unnamed American industrial interests who were trying to drive Japanese products out of the U.S. marketplace. The Greenpeace Foundation was described as an "American plot"—which was a fine irony, since it had not been many years ago when the Americans had been describing us as a "Commie plot."

Spong put on nineteen shows during his stay in Japan. Some of them played to small audiences, such as one he did especially for the scientists working for the whaling companies. Others were large, such as the one in the Mitsubishi Department Store in Osaka. Only once was he refused permission to put on a show, which was scheduled to happen in the whaling museum at the small town of Taiji, the place where whaling began in Japan some four hundred years ago. The town council voted to ban the showing of Spong's "subversive" films and slides. Otherwise, he was allowed to go where he wanted. Within two months, he had made enough good contacts with well-placed media people—mainly through members of the Japan Wildlife Club—to have launched articles in national magazines, newspapers, and television. In Osaka, on one ladies' afternoon talk show, he addressed an audience of fourteen million viewers. In all, he estimated the "Greenpeace Whale Show" reached some thirty million Japanese.

By the time Spong was ready to leave, no less than eighty journalists attended his final press conference. He was news all right—and the whales were becoming news. The Japanese were waking up to problems of pollution, Minamata disease, mercury poisoning of fish, and, of course, they were sensitive to the nuclear issue.

For her part, Linda Spong felt from the beginning that she had landed on another planet. She fell in love with the Japanese culture, hospitality, people, and style of doing things—but the feeling lingered that she was, finally, an alien.

Everywhere they went, they were met by smiles and bows and hospitality that led Spong into what he termed afterward as an "interesting misperception." One newspaper had even run an editorial titled "Dr. Spong and Commodore Perry," drawing a historical parallel between ships from the Western world with guns arriving in Tokyo Bay demanding that the economic and cultural barriers be lifted and the arrival, over a century later, of this well-mannered Western scientist armed with slides and film and tapes, requesting that the doors of a new perception be opened. Historical it might be, but in the end Spong concluded that Japanese awareness had not been affected very much—it was just the hospitality that made him think so.

While the Spong family was thus occupied in Japan, the Third Battle of Mururoa had begun several thousand miles to the southeast.

Several thousand dollars had been raised in Vancouver and sent to Greenpeace Pacific in Melbourne to ready a small Tahitian ketch named *La Flor*, whose owner, Rolf Heimann, was prepared to undertake the odyssey. Heimann was in his early thirties, a German who had survived the bombing of Dresden and who was reported to be "passionately opposed to militarism." He had plenty of sailing experience. A Canadian who happened to be on the scene, Leif Jennings, was to join *La Flor*, which would be renamed *Greenpeace IV*. Jennings had experience as a radio operator and navigator. They would sail from Australia to New Zealand and from there along the route McTaggart had taken on his second voyage.

About this time, in early April, McTaggart himself emerged from his hideaway at Buccaneer Bay, alarmed that time was running out for him to get his court case underway. If he did not take action within two months, the "transcription period" would have passed for the ramming at Mururoa in 1972. The Canadian Department of External Affairs had assured him that a thirty-year limitation period applied, but marine lawyers warned that this was just a trick to lull him into inactivity. The fact was that a two-year limitation period applied, and if he didn't get to Paris soon, he could forget about it forever. He had gotten a letter from French writer and publisher Jean-Jacques Servan Schreiber, promising to help him out by putting him in touch with sympathetic French lawyers. So there was a glimmer of hope. But of course McTaggart didn't have any money. It was up to Greenpeace.

A concert at a funky old downtown ballroom was hastily organized and a compromise worked out between McTaggart's demands for money and the demands coming from Melbourne, where, naturally enough, expenses on *La Flor* were mounting by the week. In the middle of this, it was learned that several bills had been run up by the "Greenpeace Whale Show" before it left town for Japan, and creditors were clamoring for payment immediately.

Chairman Hamish could see plainly that, as he put it, "the essence had got lost in the particular." There were so many "particulars" now that everybody seemed to have become incapable of addressing themselves to the one overriding theme to which we all rather self-consciously paid homage: the need to save the planet. Bruce himself was an "old soul," someone you intuitively felt had lived through many incarnations. He did not feel particularly comfortable in the role of chairman, since his vision of Greenpeace was fundamentally religious—in the sense that Aztec sacrifices were religious. However bloody they might be, such rituals called forth higher forces, allowing these forces to brush their great wings over human affairs. Some human beings had considered themselves lucky in the past to have their hearts cut out—such a small sacrifice!—in order to play a part in the triggering of cosmic events. Chairman Hamish was not particularly impressed with McTaggart's sacrifice, or the amount of psychological and emotional distress any of us had suffered. His was a "high" view of the situation, something that was badly needed at a time when meetings had become exercises in petty dickering over nickels and dimes.

Chairman Hamish's high view of reality generally extended to the finances as much as contemplation of the great issues themselves. He considered all Greenpeace money—that is, money that had been donated to the foundation—as sacred. It was strong medicine. It had to be used with discrimination, because there was not much of it to go around. While he did not use words like *sacred* during meetings, he did so privately so often that by the time the meetings happened, everyone present knew that there was something saintly about him—and so his integrity was never challenged. He had a very stabilizing influence on the group that way. It was only when he got angry that he tended to have the opposite effect. A powerfully built man, he exuded an aura of personal strength—there is a word for it in the Japanese language—that either backed people down or stopped them from leaping at his throat. Essentially, he rode herd on the group, but it was a time of transition from one set of basic ideas to another, from the concept of the sanctity of human life to the sanctity of all life. Chairman Hamish held the group together from September 1973, until early March 1974. On the surface, nothing much was happening. Small amounts of money were being raised and split between Spong's whale show, McTaggart's pursuit of the French navy, and the Greenpeace group in Melbourne. A few statements were issued, including a demand that the Canadian External Affairs minister be kicked out of the cabinet for telling lies during the *Greenpeace III* affair, and a few political alliances—with Indians, lawyers, hip entrepreneurs, politicians—were forged. Essentially, Greenpeace seemed to be stuck in the mud, spinning its wheels, at a junction where roads led off in a thousand different directions. That the mud would eventually harden and forward motion would again become possible went without saying. The direction, however, was still far from clear, and until that had been resolved it didn't really matter that the wheels were spinning uselessly. In fact, it was the best thing that could happen. It left a critical interlude during which many of us who were the keys to what would happen next could pull back and attend, for a while, to the business of straightening out our own individual karmic trips—our lives, if you like.

Patrick Moore had set the pace. The previous summer, while the rest of us had been tied up in the task of getting McTaggart to Mururoa for the second time and getting him back, Moore had finished off his thesis, broken up with his wife, and within virtually a matter of hours, found his new love—a lady named Eileen Chivers, a couple of years older than himself, who was just on the upswing after her own marriage to a musician had collapsed. They were an interesting polarity: a Ph.D. in ecology who had been raised in a logging camp on the west coast of Vancouver Island and a highly urban woman who had just spent close to a decade in the decaying Big City remains of the jazz underground. They fell into each other's arms with a mutual emotional hunger that left Moore's concerns with Greenpeace twitching in the dust. They lived for a while in a Kitsilano hippie crash pad called Fowler's Rest Home, ran off together to Mexico, had amazing experiences, returned, and fled up to the Moore family's logging camp at Quatsino Sound, on the northwest coast of North America's biggest offshore Pacific island, surrounded by forests and bears and cougars, fishermen, loggers, and truck drivers.

Because of his logging background and his prominent front teeth, Moore's affectionate nickname was "Beaver." He set to work during the winter of 1973–1974 living up to his name by single-handedly building a house on the edge of the ocean while Eileen went quietly squirrelly trying to cope with her new "bush" environment. Moore had no plans whatsoever for getting back into active involvement with Greenpeace. During the Metcalfe Era, he had been identified by many as Metcalfe's henchman, had exhausted himself storming about Europe and pumping the issue of Mururoa up in the media while McTaggart was undertaking his first run to the forbidden atoll. In the end of it all, Moore had been left feeling quite burned out, his marriage in shreds, and his head not feeling all that good. Now he was content for the time being to live the life of the son of the boss of a logging camp, building his own house and rebuilding his own emotional life with a new partner.

A couple of hundred miles to the south, Hamish Bruce and I were going through convulsions in our own domestic lives, and so were many of our friends. Marriage breakdowns seemed to be the theme, within our circle at least, during the spring of 1974. When your basic life-situation is changing, it is difficult to keep your mind on business. Greenpeace? How can you save a planet if you cannot even save yourself?

For me the end came on the Easter weekend. Since Bruce had taken over as chairman, I had been phasing myself out of the Greenpeace picture as much as possible; except for the long-term commitment to work with Spong to launch an antiwhaling expedition, I did not, at that point, think of it so much as a Greenpeace trip as purely a whale-saving trip. If Greenpeace wouldn't go for it, we'd use another name, like the Stop Ahab Committee. The corporate identity of the supporting organization mattered not in the least. I had high hopes that I was finally "getting off the boat," even if it meant getting into another boat.

Shortly before Easter, my wife Zoe and I packed our kids into our Volks van and headed up to Quatsino Sound to visit Patrick Moore and Eileen. As an effort to bring our poor, exhausted marriage back together, it was a disaster. The kids behaved like little barbarians, and Zoe could scarcely bring herself to talk to me. But on Easter Sunday, Moore took us out on his twenty-eight-foot homemade boat, driven by a Chrysler Crown engine, and something happened that was then still just a glint on the horizon, a possibility into which the necessary amount of energy had not yet been poured. How *do* you save the whales?

Spong and I had been worrying for over a year about the crux of the problem—namely, how would we find the whaling fleets? Getting a boat together did not seem like a tremendous challenge. It had been done before. A crew could likewise be rounded up. Money could one way or another be raised. But how, indeed, do you find a fleet of vessels in the middle of so vast an expanse as the North Pacific Ocean? We had no answers.

As we approached the mouth of the Sound, the swells from the Pacific began to lift us and roll the boat around lazily like a toy in a giant's bath. The sunlight could not have been more crackling clear. It was one of those blessed West Coast days when the beauty of the natural world around you acquires an almost unearthly luminescence. It is damn near too much to bear. Rocks and forests and hills—"spiketop country" it's called, after the spiky skeletons of old dead trees that still thrust above their green children on the slopes leading up to the sky.

In the distance, two whales surfaced.

Moore could not quite believe it. He had lived on the edge of Quatsino Sound most of his life and had never seen whales in those waters. He had spent hundreds of hours out on his boat, had walked and camped along all the beaches, and had stared out the window from the kitchen at the ocean as a child. Not once, in all that time, had he so much as glimpsed a whale in the Sound. Now, here were two of them directly ahead. He gunned the engine. There were three couples on board: Moore and Eileen, a mutual friend Jim Taylor and his girlfriend Marilyn, Zoe and myself. Not one of us failed to pick up on a kind of *heat* that came from within, a quiver, an excitement that made your bones tremble even if your hands appeared completely steady. It was some combination of the magnificence of the day, a culmination of everything we had been talking about the previous twenty-four hours—whales, whales, whales—and the meaning of the day itself, the celebration of Christ's resurrection from the tomb. Moore was perhaps the most blown out, since the miracle of the appearance of these whales had more direct meaning for him than it did for the rest of us.

As we bore in closer, they surfaced again.

"Humpbacks!" Moore cried.

If, indeed, they were humpbacks, then that was a double miracle, because there were only a handful of humpbacks left in the entire world, and they *were* the ones whose songs had been recorded and transmitted via modern stereo into the heads of each of us on that boat. We never did determine what species they were, but they had the knobs arranged like bumps on the back of a vast frog, which are characteristic of humpbacks, and they were about the right size. They never breached, so we had no opportunity to view their flippers, which would have told us conclusively.

Moore's little boat glided into one of the three-foot-high concentric rings that passed across the water when the great creatures sounded. Then they surfaced less than thirty feet away—rolling up and up and up, so that we were all holding our breath, tingling with joy and awe, mumbling thanks in our heads to God for having let such a display occur before our very eyes. The hiss of the water being displaced by their passage, the gurgle and chuckle of it pouring back in to fill the space they had vacated was like the slushing of glaciers. We felt caught up in the flows of a gargantuan presence. We felt lucky, lucky, lucky beyond belief just to come near such a huge physical expression of grace. Allowing us to come within roughly thirty feet, but never any closer, the whales moved steadily out toward the mouth of the Sound, into the foam-flecked open sea where Moore's little motorboat didn't dare venture. We had to stop. We watched them disappearing due westward.

We anchored the boat and rowed ashore in a dinghy to have a picnic. While the others were gathering driftwood, I wandered off by myself. It was time to be alone. Time to meditate. I found a stream trickling down across the beach and started to follow it upstream, into the thick of the fern-jungle coast forest. I don't know how far I crawled before finding a perfectly comfortable spot on some moss beside the stream, but it passed

151

in my head as an odyssey encompassing untold ages.

At the end of it, I sat down, yoga breathing, and had the most profound insight of my life: spring buds were bursting from the trees; the stream was icy, transparent, moving glass; the light placed a halo around everything; and instead of seeing *just* water and trees and sunbeams, I was seeing the most elegant miracle ever devised. For most of my conscious life I had assumed that miracles and God and magic and the Real Mystery were somewhere *out there*, around the corner in another realm, and all that was *here* was the leftover muck of reality. But now it was reality itself—that tangible ordinary stuff you could touch and smell and hear—that was the full-blown Primary Miracle happening right before my eyes. There was no mystery any longer—reality *was* the mystery, and there was nothing mysterious about it at all. Here it was. God and the world were one. The only mystery was the gravel I was sitting on. True magic was the bud on a tree. Not an original insight. But once you have seen the truth, what more is there to see?

The appearance of the whales—the Laws of Manifestation at work—had somehow triggered this. I had never felt happier. I was "blissed." I knew what to do now, and that was very simple: *Trust in the universe to guide you step by step. You are just an instrument in the hand of God. Begin here. Send a boat out from Quatsino Sound in the direction those whales were moving and you will have taken the first step down the long road leading to the salvation of the planet. Do not worry. Just do as you have been told. Skana has spoken to you. These humpbacks, or whatever they are, have pointed out the direction. How many more signs do you want, boy?*

Most of the meetings during the winter of 1973–1974 had been held in Chairman Hamish's living room, but his marriage, too, broke up, and he decided to move into his sailboat, devoting his time mostly to trying to figure his own head out. That left Greenpeace searching for a new chairman and a new place to hold meetings. There were some excellent women candidates, but none of them wanted to take the job. With Spong having been gone for almost two months in Japan and Hamish Bruce and myself withdrawn into our private upheavals, the main advocates of an antiwhaling campaign were in absentia. The Mururoa forces quickly named a chairman who was committed to their issue, a young teacher named Neil Hunter. Hunter had been attending meetings for close to a year. He had originally been introduced to the group by Doc Thurston, who at one time or another had dragged all his non-Greenpeace friends into active involvement, although usually not for long. Throughout the long summer of 1973, Hunter had toughed out most of the meetings, not saying much, but bringing a generally good vibration into the gestalt. He had his black belt in karate and naturally possessed an intense, restless energy. He had what you could call "good drive." Since the whale forces couldn't have cared less at the time who ran the show, he encountered no opposition. A few people found his style of chairing meetings offensive, so they dropped out. It was true that he seemed to have difficulty keeping track of the fact that he had left the classroom behind and was now dealing with a roomful of adults. It was not long before he was being accused of power tripping. Undoubtedly it was true. But then it had been true of every other Greenpeace leader too—probably because none of us had had experience with power before and did not know how to prevent ourselves from abusing it.

Chairman Neil's regime lasted into the middle of summer and ended with the testing of the last bomb in the sky over Mururoa. Modest amounts of money continued to flow from Vancouver to Melbourne to support the *Greenpeace IV*. There were scattered reports of sailing vessels being launched from Hawaii, French Polynesia, even one from Mexico, but none of them ever reached the test zone. The *Greenpeace IV* took a long time getting out of New Zealand and then simply too damn long getting to Mururoa. Rolf Heimann lacked McTaggart's ferocious drive, which had been the key ingredient in the voyages of *Greenpeace III*. Heimann's tiny ketch did finally make it into the water around Mururoa Atoll, but it was not until a week after the last bomb had been tested. Heimann had to content himself with taking water samples, hoping to prove that there had been radioactive contamination. His voyage had undoubtedly been epic—covering such a distance, it could not have been otherwise. But it captured scarcely any attention in the world media, caused no political uproars, and, in effect, might as well not have happened. With the French government committed to ending the tests anyway, neither journalists nor politicians could see any practical point to the protests.

On the other side of the world, in Paris in May 1974, McTaggart had made contact with Brice LaLonde, president of Les Amis de la Terre, the French arm of Friends of the Earth, and had found a young lawyer, Thierry Garby-Lacrouts, who was willing to handle his case, despite strong disapproval from France's legal establishment. One simply did not take the side of a foreigner against one's own government. McTaggart was still in France when the 1974 elections took place, and the Left-wing candidate came within a whisker of toppling the ruling Gaullists. Mururoa had emerged as a major election issue, and the Left had promised to dismantle the republic's entire nuclear program, reactors and bombs alike. During the campaign, the French president was forced to bow to political pressure by canceling one of the scheduled tests and affirming that only underground tests would be permitted in the future. By the time McTaggart left Paris the groundwork had been laid for a civil action under the terms of the international Law of the Sea against the French navy. His lawyer had set to work assembling the material and filing the necessary notices. It would be the first case of its kind since the Law of the Sea had been declared in the sixteenth century.

Greenpeace meetings slowly stopped happening. There was nothing to meet about. The *Greenpeace IV* had moved on to Tahiti. There were no other boats to worry about. Paul Spong had returned from Japan, pleased with the amount of exposure he'd gotten but doubtful that anything he had been able to achieve was going to put much of a dent in public attitudes in Japan. Looking almost Oriental himself after living so long among Japanese people, he only stayed in town a week, then disappeared in the direction of Hanson Island to check out the condition of his whale-watching station. McTaggart came back from France in more or less the same mood of uncertainty that he was accomplishing anything. He, too, headed up the coast, making only the rare appearance in town.

In late July, shortly after the last atmospheric test at Mururoa, I drove out to Long Beach, on the west coast of Vancouver Island, still tormented and indecisive about all the changes that were occurring in my own life. The idea of the antiwhaling expedition still blazed in my mind, and the feeling that I had been given genuine cosmic signs had not lost any of its intensity. But my emotions were drained. I had separated from my wife, Zoe; I was still alternating between the grief of leaving my family and the joy of my new-found love, Bobbi MacDonald. A little bit of time spent sitting by a camp-fire on the beach was exactly what I needed.

The slope of the beach was so gentle that at low tide the water fell back half a mile. Seal colonies clung to the rocks; sandpipers flitted along the edge of the waves. There were gulls, cormorants, puffins, eagles. Majestic pines rose above the beach, and the only sound was the deep pulsation of the sea, the zipperlike whistling of the wind. I had camped there at least once every summer for five years.

One of the attractions of the place all along had been the spouts of rainbow-trailing spray that occurred at random out in the water, launched from the breathing holes of a family of gray whales that stopped in every year to feed, sometimes staying for months at a time.

I had known about them for years, had known that they were unusual whales in that they did not remain with the rest of the grays who annually migrate from the warm waters of Scammon's Lagoon in Mexico to the icy remoteness of the Bering Sea. The whales we had seen from the deck of the *Phyllis Cormack*, which we had decided were finbacks, might indeed have been grays—might, for all I knew, have been these same grays now moving in vast circles off the shore from Long Beach. I knew, too, that the grays had been almost completely wiped out at the turn of the century, but had been protected since then and had managed to make a comeback. No one knew why this particular pod chose to break off from the rest of the herd and venture into the shallow waters at Long Beach. It was known that the same family reappeared here every year, where thousands of campers could watch them. In all the years I had been among those who watched in fascination as the great dark shapes moved leisurely through the waters roughly a mile offshore, it had never occurred to me to go out and have a close look at them. Like everyone else, I had been still essentially afraid of any "wild animal" that large.

But this year was different. I had a small plastic dinghy in the van— basically a kid's toy that had cost $9.95—and I had my wet suit. And my curiosity was more intense than it had ever been.

The waters around Long Beach are known for their fierce undercurrents. Every year, at least one person drowns. To venture out a mile from shore in a toy dinghy, especially to venture into the midst of a pod of Leviathans, was not something I would have contemplated many years before. But this summer I felt serenely secure. I had been reading a lot lately about karma, particularly the theosophical writings on the subject, and the feeling had slowly been building up in my mind that I had "good karma." It was either the kind of nonsensical idea that causes people to lose their lives prematurely, or it was a higher intuitive sense that allowed one to push out in new directions, to test the parameters of luck and reality. It might also have been plain egomania. These philosophical considerations aside, it remained that the tides and undertow were dangerous, a simple change in the wind could sweep my plastic dinghy out to sea, and a single bump from one of those massive creatures out there could finish me. I dug out my *I Ching*, threw the coins, and got a reading of "Approach," with a "line"

(Nine in the second place) that said:

> Joint approach.
> Good fortune.
> Everythings furthers.

> When the stimulus to approach comes from a high
> place, and when a man has the inner strength and
> consistency that need no admonition, good fortune
> will ensue. Nor need the future cause any concern. ...
> Everything serves to further. Therefore he will travel
> the paths of life swiftly, honestly, and valiantly.

It was not the sort of advice my Boy Scout master would have given me under the circumstances, but it was enough—given my new set of emerging beliefs about life—to overcome the last reservations. Taking a couple of bottles of beer and a package of cigarettes wrapped in cellophane as a survival kit, I set out to meet the whales on their own turf.

Getting to them proved harder than I had expected. They were probably closer to two miles out. With a slight wind against me and only two small plastic paddles to propel me through the swells, "approach" was slow and laborious. I was perspiring freely in my wet suit. The whales themselves were moving at least ten times faster than my dinghy. Coming up on the top of a swell, I could glimpse them, could hear the great hiss of air as they blew, and could calculate that the distance between the points when they surfaced was at least a quarter of a mile. The shore was already reduced to a faint silver line that vanished for moments at a time behind the swells. The whales were moving in long loops that carried them miles away, then back. The only hope I had of getting close was to intersect one of the loops, and then hope that I didn't intersect so closely that one of them came up directly under me.

Paddling furiously, I reached a point close to the path of a whale that I could see moving slowly and ponderously in my general direction, its back covered with a pattern of barnacles that reminded me of a Jersey cow. Now that I was within the boundaries of their feeding area, feelings of good karma began to rapidly melt away. God, they were *big*! Far bigger than Skana. I began to tremble. But Spong had said the thing to do was to breathe deeply, empty your mind, meditate, expect nothing.

The whale broke surface about one hundred feet away. Abstractly,

I knew that grays weighed roughly forty tons and they averaged about forty feet long, but those figures meant nothing now that I was at water level with the creature's bulk separated from me by nothing at all, a bit like finding yourself on an open plain with an elephant, except that an elephant could have stood on this monster's back without sinking him more than an inch.

Breathe deep. Empty mind. Meditate. No expectations.

I was not sure how swiftly or slowly the whale had been moving, but now it slowed to a halt. With the slightest ripple of its body, it changed course, probably about forty-five degrees, and began to move again directly toward me. Its advance seemed altogether too fast, like a submarine picking up speed.

It was slightly below the surface, with a wake of bubbles, coming straight at me. Breathe deep. Stay calm. No fear! Goddamn it, *no fear*! But when you are certain you are about to be rammed by a forty-ton monster, it's hard. Then the whale slowed down again and glided to another halt, barely two yards away. It hovered there and dimly, through the water, I could see its eyes. We stared at each other—perhaps only for seconds, but my trembling ceased, and a feeling came over me that I normally associated with spectacular sunsets. It was not that I was eyeball to eyeball with *something*, but rather with *somebody*. He gazed back at me benignly, like a giant uncle upon whom I had never laid eyes, but who seemed nevertheless to recognize me. But then my nerve failed, as it had with Skana, and the image came to mind of myself as a mouse being sniffed by a St. Bernard. Where, for a moment, there had been an interlude of astonishment and wonder, now there was the old reflex of fear, and I started splashing my paddles in the water trying to signal to him to come no closer ... *please*!

I had the impression the whale somehow shrugged. Then it sank downward and glided underneath the dinghy, coming up perhaps thirty or forty yards away, blowing and moving casually along on its feeding circuit.

I was shaking, partially from excitement and partially from relief. The light was waning, the shore was far away, and the only thought in my mind now was to get back to the beach before either the wind or the tide changed.

This particular encounter with a whale—my fourth—had little of the wild excitement of the previous experiences. Yet it served the purpose of reminding me that I had not yet transcended any of the limitations that had been etched out for me by Skana back at the Vancouver aquarium. I could not wallow forever in my own psychological and emotional dilemmas. The whales might have the patience of millions of years of waiting behind them, but they could not wait much longer. Rather than emerging from the water at Long Beach with any dazzling new insights or visions, I climbed slowly up the beach feeling very normal and real and ordinary, aware that I had been close to a normal, real, ordinary whale, and that we were both faced with the normal, real, ordinary world—one in which his kind were being hunted to death. It was time I got on with the business of doing something about it. Oddly enough, the experience had been sobering.

The final and heaviest sign that it was time for Greenpeace to change course in pursuit of the whaling fleets came within months of the cessation of atmospheric testing at Mururoa. Irving Stowe, the strongest and most consistent advocate of antinuclearism, was struck by stomach cancer and died.

He had been our patriarch. There were not a few in the group who cringed every time he stood up to make one of his speeches, but everyone respected his integrity and the primary role he had played from the beginning. He was one of our fathers. And even though he had forsaken formal Judaism for Quakerism, there was much about him that reminded everyone of an Old Testament prophet. He had usually dressed in sandals, macrame belts, blue jeans, and tie-dye shirts, and he had his hair tied in a ponytail—but one could see him in Biblical robes on the slopes of Mount Sinai, with the Ten Commandments of Ecology in his arms.

It was a rotten death. He wasted and shriveled away, bitter and angry at the unfairness of the blow. He had been a vegetarian, a health-food devotee, and he neither smoked nor drank. He had been fanatical about keeping poisons out of his system. He had waged ruthless war on cigarette smokers. To die of stomach cancer seemed the cruelest possible twist. It was not until the end, in the privacy of his home with his family, as a small skeleton of a man, that he broke through the anger into the acceptance of his death, and his great, powerful, wonderful ego laid itself down. His wife, Dorothy, said that on the deathbed his eyes held an unearthly light.

He had been an instrument of righteous wrath. He had achieved much, for without his righteousness and drive, it was unlikely that Amchitka would have been shut down and certain that the campaign against Mururoa would never have been conceived. For Irving Stowe, the ecological prophet, to die within months of the vanquishing of the bomb he hated so passionately was a coincidence of a very high order, as though his positive energy had somehow been bound up in a yin-yang relationship with the negative force of *la Bombe* itself. There was a heavy underlying suggestion in the timing of his death that his work in the world was done, that he had earned his release. There was no one left to resist any further Greenpeace's transformation from nuclear vigilantism to whale saving. And there was no one left to prevent us from dropping the hard brick-by-brick logic of the normal political world completely, seizing our *I Ching*s and allowing signs and visions to determine our course.

The older logical people, the traditional passivist and antiwar people, the legal minds and ideological types had mostly stepped out of the picture. Gone too were the middle-aged Americans like Jim Bohlen whose influence had been so dominant from the beginning.

Neil Hunter had quietly resigned as chairman. The Greenpeace Foundation had effectively ceased to function. All that remained was the four-man Stop Ahab Committee, consisting of Spong, Hamish Bruce, Rod Marining, and myself. In September we arranged for a hall—actually a large room in a sprawling gingerbread building called Kitsilano Neighborhood House—and invited everyone we knew who was remotely interested in helping us launch an expedition to save whales to come down, hear what the plan was, and pitch right in. Some seventy people showed up, ranging in age from an old man in retirement to numerous teenagers.

The plan was as old as the passivist notion of throwing yourself on the tracks in front of a troop train, except that there was more hardware involved. We'd take a boat out to sea, find the whalers, put the high-speed rubber Zodiacs in the water, and race in front of the harpoons, making a clear shot at the whales impossible without a good chance of a human being getting blasted in the process. We would become living shields. It might be the first time in history that people had deliberately set out to risk their lives to save animals.

The meeting was interrupted several times as ballet classes ended in the adjacent room and white-stockinged young ballerinas came fluttering across our floor on their way to the dressing rooms. We applauded them with gusto. They were both embarrassed and pleased. Their passage back and forth became a regular feature of the meetings.

A young broadcaster brought along a tape of sounds that had been picked up on a police radio just as a UFO was passing overhead. The sounds were exactly like those made by whales. The entire group broke up laughing, entranced by the vision of flying saucers filled with salt water and whales peering out through the portholes at the crazy little humans. Ballerinas and Space Whales—there was the tone of the meeting. The heavy atmosphere of moralistic purity that had pervaded almost every single antinuclear meeting we had had over the years was gone. There had been nothing much to celebrate so long as we were opposing bombs. It was a very negative game. But now, instead of simply fighting death, we were embracing life. It was not just that we wanted to save whales, we wanted to meet them, we wanted to engage them, encounter them, touch them, discover them. For the first time there was a transcendent element lying at the center of the undertaking. There was a Holy Grail.

The next five months were truly bizarre, blissed, blessed, lunatic, magical, karmic and so charged with energy that we burned out at least a half-dozen people's nervous systems.

A renegade Brahman from India "happened" to move in next door to Rod Marining. He walked around barefoot in the snow, carrying a hand mirror into which he stared with admiring rapture like a god gazing upon himself, twisting his head about so he could see his face from every angle. When he wasn't doing that, he played a flute and danced, rolling his eyes, usually clad only in a filthy blanket. He was so incompetent he couldn't tie shoelaces or zip up a fly. Yet he wagged his dark bony finger in our faces and raged about how he had come to North America to guide us. We were supposed to instantly fall down on our faces and worship him.

He had a habit of showing up at the oddest times and places, glaring malevolently one moment and giggling at some great cosmic secret the next, chanting to himself. It was impossible to tell whether he was nineteen years old or forty. He was by far the most spaced-out character any of us had ever met, totally twisted and useless, living in a constant state of delirium, yet ... you never *knew*. "Phenomenological psychology" was very much in vogue then, and it was fashionable to think of schizophrenics as possibly the only sane people among us. The leading exponent of this view, R. D. Laing himself, arrived in Vancouver during this time, on a lecture tour, and stated: "The light that illumines the madman is an unearthly light. He may be irradiated by light from other worlds. It may burn him out." Given this viewpoint, our skinny, cantankerous, looped-out little Holy Man from India could not be entirely dismissed. And even if he could, he hung around anyway, waiting to pounce the moment any of us gave any sign of listening to him. It was disconcerting to be haunted by a saint.

Nor was he the only Holy Man with whom we had to contend. A "shaman" came into town from his little house decorated with Zen symbols in the middle of a nearby forest to donate five acres of land with which we could do anything we wanted. The gift was wonderful—but there was a catch. It meant that our benefactor had to be allowed to stand up at meetings and roar out his poems with thunderous passion, or whenever a benefit was arranged he had to be given his turn at the microphone. And he wanted much, much more. He, too, demanded disciples. He, too, was the authentic voice of God. He had the power, he warned, to control the winds and tides. And if anybody displeased him, he would instantly hurl a lightning bolt. Yet no one laughed at him. He tended to rave every bit as much as our friend from India, but the same problem in phenomenological psychology held ... you never *knew*. Strange events did have a way of happening around him.

The presence of these two gentlemen in our midst during the winter of 1974–1975 had a definite effect on group dynamics. It was not that either of them interfered very seriously with the work of organizing the expedition—but they did tend to make life more complicated, especially life inside one's head. You couldn't climb into bed at night with any assurance that you wouldn't be awakened by the sound of chants or mantras, magical rituals or tinkling bells. What was worse was the suffocating feeling that, like it or not, these two basket cases might be genuine gurus, and they just might be casting spells, spinning webs, fiddling with events in sly, transcendent ways, despite the reluctance of any of us to worship them.

We decided we needed a musician to play to the whales. Some freaks up the coast had experimented with playing live music to orcas, and the orcas had responded with flutelike noises of their own, displaying definite interest. All right, we would conduct the same experiment in "interspecies communication." But we would need a very special type of musician, someone with a strong affinity with animals.

The decision to seek out such a musician was made by Hamish Bruce and myself while sitting around the table in the beer parlor at the Cecil Hotel, a convenient meeting place near the newspaper where I worked.

No sooner had I said: "Okay, where the hell do we find a musician with animal affinity?" than I looked up to see an impish, bushy-bearded man in his early forties threading his way through the tables toward us. I recognized him from years before, when I had written a column about how he had been beaten up by the police in an elevator. Now I noticed he had a guitar with him. His name was Melville Gregory, which, if names mean anything, suggested a strong whale connection. Herman Melville had written *Moby Dick*, and Gregory Peck had played the part of Captain Ahab in the movie version. Hm.

"Hi," said Melville Gregory, sitting down. "I just had a flash you guys wanted to talk to me."

"Melville, do you like animals?"

"Are you kidding? I've got two pet iguanas at home, three dogs, two cats. Do I like animals? Hell, I communicate with them all the time. Why?"

"Do you think you could communicate with a whale?"

"Oh sure. Easy. That'd be far out."

Melville Gregory was in fact an excellent musician and composer. At home, he frequently walked around with his favorite two-foot-long iguana, Fido, perched on his head. Like Hamish Bruce, he believed completely in telepathy, magic, clairvoyance, the *I Ching*, astrology, sorcery, the full countercultural mind-set. To him, there was nothing at all surprising that he should have materialized at our table within seconds of the thought being formed that we needed a person of exactly his qualifications—and with a name like Melville Gregory, to boot.

"That's just magic, man. Happens all the time," he said.

However much of an illusion it was or wasn't, it did seem during those months of preparation that one had only to *think* of something or somebody that was needed, and that object or person materialized within minutes, hours, or days. I had never been a believer in telepathy until then, but as the months flowed by and coincidence heaped itself on coincidence, I had to suspend all my previous prejudices.

We decided we would film the expedition. Originally, the idea was just to get footage that could be released to television stations, or used—as McTaggart's photographs had been used—as proof in the event of any hostility from the whalers. But the more I thought about it, the more it seemed that the voyage itself might be worth making into a feature-length documentary. The trouble was, you'd need an entire camera crew. That would be expensive. I was sitting in my office at the Vancouver

Sun, trying to think of somebody I knew who might tell me how you go about assembling a camera crew, when the phone rang. At the other end of the line was a young cinematography graduate from one of the local universities, Michael Chechik by name, who had heard that Greenpeace was planning an expedition to save the whales. He had a proposal. Were we willing to consider the idea of taking a camera crew along to document the trip? Within an hour, we were sitting together at the Cecil, planning the movie.

We needed a photographer; a photographer appeared. We needed a sound technician; a sound technician appeared. We needed a Zodiac expert; a Zodiac expert appeared. We decided we wanted to try making use of sophisticated modern electronic instruments such as a Moog synthesizer to reproduce whale-like sounds; a letter arrived from a synthesizer player in San Francisco who wanted to become involved. We needed a Japanese interpreter in case we intercepted a Japanese whaling fleet; several Japanese interpreters appeared. We needed a Russian interpreter too; a Czechoslovakian sailor who spoke Russian appeared. We decided we wanted to stage a massive antiwhaling demonstration at the time of the launching of the boat; a local festival producer phoned up to say he'd just had an idea that maybe we should stage a massive antiwhaling demonstration.

A computer could no doubt have made sense out of all the good karma and coincidences, showing how they were simply mathematical probabilities whose time had come. The objective truth was undoubtedly that so many human billiard balls had been set in motion that they were beginning to intersect at angles that would not have been possible, or at least would have been highly improbable, with fewer balls on the table. It was only our relative position at the center of the network that made the feedback look so magical and effortless. Be that as it may, the subjective impression was that some force greater than mere human will was at work, and this impression gave the campaign the flavor of a crusade, or *jihad*, a sacred undertaking.

One advantage of a crusade is that morale is usually high. Another is that people who become involved do so with heart and soul: the crusade becomes the most meaningful thing in their lives, infinitely more meaningful than their jobs, home life, or classes. An element of fanatic excitement enters their lives. And then, of course, there is the "contact high" that comes from interacting with new people. Certainly there were plenty of new people ... The support that developed came from just about every level of the community around us. Step by step, the establishment was joining forces with us—yet paradoxically, we had never had more freaks involved than there were now, and there had never been so much crazy mysticism in the air.

Some unknown laws of balance were doubtlessly at work. The more radicals we attracted from the counterculture, the more expensively tailored politicians we seemed to find ourselves shaking hands with. The further out the fantasies went, the more concrete the realities seemed to become. We had the support of the Lower Mainland Dealers' Association, an acid-rock group called the Cement City Cowboys, the premier of the province, the mayor of Vancouver, red-neck radio hot-liners, the Animal Liberation League, the British Columbia Teachers' Association, small Chinese grocers, yoga and meditation societies, the British Columbia Association of Indian Chiefs, the Federation of Labour, artists, weavers, carvers, plumbers, and electricians. We were also joined by highly respected old-time boat-maintenance experts like Gordy Gobel, who was himself a legend on the West Coast, and whose presence in our group signaled to all the fishermen and tugboat operators and marine-parts companies whose help we needed that, however weird some of the youngsters in Greenpeace might be, the organization had to be basically *okay*, otherwise an old pro like Gobel simply would not be involved. A former television producer named Al Clapp, who had excellent connections at city hall, managed to open political doors at parks board and city council meetings, swinging the official endorsement of the Port of Vancouver behind us. Likewise, he pulled the provincial government into the act, arranging for free buses, cooperation with the police department, and the like. A former school superintendent, Chuck Bayley, made it possible for even the Department of Education to get behind what we were doing, distributing our literature throughout the entire school system.

A critical part of the emerging gestalt was the approval of the local Japanese community. During the Second World War, the entire Japanese immigrant population had been forcibly evacuated from the West Coast and sent to concentration camps in the Canadian prairies. Civil rights had been suspended on the worst of all grounds: race. Their property was confiscated by white businessmen, who sold it at tremendous profit. For any British Columbian organization to embark on a campaign—never mind the issue—that involved criticism of Japan was to run the risk of stirring up old racist attitudes toward Japanese-Canadians, still very much a sensitive issue on the Coast.

If Greenpeace was going to proceed with its assault on Japanese whaling, we would have to be very careful how we went about it, lest little Japanese-Canadian kids started finding themselves being called "whale killers" in the school yards. There were, in fact, a few cases of that. But we headed most of it off by inviting in as many Japanese nationals and Japanese-Canadian citizens as we could find, listening closely to their advice about how to avoid even unconscious racism. Taeko Miwa, a beautiful young environmental activist from Tokyo, was one of the first to join us, along with Michiko Sakata, back in Canada after having helped out so much with the "Greenpeace Whale Show" in Japan, and another Japanese woman named Maya Koizumi.

The other major ethnic group with whom we had a special relationship was the Indians, starting primarily with the Kwakiutl of Alert Bay, who had adopted the crew of the *Phyllis Cormack* in 1971. Searching for a symbol for the whale campaign, we hit on the image of the Kwakiutl killer-whale crest, and thought of superimposing the Greenpeace ecology and peace emblems in the center of the crest. And since several of us were still officially adopted Kwakiutl, it was permissible. The "Indian connection" went further. Paul Spong's whale-watching station had been in the dead center of what had formerly been Kwakiutl lands. And no less than three of the men we planned to include in the crew had a special affinity with Indians.

Paul Watson had traveled in March 1973 to Wounded Knee, North Dakota, where the Oglala Sioux had declared the reemergence of the Sioux Nation. He had successfully sneaked past armored cars and machine-gun-toting National Guardsmen to join the besieged Indians as a medical aide. Like many Greenpeacers, Watson had been aware that the Sioux Indians had originally gone to war against General George Custer to defend the last of the buffalo from extinction. The struggle had led to both Custer's Last Stand and the subsequent massacre of the Indians at Wounded Knee. In the context of the twentieth century, with the whales and hundreds of other species faced with the same plight as the buffalo, the struggle of the Oglala Sioux had a special meaning. In a very real sense, their original "war to save the buffalo" had been the forerunner of what had now become the war to save the entire environment. Watson had been accompanied by David Garrick, alias Walrus Oakenbough, an Ontario-born archaeology graduate who had dropped out into the counterculture. During a special religious ceremony inside the fortifications at Wounded Knee, they had both been adopted as "warrior-brothers" of the Oglala Sioux.

Our third connection with the Indians was actor-singer-radio personality Don Francks, a man in his mid-forties who had performed in such movies as *Finian's Rainbow*, but who had grown fed up with the rat race in Hollywood and retreated to a Cree Indian reserve in Saskatchewan, where he had grown his hair long and braided it Indian style. It was the Cree Indians who had produced the prophecy of the coming of the Warriors of the Rainbow—and now here we were, two hundred years later, about to embark on a journey to stop at least a part of the destruction of the natural world, trying to bring people of different races and languages together—exactly as symbolized by the rainbow—and we were being joined by representatives, albeit white, of the Oglala Sioux and the Cree themselves, flying a Kwakiutl flag. It was not surprising that young Indian Red Power advocates frequented our meetings, although usually they stayed in the background and refused to commit themselves too heavily to what they perceived as still basically a "honky trip."

The question of a boat solved itself fairly quickly. Hamish Bruce, Spong, and I shopped around the docks for a few weeks, spreading the word that we were looking for a charter. We had only looked three boats over when John Cormack called up to say that he'd gotten word that the fishing looked bad for the next summer. One of the top men at his fish-packing outfit had suggested to him that he'd "better try to get one of those Greenpeace deals again." The *Phyllis Cormack* had been fixed up with a brand-new Caterpillar diesel and a total of nearly a quarter of a million dollars worth of repairs. She'd done well on the fishing grounds the previous summer and had looked good enough on paper to warrant financial backing for the work. But now ... the *situation* had changed and was saying: *Contact Greenpeace.*

"Well, yes, John, as a matter of fact I was just thinking about maybe giving you a buzz," I said. "We might just have need of your services, sir."

The original *Greenpeace* would now become the *Greenpeace V*.

At the "Greenpeace Christmas Whale Show," a man with the flash and dazzle of an Errol Flynn materialized, introducing himself as Jacques Longini, a retired law professor. His hair and moustache were silver. He had the grace of an athlete, the smile of an actor in a perpetual toothpaste commercial, and he just happened to have purchased the *Vega* from David McTaggart, who had gone off to France again to press his court case. The new skipper announced that he and his vessel were at our disposal.

The *Greenpeace III* would now become the *Greenpeace VI*.

The costs of the boats would be paid for by a lottery. The prize, of course, would be the five acres of land that had been donated to us by the self-proclaimed shaman, who still attended all our meetings, chanting and waving eagle feathers and haranguing us to publish his poems. With two boats and the means of paying for them lined up, we could now start seriously thinking about such details as maybe finding an office, so that things could get a little bit more organized. Kitsilano Neighborhood House had burned down, leaving us with nowhere to meet, and our numbers had grown too large to be accommodated in anybody's living room. Suitably, we rented three small second-floor rooms from a local antipollution group on Fourth Avenue, in the heart of the city's low-rent hippie ghetto. Now that we had telephones, a mailing address, and a gathering place where people could congregate at any time of the day or night—more a clubhouse, really, than an office—the tempo began to pick up.

By January, there were an average of at least two dozen people crowded into those three small rooms every day, feverishly painting posters, answering phones, cutting out bumper stickers, counting GREEPEACE buttons, picking up tin cans for soliciting donations outside of liquor stores, counting lottery tickets, typing letters and appeals for funds, and slowly gathering together the array of equipment we'd need for the voyage: outboard engines, Zodiacs, life jackets, radios, walkie-talkies, wet suits, diving tanks, hydrophones, boxes of dried goods. Artists began donating paintings and sculptures to be auctioned off to raise more money. Inevitably we spilled over into the offices of the antipollution group. They were tolerant and sympathetic, but obviously flustered by the mounting pile of art objects, electronic gizmos, engines, and crates, and the constant flow of Indians, gurus, fishermen, technicians, musicians, children, sea captains, and regular office workers.

Having an office meant several things; it meant that our underground period was over, that the old filter of only those who had been invited being present at a gathering was done with: anybody could walk in through the door, anybody could listen to our planning sessions. We had gone "public." But the existence of an office also stimulated a fierce territoriality. Malingerers who tried to hang around, flirting with the women, trying to engage in pointless debates, or in any way hindering activity, found themselves quickly ushered down the stairs, or, if that didn't work, being transformed into an "unperson" through which everyone looked as though he had already vanished. Such massive mental rejection invariably meant that the physical body would soon follow.

Having an office also directly affected the sexual vibrations that passed between us. During meetings, everyone is either sitting or standing in one place, and there is hardly any bumping into one another or finding yourself alone face-to-face in a back room. There is very little jiggle or bounce. But open up physical space where everyone can move around, interact, mingle, press up against one another; fill it with people whose average age is in the mid-twenties; add an extraordinarily high percentage of beautiful, intelligent, liberated, radiant, positive females, most of them single; introduce them to young men and older men, men who are alive with the excitement of embarking upon a dangerous mission—and the juices cannot help but begin to flow. "Office gossip" became a tapestry of rich and constantly changing patterns, new configurations, sometimes subtle, occasionally gross, with an untold number of meaningful looks being exchanged across the room, hands being pressed fleetingly with passion. Some couples paired off quickly, others were already paired and succeeded in staying that way; but as in any herd, there was a great deal of jostling and heavy breathing with young bucks snorting and stamping. The light of love hovered like a nimbus over all our gatherings, infusing the atmosphere with tremendous surges of primal energy.

Yet for all this natural ferment, promiscuity was not the order of the day. There were too many spiritual people. When these people did liaison sexually it was highly discriminatory and usually discreet. Most relationships remained a mystery except to the participants. There was, too, an overriding prohibition on any displays of overt male chauvinism. The rhetoric and to a large degree the actual spirit of women's liberation had penetrated to our core. Women held most of the positions of power: Katrina Halm controlled the treasury, Elizabeth Dunn chaired the critical Crew Selection Committee, Gail Meredith protected the office, and the woman with whom I lived, Bobbi MacDonald, emerged as the driving feminine energy that fused the parts of the kaleidoscope together, taking on task after task and ramming it through to completion.

The voyage had not been planned as a "rational" act, but rather as an act of faith. We were admittedly counting on a miracle. It was only to be expected, under such circumstances, that we should find ourselves automatically seeking out miracle workers, turning to prayer, consulting the *I Ching*, and hesitating to reject completely even the "gibberish" of our benefactor, the "shaman," or the mirror-crazed renegade Brahman from India. No one knew the exact route to Miracle City, so we could ill afford to ignore any directions that were offered along the way.

But there was one other path upon which no unearthly light seemed to shine at all—and it was the one we could least afford to miss. Despite our general receptivity to the occult, we had not abandoned pragmatism entirely. Nor could we see anything wrong with giving the expected miracle a nudge. To this end, we decided that our best, most practical hope for finding the whaling fleets lay in the employment of an ancient strategy: since we couldn't beg or borrow the information we needed, we'd steal it.

Blessings were fine, but exact data giving the positions of the whaling fleets was what we really wanted. The only person who stood a chance of getting it was Dr. Paul Spong, and so to him fell the task of espionage. I would shoulder the responsibility for getting the boats out on the water, but it would be entirely up to him to tell us where to go, and if he failed, well, there was prayer, but that was all.

He set out early in the year with Linda and Yashi, taking the "Greenpeace Whale Show" on its second tour. The show was to serve three real functions: firstly, it would raise the money to pay for air fares; second, it would act as a magnet to attract other interested whale experts, any one of whom might know where the data on the fleets was hidden; and third, it was an all-purpose front for Spong's secret spy activities.

They traveled to Banff, Edmonton, Calgary, Saskatoon, Medicine Hat, Winnipeg, Toronto, Hamilton, Montreal, Halifax, and St. John's. In each of these Canadian cities, they left behind a small cadre of whale lovers who formed the nucleus of what would later become Greenpeace branch offices. From Newfoundland they flew to New York, and from there to Iceland, where the real work began.

The trail took the Spongs through several schools, to the Marine Institute and finally to an Icelandic whaling station where Spong assumed the role of an "interested whale researcher." He made polite conversation with the executives of the company and obtained the name of the principal whale scientist in Norway, Dr. A. Jongsgård. Along the way, Spong found himself one afternoon in Reykjavik, chatting with the bishop of Iceland. The bishop was an elegant, middle-aged gentleman who confessed that the thought of "blowing up a whale" horrified him. For that matter, sports fishing horrified him too, because he could relate to the suffering of a fish with a hook in its mouth. But the Church of

Iceland was controlled by the state, which meant that the bishop was a civil servant, and much as he privately wished the whale hunt would be stopped, he could not officially come out against government policy.

From Iceland, the Spongs flew to Norway and tracked down Dr. Jongsgård at the University of Oslo. Spong had been given a letter of introduction by scientists back in Iceland and it was this letter that allowed him entry to the eminent whale scientist's office. After a lengthy technical discussion about behavioral responses in *Orcinus orca*, a seminar was arranged where Spong showed his slides to a couple of dozen students, fielded questions, and generally behaved as much like a detached scientist as he could. Even at that, Dr. Jongsgård took exception to some of the "subjective criteria" that Dr. Spong was allowing to creep into his thesis, and at one point rushed to the screen with a pointer to contradict something that had just been said. The "pointer" proved to be a narwhal tusk.

Apart from that incident, Dr. Jongsgård found Dr. Spong's work to be satisfactory—at any rate he was happy to sit around talking afterward, scientist-to-scientist, exchanging information. Spong was not at this point saying anything at all about his involvement with Greenpeace. He was presenting himself strictly as a whale researcher who had done field work with orcas in their natural habitats and wished to follow these studies up with similar investigations of the behavior of the larger whales, particularly sperms. The problems involved in researching deepwater mammals were obviously much greater than those involving whales that came within a few yards of the shore. And the first problem was simply where to go to observe sperms. Presumably the whaling companies would have such information, based on the logs of their whaling ships—but one could hardly expect them to release that information, since they were in competition with other whaling companies.

Perhaps the good Dr. Jongsgård would know someone who would be more sympathetic to such a request? Purely in the interests of science, of course.

It so happened that the good Dr. Jongsgård did know exactly the man—not a scientist himself, but a gentleman who had worked for the Bureau of International Whaling Statistics in Sandefjord, Norway, since 1938, which was as long as whaling statistics had been kept. His name was Mr. Vangstein. And, yes, an introduction could be arranged.

A few days later, Dr. Spong, earnest young whale researcher, and his charming wife and delightful child, were sitting in Mr. Vangstein's office in Sandefjord, on the top floor of the biggest high-rise in a town that had once been the home port of the largest whaling fleet in history, and the one that more than any other had been responsible for the near-total extinction of the blue whale. Entrance archways to large old mansions, the former homes of whaling company barons, were decorated with immense blue-whale jaws. The great days were gone—along with nine-tenths of the world's whales—and all that remained were bones and rows of neat files in the offices of the Bureau of International Whaling Statistics, watched over by an elderly clerk and his secretary.

Spong went through his story again, moving slowly lest Mr. Vangstein become suspicious. At first the old man was reluctant to part for even a minute with any of his precious files. He was impressed, indeed, by Spong's sincerity and fascinated by his work with orcas, but the information contained in the files was "company secrets," and he could hardly be expected to allow them out of his hands without a formal written request from the Scientific Committee of the International Whaling Commission.

Spong understood the delicacy of the situation, of course, but the fact was that if he was to carry out his "studies" of sperms in their natural environment, his only hope was to know where the whaling fleets had gone. And it was sad, but the "political situation" at the IWC would probably inhibit the Scientific Committee, despite the certainty that many scientists themselves would value the results of such studies. It would be a shame to miss this opportunity. His career as a scientist was at stake. And he had traveled such a long way.

By himself, it is unlikely that Spong would ever have changed the old man's mind, but Mr. Vangstein was a grandfatherly fellow, delighted by Linda's beauty and little Yashi's precocity, and in the end, he decided that coming from such nice people the request was "innocent enough." While Yashi and Linda remained in the office, engaged in jolly conversation, Spong was allowed to go through the files, picking out what he wanted, and copying it furiously into his notebook. Mr. Vangstein was in such a friendly mood by this time that he appeared at the doorway, time and again, happily displaying a new file he'd found that he was sure would help. All the information Spong had wanted now lay before him. It took most of a day, but at the end of it—while Linda's and Yashi's laughter rang merrily through the bureau—Spong had copied out thirteen pages of figures, giving the longitudes and latitudes and dates and number of kills for the Soviet factory ships *Dalniy Vostok* and *Vladivostok* and the

Japanese ships *Nissim Maru #3*, *Tonan Maru #2*, and *Kyukuyo Maru #3*. The information covered the period from August 1973 to October 1974. Mr. Vangstein even put his secretary to work typing the data out.

It was late afternoon when the Spongs bounced back to their hotel and cooked a celebratory meal over the camp stove they'd brought along because they couldn't afford to eat at restaurants.

There they were, sitting in a hotel whose walls were covered with blown-up photographs from the 1920s and 1930s of Norwegian businessmen posing proudly in front of whale ribs and jaws, men who had decimated the whales of the planet, and they had just lifted the information which was going to change the voyages of *Greenpeace V* and *Greenpeace VI* from exercises in fantasy to exercises in real power. They had done a sleek con job. The fate of the expedition had rested on their ability to carry out their roles. They had done it! The awesome fifty-thousand-to-one odds had been reduced to something approximating an equal battle. Instead of the uncertain stuff of magic and karma, Greenpeace would henceforth be able to operate on the basis of longitudes and latitudes and specific dates. There was no guarantee, of course, that the Japanese and Russian fleets would go again to the areas where they had hunted in 1973 and 1974, but there was now a realistic chance of intercepting the whaling fleets. It did not take Spong many hours to translate the rows of figures he had obtained into position fixes on a chart—and it quickly became clear that the Russians, at least, could come into range of even a vessel like the *Phyllis Cormack* off the coast of California in late June, possibly not much more than fifty miles out to sea, along a band of out-thrusting seabed known as the Mendocino Ridge.

So it was not as much information as it might have been, but it was enough. Just exactly enough.

Spong phoned me from Norway with the news right away and advised that the expedition be delayed until the beginning of June. But by this time we had already committed ourselves to staging a huge send-off party in Vancouver on April 27, both as a demonstration against whaling and as a means of attracting world-wide media attention to the expedition itself. The April date had been picked virtually out of a hat, and as Spong now made clear, it was a full two months too early for our purposes, yet the big send-off demonstration had become such a seemingly important event in itself that nobody wanted to cancel it because we might not be able to set it up again.

After a great deal of lobbying and political in-fighting between the parks board, city council, and the provincial government, permission had finally been granted for us to use an abandoned air base on the shore of English Bay as the site for the send-off. The location—near the heart of the city—had been closed off to the general public for several years while various levels of government had haggled over its ownership. It was known variously as Jericho Beach, Jericho Park, Jericho Base—but mainly just as Jericho. Its dominant feature were four massive Second World War aircraft hangars and a great tarmac stretching between them. The buildings' windows were all broken. Cracks had opened up in the tarmac through which weeds were thrusting. Surrounding the buildings were acres of rolling meadow and bush. To the north, across the bay, was a great range of mountains that marched right to the edge of the sea. In the distant past, Indians from all over the Pacific Northwest had come there to stage potlatches and religious ceremonies. In 1939, the military had taken it over, but strictly as a wartime measure. Since the end of the war, the vast hangars had been left to decay. This would be the first time they had been used for a nonmilitary purpose.

Apart from the magic of the name and the significance of the beach in coastal Indian history—the perfect place for the Warriors of the Rainbow to launch their assault—the great wooden hangars themselves were like relics from another age. For years, artists and community groups had been fighting city hall in an effort to gain entry. Now they would have their chance. The doors of Jericho had been thrown open. In itself that would have been no minor cultural event, but coupled with the launching of the Greenpeace boats, it promised to be the biggest demonstration in the city's history. And that was exactly what it became.

Transforming an abandoned air base into the stage for a festival was an awesome task in itself, coming on top of all the work that was already going into the readying of the *Phyllis Cormack* and the *Vega*, both of which had to be dry-docked and their hulls recaulked. Lifted out of the water, the two vessels looked like great wooden shells about to be fired. In the past, group energy had been exhausted just getting either one of these craft ready. Two of them meant at least twice the work. And with the frantic rush of preparations at Jericho, where the sound of hammers echoed through the hangars night after night, it was not surprising that as

April 27 approached, our work force of perhaps one hundred volunteers was thinning out rapidly as people dropped in their tracks.

There was also a certain weirdness to events around Jericho: the base was still locked up behind hurricane fences and patrolled by guards and dogs. In order to gain entry, we had to say a password. Strange, at night, to approach a military base and say "Greenpeace" to an armed guard before the rusting gates would creak open. Also, the organizational genius who had come forward to put the show together, Al Clapp, had brought with him a funky gang of bikers, carpenters, rock-concert promoters, and sound-systems technicians, most of whom looked as though they ate railroad spikes for breakfast. Mix these characters together with all the flower children who were already pouring into town, add dozens of Vancouver policemen, some on motorcycles and others on horseback, allow for representatives of several dozen Indian tribes who were already setting up their "sweat lodges" in the hills around the base, make way for the musicians and their equipment, and you have the makings of a West Coast "ecological Woodstock," which was what we were aiming for, with the exception that there would be no rock music. For in that case the City Fathers would not have allowed it to happen.

There was a full moon the night before the launch. A fifty-mile-an-hour windstorm swept over the coast, smashing power lines and telephone poles, knocking down trees, causing blackouts all over the lower mainland. But as morning dawned, the wind had spent itself, leaving a cloudless, perfect spring day. By 7:00 A.M., a dozen of us were down at the *Phyllis Cormack*, bleary-eyed and hung over from a benefit dance that had been held at a downtown ballroom the night before.

In the last-minute rush to load supplies onto the *Cormack*, a bottle of formaldehyde got knocked over. We had purchased it at the insistence of a local whale authority who wanted us to carry out a couple of obscure experiments. Now the boat was filled with deadly fumes. Paul Watson and I yanked on diving masks, wrapped wet towels around our faces, and plunged into the bunk room to start sopping the stuff up, staggering outside every few moments, gasping and vomiting over the side. The incident unexpectedly provided a useful service. A young freak who had earlier been rejected as a crew member had stowed away in the above-deck aluminum shelter around the smokestack. Now he staggered out of his hiding place, blinded by tears from the formaldehyde, and was quickly escorted to the dock.

Once the mess was cleaned up, some friends picked me up, and we drove across town toward Jericho. The boats themselves would rendezvous offshore, and my one remaining task before departure was to make sure that all the crew members would be assembled and on board by 4:00 P.M. We had to park the car two miles from the base and walk the rest of the way. As we approached the site, my stomach gave out on me. It was a combination of excitement—a fantastic *buzz* of energy emanated from the hangars, rising above the antlike heap of people who had converged there—plus a hangover, plus formaldehyde poisoning, plus fatigue from the preceding months, and especially the last few weeks, and the knowledge that this was finally it. A dozen times during the trek to the hangars, I collapsed by the sidewalk and had to sit there, blinded, with nothing left to throw up anymore, my body feeling like it was turning itself inside out, while the steady stream of healthy, happy people flowed past on their way to the big event.

By 2:00 P.M., an aerial survey confirmed that there were roughly twenty-three thousand people on hand. The hangars rang with music. Smoke puffed from the sweat lodges. Tents were pitched everywhere. Indians chanted. Hare Krishnas tinkled their bells. A snappy breeze lifted great kites everywhere. There were dogs and horses and children and middle-class couples and pensioners and hippies and a whole hangar full of artworks, most of them depicting whales. Four sound stages had been erected, and the crowds around them clapped and danced.

Somewhere in that crowd was my crew.

The next two hours were a nightmare of dashing frantically about, rounding up crew members, pulling them forcibly away from their friends and family, until we had everyone assembled on the beach. Al Clapp had signaled that within twenty minutes he would start the parade. All we had to do was somehow get across half a mile of white-flecked water. The beach sloped too gently for the boats themselves to come in close to shore without grounding. A fiasco of considerable proportions loomed.

Suddenly, to my utter astonishment, a *landing craft*—a steel Second World War landing craft—came crashing through the water toward us. It plowed right up onto the sand, and a couple of bikers in black leather jackets jumped out. In the background, I could hear the sound of the bands starting up and Al Clapp's voice over a megaphone, assembling everybody for the parade. Coming out of my paralysis, I ran to the landing craft, but before I could even open my mouth, the man at the controls grinned and said: "Hop in!" We scrambled aboard, the engine roared, the landing craft surged backward effortlessly, and we found ourselves sweeping grandly along the promenade just as the crowd led by a mock-up whale made its appearance. The sound that greeted us—twenty-three thousand cheering people—so overwhelmed me that I began to cry.

The landing craft, driven by a hip coast sailor named Dennis Feroce, who had heard from somebody that "Greenpeace might need some help," made two quick runs from the shore to the *Cormack* and *Vega* and back, delivering the crew. Externally, it looked as though everything had gone off with military precision, although, in fact, it was shithouse karma again. At exactly 4:00 P.M., Captain John sounded the horn, and we took off with the shouts and cheers still echoing and reechoing in our heads. *Vega* departed on a clean, swift tack with the wind right behind her. We were a couple of miles out, with the white hangars of Jericho and the pastel towers of the city dropping rapidly behind, when it suddenly dawned on me that we were missing several people.

The camera crew had been left behind.

By morning, the *Phyllis Cormack* was pushing out through the mouth of the Juan de Fuca Strait, her axe-bit bow cleaving huge swells left over from the gale that had lashed the coast the night before the Jericho launch.

Vega, under Jacques Longini, had headed northward with the intention of slipping over the hump of Vancouver Island and back down to Quatsino Sound, Patrick Moore's home, where the two boats would rendezvous.

The *Cormack* handled magnificently. In contrast to her first Greenpeace voyage, when she had looked like a hulking, rusty workhorse, with an old Atlas engine on the verge of laying down and dying, she rode now in a blaze of technicolor splendor. A white triangular sail had replaced the old soot-covered green canvas, and the great Kwakiutl whale symbol stood out brilliantly in the sunlight. The entire vessel had been freshly painted with glossy whites and greens. A four-man aluminum trailer had been mounted on the poop deck and converted into a sound studio crammed full of speakers, oscilloscopes, generators, six-track high-speed stereo tape decks, AKG-150 condenser microphones, transistors, capacitators, headphones, walkie-talkies, and calculators. With Zodiacs piled up like blunt rubber missiles at the stern, and a quadraphonic underwater sound system nestled near the trailer next to a half-dozen four-foot-long cylindrical hydrophones, there could be no missing the fact that she was embarked on one of the strangest maritime assignments in history—a probe designed to test the possibility of communicating with a nonhuman intelligence.

So far as Captain Cormack was concerned, of course, all this business about "interspecies communication" was nonsense. But that was all right—a deal was a deal. "If you fellas wanna go floppin' around in the water tryin' ta chit-chat with whales, that's yer business. Just point where yuh wanna go and I'll git yuh there. But I still say yer loonies."

If the skipper had had the difficulties relating to his crew on the Amchitka run, the problems had been trivial compared to the chasms that opened between him and this latest bunch. At least on the first trip, there had been men like Birmingham, Metcalfe, and Bohlen, who were all approaching fifty years of age. Now, aside from Cormack, the oldest crew member was thirty-seven. Most were in their twenties. And while, before, there had been a few hippies included—and none of them in important decision-making positions—the reverse was now true. There were a couple of "regular fellas," but the rest ... well, the rest included such "weirdos" as Melville Gregory and another musician, Will Jackson, and it was impossible to say which one outraged the skipper most.

Gregory was a graying elf, a mystic who never parted from his *I Ching* coins; short, pot-bellied, full-bearded, long-haired, a street musician who frequently played gigs in seedy downtown pubs and who had once been thrown into jail for several months. Life had not been particularly sweet for him. In Cormack's view, he was a "lush, a mattress lover and a good-for-nothing bum." All of which might have been true, but would still have been forgivable, except that Gregory went ahead one evening and "liberated" a cod that had just been caught. Gregory's explanation— and he was, after all, our animal-affinity expert—was that he had looked into the eyes of the fish as it was lying on the deck and decided that it was still alive, even though it had stopped twitching. Moreover, the fish "communicated" that it wanted to be set free back in the water. While no one was looking, "Saint Melville"—as we quickly came to call him—tossed the fish back overboard. When Cormack found out about this, he roared with absolute disbelief and rage, and Gregory had to flee into the trailer and lock himself in to avoid being thrown overboard himself. The captain was flabbergasted. "Can ya beat *that*?" he demanded. No one could.

But at least Saint Melville played music that Cormack could understand. The same could not be said of our other musician, thirty-year-old Will Jackson from San Francisco. Demonically good-looking—like Mick

Jagger of the Rolling Stones—Jackson had received a scholarship from the California Institute of the Arts in 1970 and became interested in exploring synthetic electronic music. The pursuit developed into multimedia electronics performances throughout the San Francisco Bay Area, mostly the underground circuit of art galleries, theater groups, and colleges. He'd experimented with 3-D laser imagery and holography and had worked with Serge Tcherepnin designing a synthesizer that combined the advantages of Moogs and ARP with large studio-control systems for tape engineering. His Tcherepnin synthesizer was suspended by ropes in the *Phyllis Cormack*'s hold, where it swung back and forth like a bizarre pendulum, festooned with oscillators, filters, modulators, envelope generators, programmers, sequencers, and mixers. The machine, for all its space-age sophistication, still depended, Jackson said, "on the best rock-'n-roll wah-wah boost pedal ever made." He had only to play the instrument once, sending weird warbling sound waves and whale-like screeches out of the hold, before Captain Cormack reached the conclusion that "that Will Jackson fella's no musician, he just makes a *racket*, that's all." Cormack's favorite song was "Hallelujah, I'm a Bum," from the Depression era. "Audio-processing systems" from the depths of the San Francisco avant-garde music scene were an out-and-out insult to his sensibilities. Try as he might, after his first solo performance on the Tcherepnin, Jackson could not win back Cormack's affection. The captain viewed him as a Martian sent to torment human ears.

The crew included only two experienced seamen: Paul Watson and our Czechoslovakian friend, George Korotva. Watson was on board primarily to function as "lead kamikaze" when the moment of confrontation came: he would drive the Zodiac that would place his body and mine in front of the harpoon. No one else would be asked, or allowed to risk their lives in that way. Originally, I had expected to be traveling in the Zodiac with Paul Spong, but when it became evident that Spong's place in the scheme of things was to travel to Europe to seek out the information we needed, I had been left with the thought of being alone under the harpoon. Never having driven an ordinary outboard, let alone a Zodiac, and not being particularly athletic, I did not believe I could physically handle the situation. Watson was much in disfavor with most of the Greenpeace crowd because of his impetuousness, unpredictability, and a marked tendency to brag too much. I think, too, his great physical strength intimidated many people. He was a martial-arts buff—infatuated with Japanese women, in fact with all things Japanese. He did fantastic impersonations of John Wayne that spilled over into his whole style of speech and movement. His nickname was "The Duke." He had something of a persecution complex—but his reflexes were excellent, he had great courage, and he was willing to die if necessary.

There, after all, lay the key to the whole exercise: the readiness to exchange one's life for that of a whale. Whether Watson was particularly popular or not was very much beside the point; he was a warrior, and his willingness to go with me into the valley between Leviathan and the great steel harpoon gave me strength I doubted I had on my own.

At the outset, George Korotva's role was simply that of an experienced sea dog who happened to speak some Russian. Thirty-three years old, he had skippered ships larger than the *Phyllis Cormack* along the West Coast. A Czechoslovakian, he had been imprisoned by the Russians during the 1965 student uprisings in Prague and shipped off to a camp in Siberia where he learned what he knew of the Russian language the hard way. After eighteen months of imprisonment, he escaped with two companions and walked forty-five hundred miles to the Finnish border. He'd knocked about in Europe, involved in the shadowy world of agents and counteragents, before making his way to Canada. His English was delightfully atrocious. He never had a "hangover," instead he would be "hanging over." If he was mad at somebody—which was often, his temper was fierce—he'd threaten to "drill the hole in your head and fok you in the brain." He had difficulty distinguishing between nouns and proper nouns, so that Will Jackson became "The Will," Cormack became "The John," Gregory became "The Mel," I was "The Bob," and George himself, of course, became "The George." The George inevitably had a cigarette holder clamped in his teeth, he laughed with Dostoyevskian gusto, and had an aura of strength that suggested he could tear trucks apart with his bare hands. He was an odd combination of street fighter and Pavlovian psychologist, with an actual ex-con survival reflex.

Whatever he was, The George was not at heart a sentimentalist. It was he who grabbed me one night shortly after we took off, pushed me up against the wall—we were out on deck in a light drizzle, in waning copper light—and said: "Look, brudder, it matters for nothing that you get falling-down pissed and stoopid, aye? I like it. But to *deese* people you the guru, and if you turn into a cabbage, the whole trip turns into cabbage, no shit. *Get it together, brudder!* These people counting on you. So dem whales."

I don't know what Korotva's—or anybody else's—IQ was, but it was high. He had been hiding for years behind the image of "poor dumb immigrant," yet he was a master European chess player from a family of aristocrats whose lineage went back many centuries—and he saw with pitiless clarity what was going on. He understood the games that group dynamics forced us all to play, including the need for me to play alcoholic, cynical clown in order to cope with the tremendous and mysterious energies that were swirling about us, but he could also see the lines that were drawn, over which it was unwise to tread too far.

It was evident to Korotva from the first night out that I was having a lot of trouble with my role as "expedition leader." Having always hated leaders—my father had deserted my mother when my brother and I were still young—it was close to nauseating to realize that I was now a leader and group father myself. The George, from that point onward, became virtually my personal psychiatric advisor. A Pavlovian, remember, schooled in Siberian prison camps. He also became my bodyguard. I don't know, during that voyage and the ones that were to come, how many times he saved my life, but a rough guess suggests at least a dozen.

Cormack's relationship with photographers had never been easy: on the Amchitka trip, Bob Keziere had lived in terror of him. The old man had mellowed quite a bit since then in terms of having his picture taken, so long as he knew what was happening and had time to strike the appropriate pose. But he didn't like the new-fangled habit of photographers sneaking around trying to catch him off guard. Instinctively, he was wary therefore of Rex Weyler, whose task it was to document the trip. Weyler was a longhair, too, an American refugee who had been living in Texas when the draft board had reached out for him. He'd fled into the depths of the psychedelic revolution in Los Angeles before emerging as an activist in Chicago, working with Resistance leaders Tom Hayden and David Harris, being tear-gassed several times before following the guru trail to India and Nepal, getting married in Amsterdam, and returning to the U.S. only to be chased by the FBI into Canada, where he now worked as a photographer and assistant publisher of a small Vancouver weekly. Back in 1970, in France, he'd written a letter home describing what he had learned from his round-the-world travels:

> There exists in the world today a brotherhood
> that reaches every corner of civilization. There is
> a brotherhood of people who love *life* more than
> *things*. They are not all young, they are not all stoned,

they are not all wise—but they exist. Some are very
old and wise—some are very young and not so wise.
But this brotherhood I think is unique in history, and
it gives me hope. It knows no barrier of language or
land—it is real. I can't put it into words in a letter,
but if this planet makes it out of this century without
self-destructing, miracles may happen.

In the person of Rod Marining—whom Weyler had interviewed once
because Marining had a small windmill in his front yard—the antiwar
veteran had recognized a member of his "brotherhood," and through
Marining had been introduced to Greenpeace. Now Rex Weyler braced
his legs on the deck of the *Phyllis Cormack*, a long way from Texas, his
cameras poised and waiting for miracles. The mark of once having been
an American hippie wandering through India was upon him: he was
wide open to the notion that the universe was a miracle to begin with,
and there was something about his style—a willingness to put his palms
together in prayer and bow—that suggested he had probably lived a
whole other incarnation, not just as a visitor to India, but as a native, one
who never did get a chance to visit Texas.

The cook was Walrus Oakenbough. Walrus had traveled to Wounded
Knee and had been adopted as a "warrior-brother" of the Oglala Sioux.
He toiled mightily in the galley, having to go through all the same lessons
that Bill Darnell, the cook on the first trip, had had to experience: not
opening cans upside down, not hanging cups the wrong way, making sure
the skipper had his first breakfast at 3:00 A.M. and his second at 6:00
A.M., and that he had cookies and cake and candy to keep him happy.
Walrus's food was a bit too "hippie-dippie" for Cormack at first—salads
that included kelp, mashed potatoes mixed with parsley flakes, sassafras
tea, eight-grain sprout bread.

No one ever did find out why he'd changed his name from David
Garrick to Walrus Oakenbough. With long hair and a great walruslike
moustache, he looked more like a frontier cowboy than an Indian, yet it
was to the North American Indian consciousness that he had gravitated.
His favorite expression was "Ah, shucks."

Twenty-eight years old, Walrus was a graduate from Trent University.
The road from Trent to the deck of the *Phyllis Cormack* had been long,
and somewhere along the line, David Garrick had almost completely
ceased to exist: an Electronic Age Indian-cowboy stood in his place,

humming merrily as he worked, putting off such good vibrations that everyone loved him—including, grudgingly, the captain. "I don't think much of that veggie goop he calls food," grumbled Cormack, "but he's a good worker, I'll say that."

But the person on board who was most loved and admired, of course, was Taeko Miwa, the classically beautiful young Japanese activist who had been previously involved in blocking the construction of an airport in Tokyo—by means of hundreds of people digging tunnels on the site and living underground for months while police tried to flush them out with tear gas. She had also been intensely involved in the struggle in Japan against the mercury-contaminating corporations that had unleashed Minamata disease on the world. Everybody's protective instincts about Taeko were in high gear, precisely because she was so quiet, so exquisitely beautiful, and apparently fragile, like a ceramic figurine. If anybody had tried to maul her—or even make her feel slightly uncomfortable— he would have been battered to death by the rest of the crew within minutes. Cormack warned that if anybody laid a hand on her, he'd dump them in the ocean. "No smoochin'," was his rule. That translated as "no fraternizing among crew members," and he had charged me, as "agent for the chartering firm," to make damned sure that none of the ladies who were on board yet, or who were scheduled to join us later, were molested in any way. That proved to be an easier task than it seemed: Walrus and Taeko, for instance, promptly fell head-over-heels in love, and while Cormack and I were prepared to bash heads if anyone bothered the lady, what could we do when she herself was in love?

The doctor was Myron MacDonald, a man with a prematurely graying beard, dark, almost occult eyes, and a good friend of Lyle Thurston, who had served on the Amchitka voyage. MacDonald had been married, before they broke up, to the woman I now lived with— but that didn't stop us from being the best of friends. Technically, since neither of our divorces had come through yet, Myron and Bobbi were still husband and wife, except that she was now with me and Myron was with someone else. Cormack could only shake his head. Myron had sailed around Vancouver Island in his own boat, understood navigation, understood group psychology, said little, and saw almost everything that was going on.

It was he who drew my attention the first morning out, as we were plowing triumphantly through the swells in the Juan de Fuca Strait, to the fact that one crew member, H., was showing every sign of having gone completely schizophrenic on us.

"I hate to say this," whispered Dr. MacDonald, "but we've got a problem. H. has been standing at the bow for seven hours now. He won't leave, and I'm beginning to think he might die of exposure pretty soon if we don't do something. He won't take pills. I tried."

H. was, indeed, standing at attention at the bow. The April Pacific wind was cold, and there were icicles dripping from his moustache and beard. His face was slightly blue, and his eyes, when he finally turned them on me, after several minutes of coaxing him to say something, blazed with an awesome intensity, as though he was seeing much, much more than I was.

"H, are you okay?"

A hint of a vastly superior smile played on his frozen lips. A single, slow nod was his only response, then he went back to staring straight ahead into the advancing swells.

It took about another hour before Korotva and Dr. MacDonald were able to coax H. back into the warmth of the galley. By this time everybody on board knew that he had flipped out—yet rather than having become merely lunatic or tragic, he had somehow acquired an enormous dignity. He hardly said anything. You could tell it was an effort to bring himself down into the normal world enough just to bother forming words, and you knew as well he was doubtful that it was worth the effort. He looked at us, one by one, piercingly, with a mixture of contempt and pity. He had acquired psychic X-ray vision. We felt like primitive animals being examined by a great, detached scientific consciousness that had descended into a temporary physical form. For the first time I truly understood why, in the old days, when people went mad, those around them thought the mad one had been "possessed." H.'s personality had been swallowed up by a kind of "overmind"—a higher being who now looked out at us through pupils that had expanded until they almost filled H.'s eyes.

"If you go by the book," said Dr. MacDonald, "he shows every symptom of classical schizophrenia, but on the other hand he might be an Awakening Being or some kind of Buddha. I don't think he's dangerous. Let's just act normal and see what happens."

"Acting normal" around H. was of course impossible. Except for Cormack, we all found ourselves tiptoeing about when he was present. And looking him in the eye—those supraintelligent, overviewing eyes— was out of the question for more than a second. The tension was not eased in the least when he appeared the second day out with war paint on his cheeks, or when, on the third day, he started chanting in a language that none of us had ever heard.

We stopped in at the Vancouver Island community of Tofino to pick up the camera crew we had left behind at Jericho: Fred Easton and Ron Orieux. Then we headed northward toward Quatsino Sound, the place where, over a year before, I had seen the humpback whales heading out to sea. There, at the logging and fishing village of Winter Harbour, we were to rendezvous with the *Vega*.

Slowed by rough seas, it took us a total of five days to reach Winter Harbour. Patrick Moore and Eileen Chivers were waiting at the dock, along with about a dozen other Winter Harbourites. In the confusion of hugs and kisses and jokes, I only had time to warn them: "H. is acting a bit strange."

It was not much of a warning, considering that his condition had worsened since the first morning he'd spent on the bow. Dr. MacDonald had already coined the term "bow-case" to describe him. Our bow-case ran amok that first night in Winter Harbour, stumbling through deep salal bushes, howling, chanting, raving, gibbering. The forest came alive: vines reached out and seized him, ferns spoke in semaphore. Pulsating through the entire planet beneath him was an overwhelming primal life-force. His mind went out of time and space, he visited different worlds and dimensions, his identity underwent orgasmic changes. He also terrified the hell out of quite a few of the villagers. By morning, when he found his way back to the boat, it was clear from the way the muscles on his face were writhing that he was approaching some kind of psychic critical mass and was liable to explode in front of us any minute.

We had to get him out of there. He was already dangerous to himself— so out of touch with mundane reality that he couldn't tell the difference between water and lighter fluid. Knives and forks had to be kept away from him lest he cut off a finger. Reporters—including one from *The New York Times*—were starting to appear, and the last thing the campaign needed was a story about how one of us really *was* crazy. I approached him gingerly, to tell him that Doc Thurston was on his way up from Vancouver, that a seaplane would soon arrive, and that we had arranged for a place, a hospital, for him to go.

His response was to flare up with awful, contemptuous rage. His anger seemed to communicate itself directly into the environment around him. A flock of crows broke suddenly from the beach, screeching furiously, tumbling in the air above him. A cold blast of wind abruptly flattened the swamp grass around us. I experienced a fear quite unlike anything

I'd ever felt before—as though I had, indeed, enraged a genuine sorcerer or shaman.

The seaplane with Doc Thurston arrived. He and Dr. MacDonald stalked H. around the decks, trying to talk him into accepting tranquilizers. The pilot stood around nervously, insisting "unless you get that guy on drugs or tied up, there's no way I'm putting him in my plane." It was an awkward situation, to say the least, what with well-wishers from the village wandering curiously about the boat, the man from *The New York Times* watching everything with shrewd big-city eyes, and almost everybody in our crew terrified at the thought of an open, physical confrontation with H. Somehow, his muscular strength had increased in direct proportion to the degree he had gone out of his ordinary mind. I was certain he could snap my neck like a twig.

He moved suddenly, making an effortless six-foot leap across the deck to the edge of the hold. He was down the ladder before anybody could move. Cormack and Korotva—the two strongest men on board—both grabbed the top of the ladder to prevent him from knocking it down. The ladder was a heavy wooden affair, hammered together with large nails. When H. found he couldn't wrest it from the combined strength of Cormack and Korotva, he simply broke it in half, then went and squatted at the stern of the hold, where he began chanting again.

Korotva leapt after him and I swung down on a rope. We tried every tactic—we reasoned sweetly, we poured out love and brotherhood, we threatened, we ranted, we begged and wept and howled at him. Nothing worked.

While Korotva and I were down in the hold trying to pry H. out of the corner into which he had backed, like a dangerous and badly wounded animal, the two doctors were up in the galley crushing tranquilizers into a thick powder and mixing it up with mayonnaise and pickles and cheese, putting it all into a thick sandwich which they lowered into the hold with instructions to pass it to H., who hasn't eaten yet that day. H. rejected the food with an impatient wave of his trembling hand. Korotva and I, however, were both starved. Unaware that the sandwich had been spiked, we each wolfed down half of it. I don't remember anything much after that, except a scattered recollection of being hoisted by several people out of the hold and carried to my bunk. Korotva just barely made it to the deck before he collapsed.

At last H. turned docile, even cooperating to the extent of lowering his trousers so Dr. MacDonald could give him a sedative injection. He was pulled out of the hold on a stretcher, wrapped in a sleeping bag designed to look like an American flag, taken to the plane, and flown to the psychiatric ward of a hospital in North Vancouver. He left behind an emotionally exhausted and bewildered crew. H. had been our first bow-case. Who would be next? We had all sensed that we were mixed up in the play of immense forces, maybe even some kind of archetypal clash between great evil and great good, and that there might be heavy karmic feedback that could go either way, but we had not quite braced ourselves to deal with anything so stupendous and immediate as the devastation of one of our own member's soul. And we had been utterly helpless to do anything to prevent it.

It did not seem like a coincidence that, the day after, Paul Watson came down with an attack of acute appendicitis. He had to be taken by helicopter to the town of Campbell River for an emergency operation. He would rejoin us later.

It was very much as though, in the process of being forced off the boat, H. had taken Watson's appendix with him. He had not just devastated us emotionally—he had done it physically, too.

Within days of the launch, the Japanese Whaling Association in Tokyo had attacked us for initiating a "not only regrettable, but foolish" action. At the same time, a fisheries ministry official was quoted as saying that Japanese whaling fleets were under orders to avoid a confrontation with Greenpeace. Japanese newspapers were filled with the story, but other than general advice, no firm policies or strategy had been formulated to deal with the harassing activities on the high seas. Asked by reporters what Canada would do to protect us in the event of any violence, the Canadian fisheries minister, Roméo LeBlanc, said: "It is obvious that if you are in international water, there is not very much we can do." Asked if he was saying, "They're on their own," he answered: "You said that, I didn't."

Back in Japan, indignance reached such a point that the government itself issued a statement warning that if our two little boats were somehow to succeed in interfering with legitimate Japanese whaling operations, the Land of the Rising Sun would press charges against Greenpeace in the International Court of Justice. From Russia: an Olympian silence.

In Peking, the *People's Daily* reported briefly, but approvingly, about a "Canadian attempt to challenge the hegemony of Soviet revisionism in the North Pacific." From the standpoint of a publicity campaign designed to draw attention to the issue of whaling, we had succeeded from the moment we left the dock. The Jericho launch itself had made us a media event, and while we were not exactly breaking over North America and Europe in an avalanche of chunky headlines, we *were* established as an item in the global communications network, one whose progress would have to be followed. In Japan, we had stirred up a mess of whale consciousness—and for openers, that was plenty. The trick now would be to keep us alive as an item for the long two months that really remained before we could hope to hit the Russians. Under the best of circumstances, two months is a long time for any story to survive.

As chief propagandist for the voyage, I was now fully committed to being a "flack" for the whales. In order to sustain our global media presence, I was going to have to use every trick I had ever learned from Ben Metcalfe and be ready to invent quite a few on top of that. In preparation for the trip, I had resigned as a columnist for the *Sun* in order to be on board as a reporter. The advantage of that was simple. The subjective stuff written by columnists is never picked up by the wire services. But the "objective" stuff from reporters is treated as legitimate news. Thus I could file an "objective" report from the *Phyllis Cormack* one day, and flush a reaction out of Tokyo via the Associated Press wire service within less than twenty-four hours. All I had to do was make sure never to quote myself. Instead, I invented quotes, placed them in the mouths of various agreeable crew members, then "reported" to the outside world what they had said. As a journalist, I was, of course, a traitor to my profession. As "news manager" for the expedition, I could censor any unflattering realities, control the shaping of our public image, and when things got slack, I could arrange for events to be staged that could then be reported as news. Instead of reporting the news, I was in fact in the position of inventing the news—*then* reporting it. Sooner or later, we would have to come up with the goods—a confrontation with a whaling fleet. And there was no way that could be faked, nor its failure covered up.

In the meantime, there was much to do. The Zodiacs had to be tried out. The underwater sound systems and hydrophones had to be tested and retested before our electronics expert, Al Hewitt, could get all the bugs out of them. And the crew had to be trained. That meant taking the Zodiacs out into the mouth of Quatsino Sound and opening them up to full speed, just to see what would happen. Korotva and Rex Weyler climbed in one, Patrick Moore and I into another, and our two musicians, Jackson and Gregory, into a third. The camera crew put-putted out ahead of us to get film footage from a regular outboard as we came crashing, leaping, ker-flopping wildly by. The Zodiacs were truly amazing craft. Their speed was dizzying, their maneuverability, everything we could ask for. We even dared to advance into the open sea, with the Zodiac ahead disappearing completely from view among the swells, then zipping up effortlessly like a rubber rocket that one half expected to see keep on going into the sky. The trick was to hang on tight enough to avoid being bounced clear into the water. We even practiced throwing ourselves overboard at high speeds so that in the event of an accident, we'd be familiar with the situation.

Only a few miles down the sound was a place called Coal Harbour, which had been the last land-based whaling station in Canada. It had been closed down only seven years before. This meant there were several fishermen in the village who had worked on the old whaling ships and who were familiar with all the local areas where whales had been hunted. It was a piece of real-life symbolism that lent a feeling of rightness to our presence in Quatsino Sound. It helped, too, because they were happy to tell us where to look for whatever whales had survived.

It was May 6 before the *Vega* finally arrived in Winter Harbour. Her captain, Jacques Longini, fifty-four, was a man who seemed to have done everything and been everywhere. His favorite anecdote concerned the time he claimed to have flown Madame Chiang Kai-shek's furniture out of China just before the Communists seized power. Nobody was quite sure how many wives he had had, or even how many he had at the moment, since his wallet was full of photographs of himself with various dazzlingly beautiful women. His body was athletic—yet there was a certain world-weariness about him, and a tendency to go up and down in his moods so frequently that he quickly came to be nicknamed "Captain Flapjack."

We were also joined at this point by Nicholas Desplats, a young Frenchman from Les Amis de la Terre in Paris, who had also been active in the battle of Mururoa. Desplats was never without a pipe in his mouth; he wore corduroy jackets with leather patches on the elbows, very literary, very Parisian, and he seemed to be enjoying himself immensely. It was difficult to know, for certain, because his entire English vocabulary consisted of two phrases: "It's good" and "Thank you." Nicholas Desplats had never seen a genuine North American wilderness before, nor had he ever seen such a creature as a bear. We found this out one afternoon, at a nearby garbage dump, when two black bears appeared. Automatically, the three of us who were present backed toward the car, not realizing until too late that the Frenchman had stayed behind. To my horror, I looked back to see the bear standing on its hind legs, front paws placed on each of Desplats's shoulders, eyeball-to-eyeball with our visitor, who simply stood there with his hands in his pockets, puffing calmly on his pipe. No-one dared to yell or do anything to startle the bear at that stage. After a few minutes of examining this odd phenomenon of a completely fearless human being, the bear dropped back to the ground and wandered away. Desplats returned casually to the car. "It's good," he said, beaming.

Our main task in Winter Harbour was to get all the sound systems working. In this department, Al Hewitt was our sorcerer. Tall and lean, thirty-seven years old, with bushy eyebrows, Hewitt could easily have passed for Vincent Price. He did not so much *hang around* any given scene as *lurk*. He spoke only when he had to, was shy of cameras, and was so technically competent about engines as well as electronic gear that Captain Cormack had already adopted him as chief engineer. Hewitt had been introduced to us as an "electronics genius," and a genius he was—forever perceiving mathematical patterns and relationships that eluded the rest of us completely. Now, under his instructions, the waters of Quatsino Sound came alive with strange sounds. He had rebuilt the five hydrophones to meet our specific requirements. Each hydrophone consisted of a vertical watertight tube containing recording microphones, a power source, and an FM transmitter hooked to a flexible aerial that floated above the water. Under Hewitt's direction, the Zodiacs were used to deposit the hydrophones all over the inlet so that sound waves could be transmitted to the "command module," meaning the aluminum trailer on the back deck. In theory, the hydrophones would allow us to hear whales or dolphins from miles away.

The underwater speaker system was Hewitt's other major project. The speakers were hooked up to a variable power source, monitored by an oscilloscope for sound-shape studies, which served as a visual safety valve to make sure the system didn't overamp. The total energy output was 270 watts, which gave an audio range effective over at least three hundred miles.

The final crew member to join us was Carlie Trueman, twenty-four, of Victoria, B.C., an experienced scuba diver and small-craft operator. A short, stocky young lady who wore pigtails, Carlie had loaned us her own Zodiac and was an expert at taking care of these highly sensitive air-filled craft. She didn't go for "hippie-dippie shit" herself, was practical and technically competent—definitely more a mechanic than a mystic. By May 12, we had not seen a single whale and my editors back home were getting tired of stories about appendicitis and technical problems—they wanted action.

I placed a phone call to a friend, John Allen, who managed a hotel overlooking Long Beach, from where I had paddled out in my plastic dinghy to meet the grays. There, we knew for sure, there would be whales.

"Hi, John. Bob here. Listen, we need to get some footage of whales pretty quick. Can you have a look out the window?"

"Sure. Just a sec."

There was the pitter-patter of his feet moving away across the room, then returning.

"Yeh, they're here."

"Thanks, John. Watch for a funny-looking boat showing up soon."

It was decided that Patrick Moore would take my place as coordinator of the *Phyllis Cormack* as it journeyed down to meet the grays, and at 6:00 A.M., May 13, the *Phyllis Cormack* headed south to Long Beach. Five hours later, *Vega* departed from Winter Harbour, heading due west to the edge of the continental shelf, where we planned to establish a floating listening post, monitoring the radio for any sign of whalers. On board were Jacques Longini, myself, and three Americans: Matt Herron, a photographer and navigator who had sailed across the Atlantic with his family; Ramon Falkowski, a veteran of the voyage of the *Fri*, one of the protest vessels which had sailed to Mururoa in 1973; and John Cotter, a young doctor from Kansas City, who had never been to sea in his life. There followed eight frustrating, uncomfortable days while we bounced and lurched about in winds up to forty-five miles per hour some two hundred miles out to sea, popping Gravol pills, discussing philosophy, and staring vacantly out the portholes at the waves as they came rolling over us. As early in the year as it was, we were probably the only sailing boat out this far in the eastern Pacific. The wind and water both had a damp, awful chill from the icebergs not so far to the north.

We finally picked up a clear Russian-language broadcast, but we couldn't get a fix with the radio direction finder, and in any case, we were unable to get through to shore to report because Captain Flapjack had somehow not managed to get the right matching crystals for his side-band radio. As a "spy ship," we were a flop.

But while we were busy failing to find any fleets, the *Phyllis Cormack* was having no trouble at all finding whales, and quite a bit of trouble avoiding being inundated by whale freaks all seeking a ride. Stopping at Tofino to resupply before approaching the grays, the boat was invaded by visitors from Vancouver. There were soon no less than eighteen people on board, including Paul Watson, who had swiftly recovered from his appendicitis attack and was now clamoring loudly to be accepted back on the boat immediately. Carol Brien, a nurse who was scheduled to join the expedition later, had arbitrarily decided to show up early. American musician Paul Winter, of the Paul Winter Consort, had arrived, as planned, so that he could play to the whales, but several others, including Watson's girl friend Marilyn Kaga had also chosen that moment to arrive. Patrick Moore strove mightily to keep the crew numbers down, but there were so many people straining so relentlessly to be included that by the time the boat pulled away from the dock at Tofino it was overloaded and any semblance of order had broken down.

It was a simple matter for the *Phyllis Cormack* to swing around the peninsula into the wide, shallow bay where the grays—eight of them—were feeding. In no time at all, the crew could see the whales clearly: a puff of white vapor, then a glimpse of a dark fingerlike shape against the intense blue of the sea. And off somewhere else, another puff. Another. And, occasionally, a fleeting after impression of a wing-like tail. Most of the people on board had never seen a whale before, and in their excitement, they became confused.

Zodiacs were soon buzzing about, motors were stalling, and shouted instructions from the deck were either ignored or misunderstood. There were at least eight people pointing in as many directions. Carlie Trueman and a San Francisco oceanographer, Gary Zimmerman, were soon in the water with their special cameras, paddling first in one direction, then another, but not being able to get close enough for any decent underwater footage. Because the bottom was only forty feet below and composed of gritty sand, visibility was less than eight feet, and the whales would not cooperate to the extent of remaining motionless long enough for the divers to get up beside them.

The results of the musical experiments were not too exciting at first. When Paul Winter began to play his saxophone melodiously through a recycled army loudspeaker, one of the grays did swing toward the boat, coming within one hundred yards and pausing to listen. Will Jackson's synthesizer had roughly the same effect. One of the largest of the whales pulled quietly up beside the *Phyllis Cormack* and hovered there for several minutes, as though puzzled or at least curious. Then it went back to feeding. The quadraphonic underwater sound system did not get much use. The waters off Long Beach were too shallow to allow Cormack to simply cut the engines and drift for any length of time, and with the engines running, the sound system was more or less canceled out.

Whale communication proved to be more successful when conducted from a Zodiac. Melville Gregory got what he called a "good" response when he tootled his flute at them. Paul Winter found that by resting his saxophone against the wooden floor plates of the Zodiac, he could send a vibration out through the water that led several times to the grays coming close enough, and lingering just long enough, so that the smell of their breath passed over the occupants of the Zodiac. Afterward, everyone agreed that the gray whales had the equivalent of halitosis.

The most exciting moment came when Zimmerman, Trueman, and photographer Weyler came up behind a particularly huge gray—at least fifty feet long. The creature had allowed them to motor within a few yards, so long as they motored slowly, with the engine throttled down and everyone remaining quiet. But on their third such approach, the whale altered its speed, so that when it came up, it was barely ahead of the Zodiac's bow. The backbone—a long row of saw-edged spines—rose slowly under the boat until the tail was directly below them. Carlie cut the engine immediately and Weyler snapped picture after picture, while Zimmerman simply hung on, waiting to see what would happen. The tail rose gently against the bottom of the Zodiac, then lifted, so that the craft and its occupants suddenly found themselves rising slow motion out of the sea. The whale lifted them no more than a foot before depositing them just as gently back in the water. But it was enough to impress on them forever the difference in their size and strength. There had been no sense of strain at all. As easily, the whale could have flipped them head over heels dozens of feet into the air. That it had chosen to give them a minor demonstration of its power, without harming them in the least, left everybody feeling somewhat awestruck and humbled.

It was not until the second day at Long Beach that Walrus got a chance to go out in a Zodiac, taking Takeo and Carol Brien with him. By this time, everyone except John Cormack had taken a turn at "communicating" with the grays, although, in fact, most of the would-be communicators had spent more time racing around in Zodiacs, probably annoying the whales, than getting anywhere close to establishing serious contact.

By contrast, Walrus and his crew tried a gentle approach. They cut the engine and drifted, playing on a bamboo flute and a recorder. Walrus felt they had entered "the whale's sphere of mental influence" and that a "test was brewing that required that the engine be turned off if we were to be accepted by the gray whales." It took close to half an hour, during which time it seemed to Walrus that the presence of two females in the Zodiac acted as a "powerful attractive influence that augmented the whales' response to flute music." In the end, one whale came within fifteen feet, lifting its great head out of the water to inspect them.

Walrus wrote later:

> In conclusion, we felt that the gray whales definitely responded to our presence, respecting our decision to shut off engine. We felt they enjoyed our music, especially responding to high-octave sounds—though other people felt & experienced that chanting, singing & saxophone or flute music all evoke definite decision on part of whale to take time out from munching to listen and move with people—often playfully exercising powerful sign language and body language.

Will Jackson spent several hours alone during sunset and dusk, meditating in a Zodiac, attempting to reach a Zen-like state of "no mind," whose purity in itself would attract the whales. It could, of course, have been a coincidence, but two whales did choose to linger within ten and fifteen feet of him for several minutes at a time, returning three or four times.

Jackson was never to be the same after that.

The encounter with the whales, especially the short, sweet ride Weyler, Zimmerman, and Carlie Trueman had taken on the whale's tail, had the effect of "converting" everyone involved into whale freaks, in just the way that Paul Spong and myself had been converted by coming into physical contact with whales in the first place. That the giants were, indeed, gentle, was now clear to everyone. So, even though as a "scientific experiment" the visit to Long Beach had really proven nothing, it did mean that by the time the *Phyllis Cormack* returned to Winter Harbour, which was serving as our main base, her crew was at least ten times as committed to whale saving as ever before. In sharp contrast, *Vega*'s crew was moody and depressed. And certain lines of conflict had etched themselves clearly in the collective headspace of both crews. Walrus and Taeko's attraction for each other had passed the point of cooperation in preparing meals and had reached the stage where Cormack caught them necking passionately. Now the skipper was furious and wanted one or the other of them off the boat. "No smoochin's what I said and no smoochin's what I mean." Silly hassles had developed on the *Phyllis Cormack* during its trip southward, over smoking and no smoking, and a set of "rules" had been posted dictating the hours during which one was allowed to smoke in the galley. The competition between the two skippers, Cormack and Longini, was becoming irksome, with Cormack constantly ribbing the other to the point where Longini was threatening to pull out. Between Matt Herron and Longini there was also an element of conflict, since Herron considered himself to be the more experienced sailor but Longini insisted on his prerogatives as captain. Cormack's loathing of the synthesized music played by Will Jackson had become almost pathological. One afternoon at Long Beach, Jackson had blasted an electronic concert at the whales, which could be heard on the shore, two miles away. Halfway through the presentation hailstones began to clatter down on the decks. Cormack attributed the hail to the sounds Jackson was producing.

The *Phyllis Cormack* was back in Winter Harbour by 6:00 A.M., May 22, along with the *Vega*. Several crew changes occurred immediately. Taeko Miwa was asked to leave. Doc Cotter from Kansas, Michael Chechik, the producer, and Patrick Moore also departed—Moore because he had logging work to finish off in order to be able to rejoin the boat later, Chechik because of his seasickness, and Cotter because his holiday time was over. Nicholas Desplats was reluctant to give up his bunk, almost tearful, when I told him his time on the boat was *fini*. An intense young man named Doug Dobbins was shuffled onto *Vega* to replace Doc Cotter. Paul Winter, Ron Orieux, and Marilyn Kaga left. Dr. MacDonald had by then returned to his practice in Vancouver.

Shortly before the boats set sail again, I threw the *I Ching*. I drew the first hexagram, The Creative, which was the very one I had drawn years before, when the idea of the whale trip first came to mind. That was a very good reading—but this time there was appended warning: "Hidden dragon. Do not act." My psychological relationship with the *I Ching* was still sufficiently ambiguous so that I tended to accept its advice when it went along with what I was already thinking, and ignore it otherwise. So in this case, I shrugged and decided to carry on anyway.

194

At midnight sharp, both boats pulled away from the dock.

Our plan this time was to rendezvous above an underwater range of mountains some 120 miles out to sea. The undersea mountain range—more properly a plateau—reached to within 750 fathoms of the surface, and it was here, mainly, ex-whalers who lived at Coal Harbour had told us, that the whaling ships had always gone to kill sperm whales. For some reason, sperms tended to gather above such seamounts. This one had the oddly suburban name of the Dellwood Knolls. Once out there, we would get to test our RDF equipment, our lorans, and, above all, our ability to find another ship at sea. The two boats would go at their own different speeds out to the Knolls, then track each other down. That was the practical part of the trip.

At another level, it had been over three weeks since we'd left Vancouver, and, aside from the reports and photographs of the encounter with the grays at Long Beach, we had not been generating many decent new stories. We needed something—a grabber—to boost our fading news value. Already, in the Vancouver *Sun*, my dispatches from the boat had slipped to the back pages. So now I concocted a story that we were "satisfied that no whaling ships have entered the eastern Pacific," and that the two Greenpeace boats were therefore altering course to conduct a unique scientific experiment. An eclipse of the full moon was expected shortly—our mission was to document the arrival of sperm whales on the feeding grounds during the full moon. My dispatch went on:

> "So far as we know," said Dr. Patrick Moore, an ecologist, "no one except whalers has ever witnessed the gathering of the Sperm whales at the time of the full moon beyond the continental shelf."

> Moore explained that at such times, tiny organisms in the water rise to the surface to absorb the additional lunar light. Giant squids which normally remain in the depths are known to follow these organisms, upon which they feed, to the surface.

> "In turn, giant squids are the favorite food of Sperm whales," said Moore. "We hope to film and record the gathering of the Sperms, and if we are extremely lucky, to document an encounter between a Sperm and a giant squid. The only reported sightings of such

an encounter have been at the time of the full moon." Sperm whales grow up to 80 feet long. Giant squids can reach 50 feet. "The struggle between these two immense creatures is probably the most awesome encounter to take place in the natural world."

As science, it was technically correct. As a possible reality, it was remote, to say the least. We had still not even seen any evidence that sperm whales had survived along this coast at all. But as a news story, it was fairly hot stuff. The *Sun* ran an eight-column banner:

GREENPEACE SHIPS SAIL TO SEE
"AWESOME" BATTLE OF PACIFIC GIANTS

The wire services grabbed it. In newspapers as far away as Australia, headlines appeared, saying:

MAN LANDS ON GIANT
SQUID–WHALE GROUND

Any perceptive student of media would have seen in the story the thrashings of a desperate publicist, and I was, indeed, beginning to feel desperate. More than a month remained before we could reasonably hope to make contact with a whaling fleet, and already I was having to strain to come up with news stories. If we lost the media's attention, so did the whales. Also, a month remained before the 1975 meeting of the International Whaling Commission, and it was upon this meeting that we wished to exert the maximum influence for a ten-year moratorium.

At first light the next morning—a Friday, May 23—the sea was relatively calm. A brilliant rainbow appeared off the port side to the south. But the wind was picking up, and by 9:00 A.M., Cormack had decided we were heading into a storm. True to his cautious fisherman's survival instinct, he decided we'd better duck back into the "hole" at Triangle Island, rather than take an unnecessary pounding in the open water. He had mellowed enough toward Will Jackson to be willing to give the synthesizer player a few lessons in handling the wheel, but when the skipper ordered Jackson to "come about," he did it much too abruptly. Down in the galley, Walrus Oakenbough watched in horror as one-seventh of the boat's coffee mugs were reduced to shards, one-seventh of her plates were smashed to pieces, a drinking glass and a beer bottle shattered, and milk, water jugs, rice containers, and every ashtray on the boat went over on their sides.

Reaching the shelter of Triangle Island around noon, cameraman Fred Easton expressed a desire to go ashore to get film footage of the birds, including several bald eagles who could be seen cruising above the shore. Carlie Trueman, our Zodiac expert, offered to take him to the beach, and for some peculiar reason, Will Jackson and I decided to join them. We had little trouble getting the Zodiac into the water, but as we approached the shore, it quickly became evident that landing was going to be difficult. There were few openings between the barnacle- and weed-covered rocks, and the waves were coming in three- and four-foot surges over the outcroppings. None of us had wet suits on—just rain slicks and heavy rubber boots. It seemed to me that Carlie was bringing us dangerously close to the rocks. I had just finished saying:

"Carlie, isn't this too close?"

The next moment, the Zodiac was snagged on a rock.

"Get out, get out, you guys!" Carlie screamed. "Push us out, quick, before the next wave."

Fred Easton hopped nimbly onto a nearby rock, holding his precious camera high in the air to avoid having it soaked. Jackson and I landed hip-deep in the water, and strained furiously to push the Zodiac free, but before we could budge it, the next wave had come in, lifting the Zodiac, sweeping us under it and backward off our rock perch into deeper water where we couldn't touch the bottom. Our boots filled up immediately, dragging us downward, with the Zodiac cutting off any possibility of surfacing directly above, anyway. Jackson grabbed me with one hand and clutched desperately at the Zodiac with the other, barely hanging on as we went tumbling with the wave toward shore. The only coherent thought that went through my mind was: *"Hidden dragon. Do not act."*

Getting the opening she needed, Carlie gunned the engine and rode up onto the next wave that was coming in, clearing the rocks by inches, leaving Jackson and I floundering, coughing, and shuddering from the impact of the icy water, but with loose gravel beneath our feet, enough for us to get the traction we needed to haul ourselves up onto the beach. Easton had meanwhile hopped gazelle-like from rock to rock and was standing on the shore, completely dry.

Had Jackson not grabbed me, I was certain I would have drowned.

Several cold and extremely uncomfortable hours followed before Carlie was finally able to retrieve us from the shore. The eagles, thinking we were threatening their nests, spent much of the time hissing at us and dive-bombing at our heads. My respect for warnings from the *I Ching* had increased enormously.

By the next day we were over the Dellwood Knolls. We pulled down the sail, cut the engine, and drifted, with people taking turns on the mast, watching for whales. In the past, whales had apparently gathered by the thousands over this seamount to feed and mate and generally socialize. During the day, we saw three dolphins, one sea lion, and one seal. Otherwise—nothing.

The night of the full moon and the eclipse turned out to be oddly subdued. A thin film of clouds closed in just as the eclipse began, leaving the decks bathed in a strange copper glow. Jackson sent his synthesizer sounds out gently, hauntingly into the sea. A few petrels and an albatross wafted silently through the sky, and the swells cradled the boat, making us all sleepy. No whales appeared. No giant squid. There were no awesome battles between the Titans of the deeps. There was nothing to document except the strange and lonely sounds that Will Jackson sent quivering over a lifeless, copper-colored ocean.

The next day was Sunday. We rendezvoused with *Vega* at 11:00 A.M., but the event somehow lacked focus. Our attention was concentrated on the singular fact that this area had once been bursting with life but now contained almost no life at all. "It's a biological wasteland," commented Zimmerman.

Our momentum seemed to be slowing badly.

At this point, I had still not told anyone on board that we actually knew where the Russians would be. I had the list of positions that Spong had transmitted from Norway, but had buried it at the bottom of my duffel bag, rather than risk having someone blab casually to a reporter somewhere along the line that our real game plan was to be off the coast of California during the last week in June. We knew from press reports that the Japanese were paying close attention to our activities, so it only stood to reason that the Russians—master spies that they were—would be doing the same. So long as they remained confident that we didn't have a clue as to their real whereabouts, the odds remained that they would follow their normal pattern, but any inkling that they would be jumped off California might be enough to cause them to avoid that one tiny area where we had any chance at all of intercepting them.

And it was not just that the Russians themselves might somewhere be monitoring us: George Korotva had been approached by an old acquaintance of his who worked for the Russian embassy in Ottawa and warned that "an agency of the Canadian government" was feeding the Russians regular reports on our location. It was only to be expected. Ottawa obviously did not want to see a repeat of the sort of international incident that had occurred when the *Greenpeace III* confronted France at Mururoa. Canada's voting pattern at the International Whaling Commission, of which she was still a member, was exactly the same as Russia's. Canada, no less than Japan and Russia, was an enemy of the whales, and therefore our enemy too. On top of all this, David McTaggart was in Paris during this time, patiently, persistently pushing his case against the French navy through the courts and was expecting a verdict before the end of June. Depending on the outcome of the case, Canada might well be left in a position of having to espouse his claim—in other words, take direct action against France, possibly in the form of economic sanctions. The bureaucrats and politicians in Ottawa shuddered at the thought of having to go through such an ordeal with Russia—particularly because of that damned Greenpeace outfit again.

Spong's information indicated that the earliest likely time for the Russian whalers to arrive off the coast of California was around June 20. Of course, it might be earlier—or later. The problem facing the *Phyllis Cormack* was whether to risk tipping our hand by heading down there now and possibly scaring the Russians off before we had a chance to hit them, or whether to risk missing them completely by hanging back and pretending not to know where we were going at all. The trouble with the first course of action was that there was no way we would be able to hide the *Phyllis Cormack*'s movements, not after having spent all this time and energy drawing attention to ourselves. We could not just sneak quietly down the coast. Ottawa would know about it—within no time at all, so would the Soviets. The alternative would be to head northward—as noisily and visibly as possible—making it look to whoever was keeping an eye on things back in Ottawa and the Kremlin that we were heading in the wrong direction and that we therefore posed no threat. I settled on an exercise to deceive the Canadian agency that was reporting to the Russian embassy. After we had made it abundantly clear that we didn't know where to go or what we were doing, we would slip quietly back to Winter Harbour, load on extra fuel tanks, and make a frantic run down to California, hoping to catch them by surprise.

We left Winter Harbour June 1. Carol Brien, Melville Gregory, Fred Easton, and Ramon Falkowski had departed and were replaced by cameraman Ron Precious, actor-singer Don Francks, nurse Leigh Wilkes, and Paul Watson, now fully recovered from his appendicitis attack. *Vega* headed south, down to Long Beach, where Longini was to pick up musicians Ann Mortifee and world-famous flautist Paul Horn, who were to be taken out to serenade the grays. The *Phyllis Cormack* would sweep northward to the Queen Charlotte Islands, touching in at such places as the town of Masset, where the Canadian Armed Forces had a major base. I wanted to stop in there and play dumb by appealing to the military for help, just so they would report our apparent lack of direction back to Ottawa.

The addition of Don Francks to the crew changed the whole gestalt. Forty-three years old, Francks was not only a master musician, actor, comedian, he was an avatar of tremendously positive energy. He never drank. He wore his hair in long pigtails, which Cormack delighted in pulling, and even though he had been living with his beautiful black wife, Lilli, and their daughter, Cree Summer, on a Cree Indian reservation in Saskatchewan for years, he had lost nothing of his hip, big-city sophistication and humor. His repertoire of songs went back to the Depression and included all the early-blues stuff, as well as everything modern up to the Beatles. From the first day on board, he dedicated himself to slaving away in the galley, or, when he wasn't doing that, ceaselessly polishing and cleaning the boat. His humility was breathtaking. And while there had been much talk of spirituality on the boat before, spirituality was now incarnate. Don Francks, too, had had visions and dreams, but he had also had a chance to absorb the North American Indian life-style through his pores. He was perhaps one of the wisest men I had ever met—but he never let his knowledge come out in a heavy fashion. Rather, he entertained us. Cormack in particular was entranced by this sparkling entity who had arrived in our midst and who led us in chorus after chorus of "Hallelujah, I'm a Bum!" If there was such a thing, Francks was a true Warrior of the Rainbow.

Our second day out, all my worst suspicions about the Canadian government were confirmed. A Canadian Armed Forces Argus aircraft, normally used for tracking submarines, came buzzing across the water— just like the American aircraft that had haunted us in the Gulf of Alaska back in '71. It passed about eighty feet overhead, circled, and passed again, taking pictures. Good, I thought, those bastards in Ottawa think now that we're heading the wrong way.

Late that afternoon, about fifteen miles south of Cape St. James on the southernmost tip of the Queen Charlotte Islands, we spotted three finback whales. I had been asleep in my bunk at the time, having a dream of riding on the back of a whale in Vancouver's English Bay. Then Cormack sounded the horn, signaling that he had spotted whales. Awakening to a mad hustle in the bunk room, as people grabbed for wet suits, cameras, flutes, and guitars, I had trouble for several minutes distinguishing between the dream state and the state of so-called normal consciousness. Leaping into a Zodiac with Rex Weyler, I set out after the whales, but the waves were big, our progress was none too good, and within half an hour, we had lost sight of them.

While the adrenaline was still running, we decided to have a Zodiac "kamikaze" practice run. Carlie Trueman and I would dodge in front of the bow of the *Phyllis Cormack* and the skipper would do his best to run us down.

Cormack, who loved to play rough anyway, got right into the game. With a wild gleam in his eye and a grin, he charged at us repeatedly, sometimes slowing the engine abruptly to lure us close to the bow, then jamming the throttle up to full power and veering suddenly, trying to catch us in the bow wave. Carlie quickly decided she didn't like the game. I found myself taking a perverse delight in bringing us in so close to the *Cormack*'s bow that when we looked up, it was to see the others on the deck looking almost directly down at us. The great wooden bow crashed up and out of the water, came down within arm's reach. It was a pure macho game of chicken, exactly the sort of thing Cormack liked, and in my adrenaline-fired mood, I was enjoying it too. But there was more to it, of course, than just a game. Not too many weeks ahead lay the moment when we would have to move in deadly earnest in front of a Russian bow, and the more confidence we had in the Zodiacs by then, the better. Cormack tried his damnedest to run us down. I tried my damnedest to taunt him, tease him, goad him, and to play on the very edge of the line. Once, the Zodiac actually struck the hull—and bounced immediately away. Carlie cringed with her hands over her head, and I laughed maniacally. In the midst of the game, three dolphins popped up around us to join the fun, leaping and whirling around both the bow and the Zodiac, and our musicians burst forth with a cacophony of sounds. Crazy energy seized us all. The only mishap came when one of the guy lines being used to pull the Zodiac back on board snapped, whipping a turnbuckle into Don Franck's face, bruising his jaw and bloodying his nose.

That evening we pulled into a place called Rose Harbour, where there was an abandoned whaling station. The camera crew went ashore to film it while Watson, Zimmerman, and Trueman went diving for artifacts: rusted pre-World War II harpoons, flensing knives, and the like. An unbearable silence seeped out from the moss- and vine-covered ruins. Immense bones littered the beach. Having visited such a place before, I felt oddly unmoved, but some of the others were on the edge of tears. Cast-iron and sheet-iron vats lay scattered over the three-acre site. Only the furnace where the oil had been melted down was intact. On its door, barely visible, were the words: CONSOLIDATED WHALING CORP. LTD. Reverting to his role as archaeologist, Walrus Oakenbough picked through whale teeth and metal fragments for hours before announcing the conclusion that the place had been abandoned sometime between 1925 and 1930. Several trees at least fifteen years old—and one perhaps twenty-five—were growing up through the furnace. Parts of the beach area were built up over three feet thick with deposits of whalebone cinders. *Rose* Harbour! The color of the blood.

The next day, at 3:00 P.M., we pulled into the small mining community of Tasu, on the west side of Moresby Island. Only 350 people lived there, almost every one of them working for the Falconbridge Corporation. Across the inlet from the dock where we tied up, there was a tremendous pile of rocks that went halfway up a steep mountain from which they had been dumped after being clawed from the insides of the earth. Every few minutes, a minor avalanche would occur. The sound of the rocks as they fell and bounced prompted us to name the slope "Thunder Mountain." The locals apparently had no name for it.

It was at Tasu that things really began to get strange. ...

Two Buddhist monks showed up to give us a flag from a Tibetan monastery. It had been left behind in Vancouver several months before by the Venerable Trungpa Rinpoche, one of the handful of Tibetan heavies who had fled from the Communists to North America and who were now tromping about the West, spreading the teachings of the Buddha. The two monks—young whites who had adopted the Tibetan names of Karma Shen Ten Pay Yay and Cho Chin Yi Ma—had been involved in the ceremonies during the Trungpa's visit to Vancouver and had been astonished when he left this particular flag behind. The Venerable Trungpa Rinpoche never left *anything* behind by accident. Realizing there had to be a purpose to the leaving behind of the flag, the two monks had picked it up and carried it with them ever since. They now found themselves in Tasu, working temporarily in the mines. And then they had seen the *Greenpeace V* coming into dock. To them, it was instantly evident what to do with the flag: give it to the boat that was out to save the whales!

They both knew that the entire voyage had been given a blessing the previous autumn in Vancouver by none other than His Holiness Gyalwa Karmapa XVI, the Grand Lama of the Oral Tradition of Tibetan Buddhism. He had liked our project and had given me a strip of red cloth that I still wore around my neck and had every intention of continuing to wear, especially when we approached the Soviet fleets.

The flag, which was covered with drawings of animals, had printed on it Tibetan mantras, including a verse "asking for blessings and prosperity for all sentient beings and emancipation for all beings from the cycle of suffering."

A few minutes after the monks had brought the flag down to the boat, two astonishing things happened.

First, a wizened old man hobbled painfully across the dock directly toward me while one of the monks tugged at my sleeve giggling and pointing. "This is someone you know," the monk said. I stared and stared—then almost fell over from a burst of emotion. It was a long-lost uncle of mine, Uncle Ernie, who had disappeared from the family when I was a child and whom I had long since given up for dead. He had been my father's best friend during the war years and was so much like my father, who was now dead, that I fell into Uncle Ernie's arms, weeping. The two of us stood there on the dock, hugging and sobbing—for I was the first of his family to see him in over ten years—and for a great moment the alienation that had torn the whole family to shreds and scattered us all over the continent, and the pain of separation with it, was gone, and my dear, beloved, favorite Uncle Ernie and I were together here in a place so far from home it made my heart tremble.

We had just managed to pull ourselves together enough to go up to his room in a nearby bunkhouse—he was the mining camp's general handyman—when the other monk suddenly tapped at the window, urging me to stick my head outside and look at something. He was hopping up and down gleefully, pointing at the sky. Directly above was a stunningly vivid, complete double-ringed rainbow.

"An auspicious sign!" cried the monk.

At that moment, Rex Weyler was up on the road on Thunder Mountain, where he had gone to find a vantage point to shoot color photographs of the *Phyllis Cormack*. From where he stood, when the rainbow blazed up into existence, it curved flawlessly down to land like a sheet of rippling color on the decks of the *Cormack* herself.

The townspeople loved us. We were invited that night down to the union hall so I could give a quick speech before turning the show over to Don Francks, backed up by Jackson and Weyler. Soon, we had everybody singing "Hallelujah, I'm a Bum." Uncle Ernie and I sat together, and when the folks in the camp learned that one of the *Greenpeace* crewmen was Ernie's long-lost nephew, the drinks wouldn't stop flowing.

The next day, Francks took his "Traveling Medicine Show" to the local school, at the request of the teachers, and from noon onward, children and their parents streamed onto the boat, buying up our entire stock of T-shirts, buttons, and posters. Donations of money and food poured in, which was a good thing, since the Greenpeace coffers back home were exhausted, and there was no money at the moment to restock the boat. If it were true that, for the next short while at least, we were literally going to have to sing for our suppers, we had made a good start.

In the middle of all this, I had had to tell Uncle Ernie that my father— his oldest and closest buddy—had died four years previously. The two of us spent several hours together, sharing our misery.

And then, at dawn, June 5, we were in motion again. Cormack found me on the deck just before we left, so overwhelmed—both emotionally and intellectually—by the events of the past two days, and having consumed so much free liquor, that he told me sternly: "Yer not fit. Git t'bed." Which I did.

Dolphins appeared the next day as we proceeded around the north end of Graham Island. A strange mood of fuzziness or confusion seemed to grip the whole crew, with the exception, of course, of Cormack, who continued to bustle about the boat, poking fun at Don Francks's pigtails, gossiping loudly about the discovery that Leigh Wilkes had brought along a couple of stuffed toy animals that she slept with, complaining about Walrus's food, and generally bumping heavily up against any male crew member who made the mistake of getting in his way.

Our visit to Prince Rupert was short, spicy—and bizarre. To raise money to buy food, we staged two hurried benefits, the first on the dock during the annual bathtub race, while hundreds of people milled about and dozens of assorted boats zipped back and forth in the water, and the second at a local Bavarian beer festival. Greenpeace had been popular in

Prince Rupert since the time of Amchitka, so several parties were held in our honor. Lechery and ecology had come together so very pleasantly that it took two days to straighten ourselves out.

Back home rumors of orgies and booze-ups were circulating. The Vancouver office was buzzing with righteous wrath. "What are you doing having parties in Prince Rupert," the message read, "when you're supposed to be out saving the whales?" There had even been a move to fire me as expedition leader. Of the people back home, only Rod Marining and Bobbi knew that we were deliberately stalling—the rest had been expecting a confrontation with the whalers to occur any day for well over a month now.

On Monday, June 10, we slipped out of Prince Rupert and headed back to the Queen Charlottes to the town of Masset, where I arranged for a private interview with the major in charge of the military base. I pretended to try to con him into giving us information on the position of any whaling fleets that might be in the area and allowed myself to appear desperate. The major was sympathetic but firm: there was no way he could give us the information we wanted without authorization from Ottawa. Faking dejection, I left, certain that a message would be going out either by telex or phone to the effect that there was no danger of the Greenpeace boat touching off another international incident, since they were hopelessly lost and didn't have a clue where to look for whalers. Our latest "counterintelligence operation" was complete. Now we were free to head back to Winter Harbour and get ready for the final leg of the journey.

On the way back south along the Inside Passage, we encountered a pod of seven orcas, probably some of the ones that Paul Spong had studied during the time he maintained a coastal whale-watching station. Several of us clambered into Zodiacs, and we set out, armed with flutes and cameras, to meet the so-called "killer" whales, not entirely without a feeling of respectful nervousness. Skana's free cousins could easily chomp through our Zodiacs and swallow us whole.

Instead, they responded by allowing the Zodiac with Don Francks to come within twenty feet. Two adult orcas, who had been riding on either side of a five-foot baby, stopped for just a minute, falling back a few feet from the small one exactly like two proud parents showing off their kid to the cameras. Then, with blasts of air through their blowholes, they continued their leisurely movements through the water.

Back on the deck of the *Phyllis Cormack*, cameraman Ron Precious was aiming his camera randomly at the water while making adjustments to his lens. Suddenly, two orcas broke the surface directly in front of him. Precious instantly started filming—then saw something that made his head jerk back from the camera.

"Hey! One of them's *hurt*!"

The whales were splashing furiously about—and it looked as though one of them had a long intestine hanging out from a wound on its side. Then it dawned on several of us at once what was happening—including Precious who had just realized he probably had historical whale footage on his hands. No one, so far as any of us knew, had ever filmed whales in the act of making love. The whale's penis looked to be five or six feet long. It was pink.

"This will be a whale of a sex movie," quipped Precious afterward. A lot of jokes went back and forth about how our documentary would get an X rating now and what the Canada Council—which had donated ten thousand dollars for the making of the film—would think about a porno whale flick.

By Saturday, June 14, we were back in Winter Harbour, just as Rod Marining arrived from Vancouver, leading a five-vehicle convoy of trucks and cars, carrying eight 250-gallon fuel tanks, which, once filled, would increase our range by one thousand miles. We had a quick, somewhat painful, crew change. Don Francks, Leigh Wilkes, and Gary Zimmerman departed—Francks because his *I Ching* reading told him to, Zimmerman to act as our contact with U.S. ecology groups whom we hoped would organize flights by private planes from the coast of California to try to spot the fleets, and Leigh Wilkes because her scheduled time on the boat was up. In the place of these three, we brought back Melville Gregory, Patrick Moore, Fred Easton, and George Korotva, giving us a total "confrontation crew" of twelve men and one woman, Carlie Trueman.

A tremendous work binge followed while we installed the extra fuel tanks. Friends and lovers and wives showed up from Vancouver. Everybody sensed now that the moment had finally come. The last of our money had been plowed into the extra tanks and one last provisioning and fueling of the boat. If we failed to hit the Russians on this run, we wouldn't be able to afford a second chance. Everything depended on this one last long shot. A fine, tingly feeling was in the air.

The day before we set out, news came through from Paris that David McTaggart had won a victory in the French courts. A three-man tribunal had ruled that a French navy minesweeper had, indeed, rammed *Vega* at Mururoa in 1972. The court ordered the French navy to pay damages.

On the second part of McTaggart's case—a charge of armed piracy—the court declared it was "not competent to judge," which meant that McTaggart had only won half his battle, and he was reported "not happy" with the decision. Yet it remained that the court's decision was a blow to the French military program and to the prestige of both the government and its officers. *Time* magazine described it as a "moral victory" for McTaggart.

We had a party in Winter Harbour that night, a great gleeful celebration that ended up with several couples going off into different hiding spots in the forest to make love with that special intensity that comes when deep inside you are very afraid. At 5:00 A.M., June 18, the *Phyllis Cormack* pulled away, leaving a handful of tearful, waving figures on the dock. Our *I Ching* reading that morning was hexagram 49, Revolution, which says, in part,

> Political revolutions are extremely grave matters. They should be undertaken only under stress of direst necessity, when there is no other way out. ...

> Times change, and with them their demands. ...

> Fire below and the lake above combat and destroy each other. So too in the course of the year a combat takes place between the forces of light and the forces of darkness, eventuating in the revolution of the seasons. ...

The footnotes to the text referred to "Goethe's tale, 'Das Märchen,' in which the phrase, 'The hour has come!' is repeated three times before the 'great transformation' begins."

Our final crew was our best yet. If the previous weeks had done nothing else for us, they had done one thing: given us a collection of people on board who could whip Zodiacs in and out of the water as though they had been doing it all their lives; all of whom were intelligent, with strong stomachs and plenty of muscle. We had all the navigational, electronic, engineering, and seamanship skills we needed.

As we headed out on the final run toward the fleets, it seemed to me that we were like a seagoing gang of ecological bikers, bikers who had adopted the Satyagraha philosophy of Mahatma Gandhi, but who rode high-powered roaring machines across the waves and whose collective aggressive energy was more in tune at times with the mood of pillagers descending on a helpless village than 1960s hippies coming to tuck flowers into the rifles of National Guardsmen. If there was any other basic sense of it, it was the dreamlike sensations that we *were* reincarnated Indian warriors whooping and hollering as we surged down out of the hills toward the wagon train. It was not that we lacked fear or that we were not worried to the point of nausea about what would happen when we finally came face-to-face with the whalers. Rather it was that we had psyched ourselves, both individually and collectively, to the point where we enjoyed an almost breathtaking confidence in ourselves, a sense of having picked up, like electrical discharge, some shimmering, sparkling power; as though the *Phyllis Cormack* were a lightning rod pulling in tremendous pent-up energies from the sea and the sky and all things dwelling in them.

Night after night in the galley, at the top of our lungs, we sang:

> We are whales, living in the sea
> Come on now, why can't we live in harmony?
> We'll make love above the ocean floor
> Waves of love crashing on the shore.
>
> I'm fifty tons of blubber
> Spouting rainbows to the sun.
> I'd just love to go on cruising,
> Making love and having fun.
> I'd love to go on living, yes,
> And raise my children too.
> Can't you see dear man, dear woman,
> That we're just the same as you?
> For fifty million years or more
> I've watched the ocean shore,
> Seen the mastodons of yesteryear
> And the mighty buffalo.
> Watched man emerge from mountain caves,
> Pursue me with a spear,
> While building bloody empires
> Out of greed and lust and fear.

Our great race is vanishing,
Our lives are put to waste.
Make haste, men of compassion,
Least you follow in our wake.
Our children die to feed your dogs
And paint your lady's face.
We are guardians of the ocean
And our lives not yours to take.

The excitement was not just contagious—it was all-pervasive. Each and every one of us hummed with energy. We *were* the whales, damn it! Our righteousness was theirs. We *were* the sea and the earth come alive in some mysterious but perfectly natural way! The mechanics—Cormack, Hewitt, Carlie Trueman, Moore, and Korotva—held back, watching the rest of us very carefully, as though deeply worried that we might, in fact, *all* be turning into "bow-cases," as H. had done at the beginning. Yet their nervous systems, too, clearly trembled with a sense of hovering on the verge of some amazing experience. The waves came at us—nasty weather was the rule—and we whooped and raced to be the first on deck, clinging to the mast or the guy lines, exulting in the wind and the stinging lash of spray across the face. All the old clichés of the ecstasy of being at sea now took us over, one by one, as our ordinary belief systems collapsed like burned-out light bulb filaments.

Our "normal" sense of reality would have told us to hide out in our bunks, quivering with terror as the wind rose and the sea rose and the whaling fleets became a palpable presence, their harpoons looming like ICBMs in our minds. But now, as though afterburners had been attached to our adrenaline pumps, we thought we heard the waves as the rumble of a vast, perhaps infinite audience beginning to signal its approval. Crazy? Were we crazy? I think most of us were. At least, most of us were sharing what would otherwise have to be described as a collective hallucination—the hallucination being that we were on a brightly painted old boat, as glossy and solid as a horsey on a merry-go-round, charging to the accompaniment of music across the waves to the rescue of giant sentient beings, almost like the reverse of the legend of St. George and the dragon. It was to the *dragon's* rescue we went!

In this fantastic collective hallucination, we perceived ourselves as having been blessed by one of the agents of the Buddha, of possessing a great electronic sword that could reach into millions of minds at once, of being involved in an archetypal battle between the forces of darkness and the forces of light, a battle that was taking place near what might well prove to be the end of time, somewhere in the last days before Armageddon. Many of us, moreover, perceived ourselves as being guided by two forms of magic: an electronic magic that came out of little metal boxes, picking the voices of our prey out of the air just as the cawing of crows attracts the stalker, and another kind of magic, as ancient as the electronic stuff was modern—magic that came in the form of hexagrams in an old Chinese book.

The excitement was so intense that I threw up the first night out, having digested a bunch of Quaaludes, yeast pills, desiccated-liver pills, cough syrup, beer, cigarettes, vitamin C, and herbal teas. Nothing else.

We'd left on a Wednesday, heading in a course that would normally have taken us straight to Hawaii, except that we intended to go no further than Cobb Seamount, some 120 miles off the coast of Washington State, and then head due south along the edge of the continental shelf into the upper reaches of the area where Paul Spong's information said the whalers had been for three years in a row during the last two weeks of June and the first two weeks of July. By Thursday, we were already beginning to pick up Russian voices on the radio.

June 22 was the night of the full moon, also the day the annual meeting of the International Whaling Commission opened in London—the day we had hoped to be confronting the Russians. Their voices crackled by the hour through the radio and the dial on the radio direction finder twitched hysterically in every direction. The voices were all around us but none close enough so that we could get a solid fix. We had mounted a speaker down in the galley so that Korotva wouldn't have to spend all his time in the radio room, listening. The result was that the whole boat blared with static and incomprehensible Russian voices, haunting us day and night. We had three tape recorders on standby to be flipped on the moment the voices started again at any time when Korotva was either in his bunk or otherwise not right beside the speaker, lest some key phrase be missed. So far as Korotva could make out, the ghostly voices belonged to Russian fishermen on huge draggers and seiners, not whalers. The need to keep the radio on full blast meant that no one was going to get a proper sleep until we finally found what we were looking for. Amazing wrinkles appeared in all our faces.

And it was not just exhaustion setting in: we were all also suffering the "diesel fuel burps." There had been a spill from one of the extra tanks in the hold. Walrus had gone below to get more food only to discover all the tinned goods floating about in oil with their labels gone, dozens of cardboard boxes filled with fresh fruit collapsed soggily in a blend of salt water, bilge gunk, and diesel. The wrappers on the margarine had turned a gray-yellow color. It took several hours for the food to be salvaged, each orange washed by hand on the deck, but even then we were stuck with the fact that everything we ate from there on reeked of diesel fumes. It got into our bodies, like garlic, causing us to pass the strangest-smelling gas and making most of us nauseous. Between the diesel burps and the eerie crackling of Russian voices through the air, life on board the *Phyllis Cormack* was becoming distinctly nightmarish.

By nightfall, the skies were perfectly clear, and the full moon lay almost dead ahead, its light illuminating the decks as though we were truly standing on a stage. A "mass bow-case situation" developed with no less than seven of us all clustered at the bow, howling, baying, whistling, roaring, chanting, blasting on flutes and recorders, with Melville Gregory naked at the wheel on the upper deck, laughing maniacally. Two dolphins and two porpoises showed up to join us, zipping like silver bullets through the moonlight and phosphorescence and cheers. The party came to an abrupt end near midnight when a sleepy John Cormack emerged from his cabin to discover that Gregory was steering straight toward the moon, which meant that he was about thirty degrees off course. Surprisingly gently, the old skipper grabbed the naked Gregory by the beard and hauled him away from the wheel, saying: "That's it Mel, yer fired." To the rest of us he yelled: "What're you guys, a buncha *werewolves*?"

At 1:00 A.M., a rainbow appeared on the horizon off the port bow. A rainbow at night? None of us had ever heard of such a thing. Perhaps it was only possible because the moon was so bright. Weyler managed to adjust his camera sufficiently to capture it on film. The Warriors of the Rainbow stared, awestruck, at the phenomenon—which lay, oddly enough, in the direction from which the strongest Russian radio impulses had been coming. Goosebumps crawled on our flesh, and Weyler whispered: "Get ready, the miracle is near."

The next day, at 4:30 P.M., as we were steaming at full throttle southward, nine finback whales appeared. We had just heard over the radio that a ban on the killing of finbacks had been recommended at the IWC meeting in London, now going into its second day. Another coincidence, oh, yes indeed. Within moments, two Zodiacs were in the water and the *Cormack* was chugging around in circles and figure eights with Walrus up on the mast, pointing to the locations where the whales were emerging. Walrus had by this time assumed the role of "whale scout" and he demonstrated an uncanny ability to anticipate the movements of the great creatures. Later, he explained it this way:

"It's a most delicate act of creation. While I was up on the mast after the initial sighting, before locating the other five or more, by focusing on the two that were nearby, I flashed that they were feeling concern for themselves ... and this feeling came to me as my own bodily agitation and emotional changes. I felt these whales were connected mentally or were in some form of communion with others that weren't visible yet. ... The direction of these others seemed to hinge on a feeling of 'knowing their whereabouts,' so I looked along this line of *feeling*, almost unconsciously sweeping the horizon to the west, and sure enough, just as I 'thought' or felt the whales were communicating an aura of presence about themselves, which I could pick up telepathically, several spouts appeared over two miles away to the west. Then, beyond these, three or four faint spouts almost on the horizon it seemed."

George Korotva gunned the big thirty-five-horsepower engine on his Zodiac and was coming up on the nearest two whales within moments. One was much smaller than the other, so Korotva made sure he came up beside the mother, rather than frighten her offspring. She just kept cruising at five to eight knots, apparently unconcerned. It was only when Paul Watson poised himself to attempt a leap on her back—imagine if we could get footage of a man riding on a wild whale's back!—that she suddenly lifted her flukes and slammed them down in the water to signal very clearly to us that it would be unwise. Watson froze. Korotva let up on the engine. We changed our minds about trying that one.

A few minutes later, with Rex Weyler at the bow of the Zodiac, his camera aimed, Korotva brought us up over a wave, only to discover a whale surfacing a few yards ahead. Korotva cut the engine so hard to avoid a collision that Weyler was unable to stop himself from a beautiful slow-motion forward tumble into the water, with all his camera equipment. I grabbed his belt to prevent him from going over, only to end up traveling with him in an amazing head-over-heels arc into the sea right next to the whale. Weyler burst to the surface, raging because his equipment was probably all ruined, and we raced back to the *Phyllis Cormack* like ambulance attendants bearing a mortally wounded accident victim. By soaking everything in fresh water, Weyler was able to salvage the equipment, and somehow most of the shots he got of the finbacks were saved, too.

Russian voices continued to shriek at us maddeningly from all sides, as though we were in the midst of a fleet of ghost ships or maybe in different dimensions, or perhaps the Russians had invented a way to become invisible. From the point of view of hoping to have a political impact on the IWC meeting in London, the time was running out on us. The Russian ships might be anywhere just over the horizon, but however far away they were, it was just far enough so that our RDF continued to fail to hold on to a radio signal long enough for us to get a positive fix. The needle leapt about the dial like a wonky compass in an electric storm. Our only definite course of action—short of darting every which way each time the RDF needle flickered—was to proceed doggedly to the exact positions where the Russians had gone on previous years, aiming primarily for three spots just offshore from Oregon and northern California which were the only ones, out of hundreds of positions they'd visited, where they had hunted whales on the same day for three years in a row. These would be the "hot" whaling grounds. If the Russians would go anywhere again this year for sure it would be to these choice locations. One was at Jackson Seamount at the edge of the continental shelf, and the two others were along the Mendocino Ridge, a narrow finger of high sea bottom thrusting out some three hundred miles west from Cape Mendocino, California.

The great theological debate centered, of course, on the validity of the *I Ching* as a workable and trustworthy oracular device. The technical arguments of the time paralleled the theological debate closely, except that they concerned themselves with the workability and trustworthiness of the RDF. The hard-core *I Ching* devotees on board—mainly Gregory, Walrus, Will Jackson, Weyler, and myself—were pitted against the Mechanic Gang, led naturally by Cormack, but with Korotva, Moore, Hewitt, and Trueman right behind. The rest stayed more or less on the fence. The question came down to this: If all else failed—the RDF, the marked charts, even the overflights by planes from California, which had been promised—and it proved utterly impossible with the technical capabilities currently at our disposal to actually catch up to one of the Russian boats, could we count on the *I Ching* to guide us to victory long after Western rationality and industrial genius had failed? If you believed we could, you were a bona fide mystic. Not surprisingly, the mystics all also expressed the opinion that even without the *I Ching*, we'd succeed, because there would still be dolphins to guide us, along with the rainbows and moons. ... Hadn't our coming been prophesied by the Cree Indians? Had we not been blessed and blessed again? Had there not been signs and portents for a long time now? This was no joking matter any longer—if it ever had been. Jackson, Weyler, Gregory, Walrus, and I were true believers. Our usual greeting to each other was just like that of monks: a quick prayer gesture and a bow, always done with great good humor rather than solemnity. Others, like Watson and Moore and Precious, seemed to have their own mystical feelings and impulses, but they were careful about what they said and were not willing to reveal so much about themselves, except at certain high moments. Korotva and his European mind was something else, hair-trigger fast but absolutely elusive. You could never pin him down. Fred Easton was positively inscrutable. Carlie Trueman loved to debate everything, but was not considered by the mystics to be one of the heavyweights in the Mechanic Gang.

It was on Saturday, June 21, in mid-afternoon, that Cormack issued his big challenge to the mystics.

Squinting at me eyeball-to-eyeball, he leaned across the galley table and said with elaborate politeness:

"Now then, Mr. Hunter, sir, you and I've always had a *gentlemanly* acquaintance, and what I'd like to know is this: If I was t'ask ya man-to-man t'tell me the honest-t'-God truth about somethin', could I trust ya t'do it?"

"John. Sir. Captain. Skipper." I replied just as elaborately, "I swear t'God I'd tell you the truth."

He held out his huge hand. We shook, looking each other right in the eye.

"Well then," he said, sitting back like a chess player bracing himself for a major move and keeping his unwavering eye fixed on me, "tell me this: D'ya really believe in this here book from abracadabra two-humps land?" Meaning the *I Ching*.

"Yes sir, I do."

He contemplated me thoughtfully for a full moment, rubbing at the silver bristles on his chin.

"Okay then," he said, reaching for the three Chinese coins on the table, with the small square holes in the middle. "What I wanna ask yer hocus-pocus diddily-ocus book is this: *Are we gonna find these here Russians? What do I do? Just toss these things on the table?"

"That's right, John. All three coins. Six times."

An intense audience of both mechanics and mystics had gathered.

214

Pens came out all around the table and notebooks were flopped down to record the results.

Cormack got hexagram 19, Approach, which also translates as "becoming great," and which refers, among other things, to "the approach of what is strong and highly placed in relation to what is lower."

The skipper grunted. In fact, there were a series of grunts around the table.

Shortly after dark that day—perhaps less than four or five hours later—the first Russian voices had come to life over the radio in the trailer. "Huh," said Cormack. And we didn't discuss the matter anymore.

Even so, whatever the *I Ching* said, short of giving us precise compass bearings to follow, the RDF was the tool upon which the success of the whole venture now depended. In this department, the mystics had little to offer. The guru was Al Hewitt, and it was entirely up to him now to provide the last link in the chain. Operating on the kind of budget we'd been operating on, we had of course been forced to buy the cheapest equipment, mostly secondhand. The RDF was no exception. Its range was supposed to be 150–200 miles, but after we'd been traveling among the Russian boats for a couple of days, not able to get a proper fix on anybody, Hewitt began to suspect that the range was much smaller than he'd been led to believe. Back in his trailer, amid boxes of tools, coils of extra wire, spare transistors, capacitators, and diodes, he hunkered for hours on end, probably exactly as he had as a child playing with dozens of Meccano sets, brandishing his soldering iron and tiny screwdrivers, trying to rebuild the entire RDF, tripling the length of its antenna—bouncing and flopping about in violent, choppy sea—to give us that extra range we so desperately needed. Between bouts of complete concentration on this, he'd wander out on deck with a detached, indifferent air—seemingly oblivious to the mood of panic and desperation that was building up around him—and do things like hang his socks off the side to the boat to get them washed, while we all waited with a barely suppressed frenzy of expectation for him to emerge finally with the "new RDF." Considering he had to improvise the whole thing from odds and ends he happened to have around, the thing, when it was finished, was a wonder of ingenious engineering.

There was also a lively political debate taking place—centered around the fact that within hours of Korotva having heard the first Russian voice, I was on the radio to Vancouver—indeed, I had a "radio fetish"—blabbing the news that *Greenpeace V* was "in the midst of a Russian fleet." After all our previous counterintelligence efforts and elaborate efforts to play dumb so that we could retain the presumably vital element of surprise, I had gone and blown the whole gig. Now everybody, including the Russian embassy in Ottawa, would know where we were. And since we could be certain our radio transmissions to Vancouver were being monitored anyway, the mere act of getting on the radio itself was enough to have blown our cover. So now, thanks to me, it was twice blown. And, sure enough, the very next day, a change came over the transmissions from the Russian ships. Whereas they had been loquacious at first, obviously carrying on with friendly chatting back and forth, now their messages to one another seemed to be cryptic. It was reduced to information of a purely factual nature: tonnages caught, fuel-consumption levels, requests for servicing. Everyone glared at me furiously. By the single, impulsive act of getting on the radio, I had not only made our task that much harder, increasing the odds against us enormously, but had possibly ruined any chance at all of us getting to save any whales. If that proved to be the case, what a guilt burden that would be to bear! The fact was I had buckled under the pressure and was so desperate to make some kind of impact, any kind of impact at all, on the proceedings at the IWC meeting, that my tactical judgment had gone all to hell. Maybe I *was* getting to be too much of a mystic to be left in any central position of authority.

It was a mistake I wasn't allowed to compound. As though an angry lightning bolt had come down from the sky, the radio suddenly ceased to function at our end. I sat for the next few days in the radio room, screaming myself hoarse over the microphone, unable to get anything more than a rare partial phrase through to Vancouver, while Rod Marining's voice boomed loud and clear into my ears. We could hear the shore perfectly, but they couldn't hear us. They were desperate for news, because they had interpreted my report that we were "in the midst of the Russian fleet" to mean the *whaling* fleet, and since then they hadn't been able to get anything but static-wrecked snatches of words out of us. For all they knew, we might have been massacred on the decks. A fine communications disaster had developed—and either way you cut it, it looked to be mostly my fault. My *I Ching* reading on Wednesday, June 25, was hexagram 18, Work on What Has Been Spoiled (decay). It states that "What has been spoiled through man's fault can be made good again through man's work," and advises that "It furthers one to cross the great water."

That day we came upon a massive Russian dragger fleet, dozens of huge rusting steel ships moving ponderously along the American coast, staying barely outside the twelve-mile limit. It was an astonishing sight, particularly since all the large ships in sight were either Russian or Polish, and all the little wooden boats—the size of the ones we were used to on the coast of B.C.—were American. There were no large American vessels to be seen at all. Cormack was incensed immediately. The Russians and Poles, he explained, cruised back and forth along the continental shelf, dragging the bottom with gigantic nets, scooping up everything that lived, throwing nothing back. It was very much like an undersea strip-mining operation. Yet because of their own conservation laws, the American fishermen were forced to take only certain species and to throw back everything under a minimum size. As a fisherman bound by similar restrictions, it galled Cormack no less than the men in the little wooden boats, to stand by helplessly as rusting metal monsters from Europe and Asia gouged the shelf bottom without even a pretense at preserving anything for later, grazing, exactly, like the dinosaurs who had rendered themselves extinct.

Taking a Zodiac over to one of the American boats, a forty-foot trawler called the *Tilko*, Patrick Moore and I managed to make clear radio contact with Eureka, California, and from there got a line patched through to Rod Marining in Vancouver. The news was all bad. Canada had voted along with Russia and Japan at the IWC meeting to increase the whale quotas. The media back home was extremely suspicious about us because of the cross-up that I had precipitated with the "amidst Russian fleet" story. The IWC meeting had only two more days to run before it was over for another year, and the delegates from the whaling countries were making jokes to the press about how *Greenpeace* was hopelessly lost and would never be able to interfere with them. Gary Zimmerman had several "leads" on people who might be willing to fly out along the Mendocino Ridge to look for whalers, but nothing had materialized yet for lack of money.

In desperation, we raced back to the *Phyllis Cormack* and set a course taking us straight out along the ridge as fast as we could, having nothing more to go on than the tiny markings on the chart that indicated that it was still somewhere out there that the whalers were most likely to be. Exhausted and sick, I collapsed into my bunk.

By Thursday, we were one hundred miles out along the ridge, and there was still no trace of whalers. The pancakes we had that morning were particularly thick with diesel oil, and at least four people were writhing in their bunks, moaning from stomach pains. The tobacco was almost gone—leaving the five chain smokers on board feeling edgy and nervous. Only seven beer cans remained. Dark clouds swept in bands across the sky. The waves were ten feet high, which meant that the decks were awash most of the time. One of the stabilizer lines broke, leaving the boat to roll and bounce without restraint. When I tried again to get through on the radio, a pitiless wall of static interposed itself, making communication impossible.

At dusk that night, Al Hewitt hooked up the new RDF system and turned the switch. Immediately, a loud Russian broadcast came through from the northeast. Whalers or draggers? By this time, George Korotva was punchy from listening to the radio, sick himself from diesel fuel, and quite unable to tell anything from what the Russians were saying. Still, it was a signal, and the expanded RDF unit with its new antenna was at least giving us a definite fix. The needle was rigid, pointing in a direction, and that was at least something to go on, even though it meant turning around and heading back at an angle toward the shore. Hewitt had done his job. Now he went wearily back to his trailer. The next day, Friday, June 27, would be the last day of the IWC meeting.

By midnight, we were forty miles SSW of Cape Mendocino. Twice that evening, Korotva swore he'd heard the word *Vostok*, which, if it was true, would mean that we were bearing toward the giant floating factory that was the flagship of one of the Russian whaling fleets. All night, Russian voices squawked and screeched loudly over the radio. The seas stayed at about ten feet high, and since we were punching directly into them, the boat shuddered and thudded as it moved at full speed. Nearly everybody had fantastic dreams that night, as though the moment we closed our eyes, our minds all rolled helplessly down into some immense subconscious cavern. Korotva, Moore, and I, in particular, were so strung out by then that we found it impossible to wind down enough to get to sleep without taking Quaaludes. In my own dreams, I saw Vancouver laid to waste by a nuclear war, and then had a series of exceptionally vivid flashbacks of my childhood, youth, and early travels, as though some instinctual lever had been triggered by all the tensions, bringing up material that had been lost to my conscious mind for years.

All through the morning, while fully half the crew remained zonked out in their bunks, the RDF finger grew steadier and steadier; then, around ten o'clock in the morning, Russian broadcasts suddenly stopped, and the needle went limp. Cormack ordered a steady course forward—leaving a presumably "reformed" Melville Gregory at the wheel. Around noon,

Gregory spotted a rainbow at 125 degrees on the compass. To himself, he whispered: "That's the sign." And without anybody else knowing about it, he altered course, perhaps 15 degrees, heading directly toward the rainbow. Cormack himself was taking a quick nap and didn't notice. The rest of us were either asleep, or slumped in the galley, burping up diesel fumes.

Thirty minutes later, the Russian whaling fleet appeared on the horizon—dead ahead.

Brilliant midday light poured down on the scene. All the *Phyllis Cormack*'s flags were flying, United Nations flag, Greenpeace flag, British Columbia flag, Canadian flag, Oceanic Society flag, Buddhist monastery flag, and half a dozen pennants embroidered with whale symbols, ecology symbols, peace symbols. Cameras clicked and shutter lenses whirred. The sky was so blindingly bright that one had to squint looking upward. The sea was whitecap-flecked monochromatic aquamarine blue. Everything had a quality of *realness* that brought you up sharp at odd moments, as though you had just awakened from a long fuzzy slumber and were perceiving the vitality of the world for the very first time. Strangely, there was no running about on the decks, no chaotic bumping into one another. Everybody's voice had changed octaves and no one said anything unnecessary. We appeared perceptibly more calm than we had been at any previous point in the voyage, as though we had already been delivered. A feeling prevailed that we had come upon a vast stage that had been set up precisely for this moment.

And a vast stage it was: on the horizon was a towering gray shape, the factory ship *Dalniy Vostok*, some 750 feet long, more like a floating apartment complex than a ship, and spread out, moving in circles like a pack of mechanized wolves, were at least half a dozen white-topped harpoon boats with red-on-black hammer-and-sickle symbols on the smokestacks. They were badly rusted—patchworks of orange anticorrosive splashed on blunted black metal hulls. They moved jerkily, swinging around with bursts of smoke, darting forward, stopping, making abrupt course changes: hunting.

Moore had been the first one into his wet suit. Now he was up at the wheel with Cormack, peering ahead through binoculars, while Carlie Trueman did a final check of the outboards and Zodiacs. Down in the bunk room, Weyler and Watson and myself passed the talcum powder back and forth as we pulled our lightly shivering bodies into wet suits, not finding anything at all to say to each other.

The nearest harpoon boat was still a good two miles away when Cormack yelled that there was something in the water almost directly ahead, just slightly off to the starboard. It was a small red triangular flag, and there was something attached to it: a beacon. Otherwise it was a bit as though we had stumbled across a watery golf course. Then we saw a shape beneath the marker, a glistening gray-blue slug-like something that wallowed heavily in the waves. Several voices cried at once: "It's a whale!"

Cormack throttled down, coming to a halt less than thirty feet from the corpse. We had expected it to be titanic, for this was the first sperm whale we had sighted, the descendant of Moby Dick, the eighty-foot giant who laid waste to the puny men who pursued him. But this was no Moby Dick here in the water a century later. It was small, far too small, and until we saw the toothed under jaw wagging slackly as the body rolled, we did not believe it *could* be a sperm. But it was.

"My God!" Carlie Trueman screamed. "It's just a baby!"

The wave of emotion that hit us then—revulsion, rage—was so concrete that it staggered us all in our tracks for several seconds, so that there was a sharply focused instant when nothing seemed to move, neither the boat, the people on the decks, nor the dead whale-child in the water. From Walrus Oakenbough came a short howl or wail such as an Indian might have made coming back to his camp to find his children massacred and his world in ruin. And then everyone was in motion. "All right," I yelled, "this is a film thing, so let's get some shots."

"Zodiac launching crew!" Cormack bellowed.

Within minutes a Zodiac was in the water, manned by Korotva, Weyler, Watson, and Fred Easton. It surged over to the side of the whale, the rubber boat itself at least as big as the part of the whale that showed above the surface, one flipper outthrust like the cup of a thin, tilted mushroom. Watson flopped out of the Zodiac and scrambled onto the whale. It was lying on its side, one eye the size of a teacup, still open and staring skyward. The eye was fogging over, but still retained a trace of the crystal sparkle of life. The body was still warm. Gently, Watson leaned forward and pulled the eyelids shut. While he bent over, cameramen on the Zodiac and the deck of the *Cormack* shot film so that, later, we would be able to show the size of the whale by measuring it against the size of the man perched upon its body.

"How long is it, Paul?" Easton shouted.

"Oh, seventeen, maybe eighteen feet," he yelled back. "It's small."

Later, from the photographs and film, we were able to calculate its exact length, which proved to be twenty-five feet. The legal minimum size, according to the rules of the IWC itself, was thirty feet. We knew, at least, that the whale was definitely undersized and that in the first moments of our encounter, we had stumbled across a case of an illegal kill, a violation of all the official rules of whaling. If the very first whale we spotted proved to be undersized, it meant only one thing: the Russians routinely ignored the rules and slaughtered whole pods of whales en masse, right down to the children. "Whale management" could therefore only mean, in truth, whale massacre. The whole superstructure of "controls" over whaling was a farce—the evidence lay before us, a whale that, in terms of a human lifespan, would have been the equivalent of nine or ten years old. Somewhere close by, its parents were presumably also floating lifelessly with beacons and markers attached through perforations in their flukes.

Meanwhile, from the upper deck of the *Phyllis Cormack*, we could see that the nearest harpoon boat had wheeled around with a burst of smoke, startled by this unheard-of event, the arrival of a strange vessel in the midst of a hunt. Moreover, one that had pulled up beside a dead, marked whale, and whose crew was actually climbing on the whale. What the Russians thought then is anybody's guess, but their reaction was what you would have expected if their first impulse had been that we were somehow out to steal the whale's body. As the ship roared directly toward us, pushing up great white wings of foam, a man appeared at the bow, next to the harpoon cannon, with a hose in his hands, water spurting out of it.

"Look out, you guys," Carlie yelled to the Zodiac crew, "they've got a hose."

It was tempting to simply have Watson remain on the whale to see what would happen, but we could ill afford to allow all the camera equipment on the Zodiac to be doused and ruined. Watson scrambled off the whale, and the Zodiac came back to the *Cormack* so fast it looked like it was hopping.

The Russian ship sloughed past us, some 50 yards away. We could see the crew lining the decks, staring at us. Up on the flying bridge, armed with binoculars, was a thick man in a Bolshevik hat, presumably the captain. The harpoon boat's name, barely visible for all the rust, looked something like: *СВЕРНIIЫН*. Its ID code was NK-2052. It was at least

150 feet long. The boat swung around our stern, almost skidding to a halt beside the dead whale. Few of the Russians looked at us for long. They acted quickly, almost furtively, lashing the whale with cables to the side of the boat, as though intensely aware that they had been caught red-handed in a violation of the rules. In their haste, they pulled too quickly on the lines and, even across the 100-yard gap of water that separated us, we could hear the sickening muffled crunch as the whale's jawbone snapped. Then the Russian boat was in motion, cranking quickly up to full speed, and heading directly toward the still-distant *Vostok*.

We followed, spotting one more small dead whale, but leaving it to keep on track for the *Vostok*. The engine of the *Phyllis Cormack* hummed and vibrated through the decks into the soles of our feet in perfect harmony with the humming vibrations that were passing through us anyway, so that it was hard to say whether this delicious, teasing, trembling sensation had originated from the engine room or had its real origin in our own flesh or maybe our souls. Within a few minutes, we were surrounded by metal boats, some of them as much as four miles away, others only half a mile or less. There was an odd manic edge to their movements, as though they were slightly confused. There were no other signs of whales, neither markers nor spouts. Will Jackson clung to the mast, watching for any sign of a living whale to whose rescue we could immediately turn. In the bursts of cold wind that hit us, there were traces of a musky odor, possibly the stink of fresh blood, but in the water itself, no sign that anything had died this day other than one small whale. The Russian ship pulled away from us swiftly as it steamed with its prize toward the factory ship, looming larger and larger as the waves steamed by, spray driving Melville Gregory and Ron Precious back from the bow. Rainbows flared in the spray. Down in the galley, Walrus Oakenbough, warrior-brother of the Oglala Sioux, struggled, still full of pain and anger triggered by the sight of the dead whale-child, to put his emotions into words in his logbook. Finally, in heavy furious letters, he wrote:

THE WARRIOR OF LIGHT IS HERE.

It took two hours to close the distance that separated us from the gray silhouette of the *Vostok*, moving with leisurely unchallengeable power through the water. Two harpoon boats flanked either side of its stern, where a black rectangular opening was evident: the chute up which the bodies of the whales were hauled to the flensing deck, several stories above. The people on the upper deck looked down at us as though from the roof of an apartment block. To our astonishment, the stern deck was covered with a volleyball net, and among the men crowding the edge there were several women, one of them in a bikini.

The factory ship was cruising at about four knots, while lines were transferred from the harpoon boats to the slipway. The whales were tugged into the gurgling wake and up the chute, moving slowly, swollen, their tongues lolling out, unbeautiful, the chain moving them jerkily, like frames in an early silent-screen movie. With each yank of the chain, their dull, blubbery bodies wobbled and sometimes even flip-flopped in a parody of life. The great toothed lower jaw changed position under the weight of the head, like the rudder of a boat under tow.

There were openings in the flat wall of the stern above the slipway on either side of the ship, so that as we came up on the *Vostok* from behind, looking into the huge mouth-like opening with the dark narrow slits above, it seemed we were staring into the face of a giant robot. If the factory ship were to move backward, it would look almost whale-like itself, its mouth agape, gobbling up everything in its path, a steel twentieth-century sea monster deadlier than anything the ocean had ever known. Into its mouth the whales disappeared, looking no bigger than sardines being swallowed by a large bear. From an opening the size of a sewer outlet halfway along the length of the massive hull, blood flowed as casually as oil from the bilges of ordinary boats, fountains of thick red blood that poured out and kept pouring out, enough blood every minute to fill a bathtub. Who had ever seen so much blood flowing before? The smell, as we came downwind, left half the *Phyllis Cormack*'s crew retching over the sides. The peculiar obscenity of the *Vostok* came into focus the moment we realized that here was a beast that fed itself through its anus, and it was into this inglorious hole that the last of the world's whales were vanishing—before our eyes.

Two Zodiacs were dispatched to allow Fred Easton and Rex Weyler to get film footage and stills, both black-and-white and color. With Moore and Korotva handling the outboards, the little bullet-like boats buzzed like bees toward the immense floating fortress of death. Ron Precious remained on the deck of the *Phyllis Cormack*, with a long-distance lens, filming the film makers. At that point, we were more like reporters on assignment than a troop of eco-guerrillas.

The Russians—at least the deckhands—were plainly amazed. They had been at sea probably at least a couple of months, routinely going about their business in a battered giant rust-bucket that had been crossing and crisscrossing the North Pacific every summer since the end of the Second World War. There was no drama left in whaling other than the normal dramas of being at sea. Not for more than a generation had any men gone into the water in little wooden skiffs to confront Leviathan, and so none of the men and women lining the decks could really be described as "whalers." They were common laborers on the kill-floor of a seagoing slaughterhouse. It was dull, stinking work, with few breaks in the monotony, duller even than ordinary land-based abattoirs where, on certain occasions, a pig or a cow might get loose. By the time the whales arrived at the *Vostok*'s dark passageway, they had been dead for hours, sometimes days. There was no excitement at all. But now—here was something to break the monotony: a little wooden fishing boat appearing from out of nowhere with a strange brightly painted sail, and two tiny space-age bullet boats that bucked and lurched in the blood-and-foam turbulence generated by the big ship's propellers while hippie-type characters aimed cameras at whales being hauled up the slipway. Was it an American plot? Could this be the legendary CIA up to some clever capitalist trick? Or were these maybe crazies from Hollywood, making a movie?

The Zodiacs darted all around the factory ship while Easton and Weyler squinted through their bouncing lenses. Much closer to water level, the men on the harpoon boats shouted questions in Russian, and Korotva shouted brief replies, but over the noise of the outboards it was difficult for anyone to hear anything. A black dog on the nearest harpoon boat barked wildly. On the *Vostok*'s flensing deck, work went on as usual. From the bridge of the *Cormack*, we could see automobile-sized slabs of whale flesh being hoisted into the air, but from our angle, so far below, we could not make out the individual workers. The sounds we could hear from above were the creaking of chains, the high-pitched whine of winches, the shrill buzzing of electric saws.

Satisfied that they had documented everything they could, the camera crew returned. Weyler was ecstatic with the photos he'd gotten, but Easton had a worried, almost panicky look on his face. He and Precious consulted hurriedly, both of them tense and uneasy, even more so than the rest of us. "Look, guys," Easton said, "we're running outta film. I mean, *really* running out."

"Have you got enough to cover us?"

"Jesus, Bob, I dunno. We'll be lucky if we've got ten minutes worth of stock left, and half of that is shit-poor quality stuff that I'm not even sure'll work. Whatever we're gonna do, we'd better do it quick while we've still got this incredible light."

While the documentation crew had been out on the water, the musical crew had not been idle. Gregory and Jackson had mounted our four-foot-high speakers on the upper deck and assembled their instruments, including Jackson's Tcherepnin. Microphones had been set up.

"What now?" Cormack demanded.

"Well, John, we're gonna put us on a little show for them, play some music, and George'll give them a little pep talk in Russian. Maybe they'll surrender."

"Huh," said Cormack. "This's a goddamn circus, that's what."

As we pulled alongside the *Vostok*'s midsection, perhaps fifty yards off her port, Gregory and Jackson cut in their guitars and sang:

> We are whales, living in the sea
> Come on now, why can't we live in harmony?
> We'll make love above the ocean floor
> Waves of love come crashing on the shore.

Now the Russians were not merely amazed, they were astonished. We counted well over one hundred of them lining the decks, one out of three with a camera. As Gregory and Jackson went into their chorus, several dozen of the Russians started clapping in tune with the music. The lady in the bikini waved to us. Rex Weyler waved back madly.

When the song was done, we plugged in tapes of humpback whales, whose eerie, alien voices, played at full volume, went echoing and reechoing off the vast metal hull, even as slabs of whale meat continued to swing on chains above the flensing deck and the fountain of whale blood continued to gush from the hole in the hull below. The expressions on the faces of the Russians changed to bewilderment, then froze as they realized what they were hearing. The clapping stopped. No one waved anymore. Several men shook their fists.

"Dat fok der brains, aye?" said Korotva savagely, taking the microphone as the whale songs ended and the Russians—those who had not walked away—glared down at us with the look of people who have just been slapped in the face.

"Hey, *Vostok*!" Korotva yelled, his voice booming and echoing. Then he launched into a brief burst of Russian speech, saying, in effect: "We are Greenpeace, we represent the fifty-two nations that voted at the United

Nations for a ten-year moratorium on whaling, and we are here to stop you from killing the whales."

When he had finished, one Russian yelled something at us.

"What's he say, George?"

"He says get fokked."

Will Jackson hit them next with his synthesizer. The feedback off the *Vostok*'s hull whanged and ricocheted so loudly that it pierced the eardrums and stabbed at the brain, sounding at moments a bit like the whales, but mostly like a crashing jet full of terror-crazed cats. Cormack became furious. "Turn that goddamn thing off!"

By now it was 4:00 P.M. The two harpoon boats that had been transferring whales to the factory ship when we arrived had completed their operations and moved off to join the other boats that were spread out almost to the horizon, still hunting. A third harpoon boat, the *Vlastny*, NK-2003, had arrived with a load of half a dozen whales lashed to its side, these ones larger—but not much—than the previous batch.

Our next move was now clear.

"Okay, John," I said, "when that guy finishes unloading, we'll just follow him out, I guess, and see what we can do."

The last leg of the voyage now began. It was a bleak moment because this was the point where all the numbers went against us. Having overcome all sorts of odds to get this far, the likeliest outcome at this stage was still that we would get no further. Neither charts, RDF, *I Ching*, nor rainbows could help us now. It would be a straight race between the nine-knot *Phyllis Cormack* and a twenty-to-twenty-five-knot Russian harpoon boat. This was the weakest link in the plan, the stage to which Will Jones, who had served on the *Greenpeace Too*, had pointed, drawing on his training in the U.S. Navy and shaking his head, so many years before. His exact words had been: "Nice, but completely impractical, I'm afraid." If ever there was a tortoise-and-the-hare contest, this was it. The *Vlastny* pulled ahead of us swiftly, moving at twice our speed. And as she dwindled in size in front of us, our hopes shrank down to almost nothing.

Fred Easton asked Cormack: "Is she running from us, John?"

Cormack snorted. "Hah! She's not runnin' from us. She's just lookin' fer whales, that's all."

Our hopes shrank as the harpoon boat got farther and farther ahead. Knowing that we had only a few minutes of film footage left contributed greatly to the general feeling that perhaps we had just about reached our limit, and it had been close, but ... just not quite enough. "It's like that kid's nightmare of running slow-motion to catch up with somebody, but you can't," remarked Carlie Trueman. "Isn't it?" At one point in what had been a two-hour chase, we had three harpoon boats strung out like beads halfway to the horizon ahead of us. A fourth overtook us from behind, sweeping past on our port side, less than twenty yards away. It paced us for a few moments while the crewmen lined up on the deck and jeered. At the bow, the gunner toyed with his harpoon cannon, swinging it so that it was aimed directly at us a couple of times. The display was intended to humiliate us, perhaps to intimidate us. All it brought forth were a couple of outbursts from Walrus and someone else. Then the Russian boat surged ahead and swung deliberately in front of our bow, leaving the *Cormack* to bounce corklike in its wake.

Each and every one of us was aching from the tension of straining forward psychologically, trying to drag our old halibut seiner across the water by sheer ESP. We hardly spoke—just exchanged quick, noncommittal glances. Several of us had been in our wet suits since back around noon. Our bodies itched and chafed. A dinner of sorts had come together in the galley, but no one except Cormack and Hewitt could eat. Hewitt was otherwise busy hanging his socks by a fishline off the side of the boat to get them clean. I hid out for the most part behind the trailer, eyes closed, trying desperately to meditate—but my nervous system was so jangled that a mishmash of images, half-formed prayers, glimpses of Gyalwa Karmapa XVI, and flashes of scenes on the *Vostok* whirled through my head, making concentration almost impossible. Korotva and Moore were chain smoking furiously. The others either paced the decks or lay in their bunks, staring through the portholes. The only sounds were the whacking of the sail in the wind and the underground-railway thunder of the engine at full speed. Weyler, Easton, and Precious nervously checked and rechecked their light gauges, eyes scanning the horizon, mentally calculating how much time they had left before it would be too dark to shoot.

The only break in the tension came about five-thirty when a half dozen Dall's porpoises appeared around our bow, flashing and glittering in the air. The water was such a clear royal blue that we could see them—ghostly writhing submarine missiles—down to a depth of twenty or thirty feet.

The *Vlastny* was now the only boat in sight. It was roughly four miles ahead.

When no one was looking, I gave up meditating behind the trailer and instead got down on my knees in my wet suit, bowed my head, and prayed in a very old-fashioned, traditional style. "Dear Lord …" Gregory reported later that he, too, had been praying steadily. So had Will Jackson. So had Walrus—except that Walrus was doing his praying up on the mast, with a pair of binoculars strapped over his shoulder. It was roughly six o'clock when Walrus saw the *Vlastny* change course, darting first northward, then, with a burst of smoke, coming about so that her bow was pointed directly toward us.

Within minutes, everyone was on deck, holding our breath while we stared. The *Vlastny* made two or three more darting-about movements before settling on a course that was bringing her directly toward us. We stood there for long moments, too astonished, too filled with sudden eye-watering hope, to speak. No one dared yet to come out and say, "This is it." It was still too good to be true.

Walrus screamed from the mast:

"Whales!"

Then we all saw it at once—spouts in the water directly ahead of the *Vlastny*, maybe a dozen of them, white puffs of vapor. The *Vlastny* was in pursuit of a pod of whales, and the pod of whales was coming directly, unerringly, straight as an arrow, toward the *Phyllis Cormack*. Out of the 360 degrees on the compass that the whales had to choose from in their flight, they had somehow picked the one particular bearing that would bring them to our side, dragging the whalers in their wake, making our whole effort to protect them possible. That moment on deck, as the significance of what was happening hit us, was worth enduring a whole lifetime of meaningless struggle just to experience once. There was nothing rational to be said about it—the mechanics had no explanation and the mystics were too awed to attempt an interpretation. Cormack roared: "Zodiac launching crew!"

The months of preparation paid off. Within less than three minutes, three Zodiacs were in the water, Watson and I in one, Moore and Easton in the second, Weyler and Korotva in the third. But something was wrong with the third one's engine. The propeller wasn't working. There were shouts of rage from Korotva and Weyler, while up on the mast Walrus yelled:

"They're turning, they're turning! Hurry up!"

Watson looked at me. I nodded. He immediately opened up the engine full blast, and without a backward glance we screamed away from the *Phyllis Cormack*, hearing the sound of Moore's engine roaring as he came out directly in our wake. All thought of waiting for the photographer's

228

Zodiac to be fixed was abandoned immediately.

Watson had tied a white cloth over his forehead with a long white plume that snapped in the wind behind him. It was a tradition that had existed among Japanese kamikaze pilots. The word *kamikaze* meant literally the "divine wind." We were undoubtedly a strange pair to be flying across the waves toward a harpoon boat—Watson in his blue wet suit and white kamikaze scarf, I in my black wet suit, wearing a multicolored Peruvian cap that had been given to me years before, with the leather pouch containing the Karmapa's red cloth bouncing against my chest. When we were about a mile out across the water, I threw my hand back to Watson, grasping his in a revolutionary handshake and yelling over the baying of the engine:

"We're doing it, Paul! We're doing it!"

Then it was all I could do to hang on with both hands to the bow rope, standing legs apart and braced on the front floor plate of the Zodiac, watching the *Vlastny* draw closer and closer.

Back at the *Cormack*, the problems of the third Zodiac were being overcome by an adrenaline-filled burst of sheer muscle power. Will Jackson had flung himself into the hold, where a spare fifty-horsepower outboard had been stored. The thing weighed at least two hundred pounds. Jackson had seized it, hurled it at least six feet straight up in the air, where Cormack's treelike arms shot out to grasp it then haul it with a bump and a clank across the deck and throw it over the side to Korotva, who somehow caught it in his arms and jammed it into place to replace the other engine. The whole operation, which would normally have involved pulleys and winches and rope and taken at least half an hour, was accomplished with tendons, muscles, and heart in less than eight minutes—and Korotva and Weyler were blasting away at twice the speed of the other Zodiacs across to the point a mile ahead where they could see two small splashes of foam approaching the great bow wave of the Russian boat.

By the time Watson and I pulled abeam of the *Vlastny*, my legs were buckling from the strain of standing up. Something of the exhaustion of a bronco rider had come over me, but it seemed important to remain upright in defiance, if nothing else, of the men who lined the rust-corroded sides of the boat. The Zodiac lifted and flew clear into the air twice as we swept past. Then we were abreast of the harpoon gun. It was painted a bright baby blue, as though it were some kind of kiddy's toy. Its operator, a middle-aged man with a tweed cap, stood back from the gun, where he had been crouched smoking a cigarette until our arrival. He spat over the edge of the deck at us. Up on the flying bridge, one man was shaking his fist and howling.

Taking us no more than thirty or forty feet ahead of the ship's bow, Watson cut abruptly to the starboard, swinging directly in front of the harpoon. By this time, my legs buckled and I was forced to slump to the floorboards, facing backward so that I was looking over Watson's shoulder at the immense metal wedge rising and falling behind us, sending heavy bursts of foam ten feet into the air, rising so steeply that the curve of the bow obscured the rest of the superstructure. The harpoon wagged against the early-evening sky like a horn or proboscis or dentist's drill. Even the gunner was obscured at times as the ship came up over a wave. It seemed as though we might as well have placed ourselves in front of a machine set on automatic. It was all metal and cable, as impervious to appeal as a steel trap.

Then—ahead—we got our first clear glimpse of the whales.

They surfaced not much more than sixty or seventy feet in front of our bow. Their heads broke out of the sea, glittering and black, waves unfolding like lenses around them as they surfaced. It seemed as though a giant surreal mouth was opening in the middle of each of their backs as they exhaled a blast of air and spray like a jet whistle. There was a tremendous explosion as the water closed back in around them, leaving an upswell of whirlpools and eddies. And as they surfaced, in a brief, desperate effort to gasp for air before plunging below again, small perfect rainbows formed in the spray from their blowholes. In a single glance, I could see what seemed to be about seven rainbows, like multicolored halos, over seven thrashing whales. And then they were gone—the whales into the hoped-for security of the sea and the rainbows into nothingness.

Our Zodiac skittered across the turbulence of the water where the whales had been only seconds before, with the *Vlastny* pounding along immediately in our wake. The sun edged down to the horizon somewhere behind the ship so that the wedge-like portion we could see was in shadow, the gunner was a dark silhouette, and it seemed for a moment that we had arrived at a pivot point on a celestial scale that weighed light against darkness, rainbows against shadow, flesh against metal, life against death.

At moments, we hung back, less than ten feet in front of the harpoon boat's bow, lest we overrun the whales themselves and have them come up under us, flipping us into the water to be chopped to pieces by the

Vlastny's props. At other moments we moved forty or sixty feet ahead, trying to keep ourselves directly in front of the harpoon, which was being swiveled this way and that, searching, as though it had the power to sniff the wind, seeking out the exact spot where the whales would break wildly and hopelessly to the surface again. For the men on the ship, there was no guessing involved. On their sonar they could see the whales clearly, underwater or above. The ancient refuge of the sea had been penetrated by the equivalent of X rays. Down at the water level, without such equipment, it was impossible for Watson or me to know where the whales would come up—all we could do was attempt to stay directly in front of the harpoon, knowing that it *knew*.

Rather than being terrified at this point, we were exultant. We did not believe they could fire. We did not believe they would take the political risk of killing two human beings in international water. Several times, as though to enforce our belief, the gunner walked away from the harpoon, leaving it without an operator and therefore harmless.

"Gotcha, you bastards! Gotcha!" was the only clear thought I remember having.

Then our outboard engine died.

One instant it was going full blast, the next it was making a gurgling sound, and then it was silent, and we were gliding rapidly to a halt with the *Vlastny*—some thirty or forty feet behind—bearing directly down on us. A heavily built bearded man was hanging over the bow, laughing wildly and slashing his forefinger back and forth across his throat. I took him to be the captain. There could be no doubt he was signaling that he had us now and fully intended to run us down. Feebly, I flashed him a peace sign.

With a classic comical look of dumbfoundment, Watson stared at the engine.

"Oh oh," he said.

The *Vlastny* was coming down on us like an express train. "I think we're going to have to jump, Paul."

But Watson was clawing desperately at the engine, making no move to leap. The fuel pump had bounced into the air and come down on the fuel line, choking off the supply, but neither of us knew that then. I remained frozen at the bow, unable to jump unless he was jumping, staring with a feeling of great detachment—as though I was merely watching a 3-D movie—as the harpoon ship reared over a wave, flung its whole weight in a mighty swing down into the water scarcely yards away, and was about to grind us under, when the bow wave it was pushing before it seized us, our outboard engine coughed briefly, biting into the wave, and we were lifted and swept lightly aside, the *Vlastny* passing so close at full speed I could have reached out and touched it. The feeling was that guardian angels had swooped down at the last second and given the Zodiac a gentle nudge. Others might say it was the fact that we had an air-filled rubber boat that saved us.

Whatever, we were left bobbing in the harpoon boat's wake while it steamed steadily after the whales, and there was nothing any longer to prevent the gunner from firing. While Watson continued to tear at the engine, I pounded the side of the Zodiac, screaming obscenities aloud. Then the two other Zodiacs were flanking us on either side and George Korotva was yelling: "Get in! Get in!" Rex Weyler and I changed places without a single comment, and there ensued a wild shuffle leaving Easton with Watson, while Korotva and I, powered by the big fifty-horsepower engine, surged after the *Vlastny*, Moore and Weyler close behind.

Within minutes, we were back in position.

I now had time to see that an amazing melee was taking place in the water around us. We had stirred up a hornet's nest. The *Vostok* itself was steaming at full speed toward the scene. From all over the horizon, harpoon boats had changed course and were likewise throwing gusts of smoke into the sky as they converged toward us. Directly ahead were not just seven, but more like a dozen whales. And it was clear now they were taking advantage of the *Phyllis Cormack*'s presence in an effort to shake their pursuers. With a shock, I looked up to see the old halibut seiner directly ahead of us, with Cormack hunched at the wheel, bringing his bow into line with the onrushing harpoon boat so that if there was to be a collision, it would be head-to-head. The whales were diving under the *Cormack*, so that, as we followed them, we were forced to cut in an arc around our own boat, and the Russians were forced to do the same. Everything was in kaleidoscopic motion: twelve whales, three Zodiacs, one halibut seiner, ten harpoon boats, and a looming, full-steaming factory ship like a startled rhinoceros charging to the center of a disturbance.

Moore's Zodiac was flanking Korotva and me on the starboard side, while Weyler frantically snapped pictures. To our port and slightly ahead, so that the whales several times seemed to come up around them, were Watson—his engine running again—and Easton, who by this time had only thirty seconds worth of film left in his camera; his battery pack had gone dead on him so that it seemed we would get no film footage.

The whales were staying underwater for shorter and shorter periods as their aching hearts began to lose the race against the diesel and pistons and camshafts. The moment was approaching when the Russians would either fire or not fire.

Several times, we had glimpsed a man running along the *Vlastny's* catwalk from the bridge to the bow—the catwalks being the most noticeable feature of the ships other than the harpoon itself. In the wheelhouse, the radio was clearly crackling with noise, as orders and queries flew back and forth to the commander on board the rapidly approaching *Vostok*. Had anyone put a call through to the Kremlin? Then, the runner appeared on the flying bridge, spoke with the captain, and walked—did not run—along the catwalk to the gunner. From the way he walked, purposefully, almost thoughtfully, it was easy to see that the order—whatever it was—had come.

"George, get ready," I said.

Korotva said later he could tell the moment the gunner was about to fire just by watching my face, which suddenly went ashen and sick-looking. "Twisty," he said.

Some thirty yards away, bouncing across the water, Fred Easton could not make out enough details to know himself what was about to happen. All he knew was that the chase had been going on for close to three-quarters of an hour, the light was failing, and that with thirty seconds of film left and a dead battery, he had nothing to lose, so he told Watson to stop while he hoisted his camera once more in his arms, swung it toward the *Vlastny's* bow, and pressed the trigger. To his astonishment, the battery inexplicably came back to life. He panned quickly forward from the rusty bow that he could see so clearly through his lens, caught a clear image of Korotva and me in the Zodiac, and kept panning until he caught another glimpse of the whales just as they broke the surface. He had just started to pan back when he heard the loud firecrackerlike *whack* of the harpoon being fired, and saw it whispering through the air, cable unraveling behind it. Reflexively, he jerked the camera around, managing to follow the harpoon down into the whale's back where an explosion of foam occurred, then panned back once more, catching our Zodiac as it rode up on a wave, almost directly on top of the whale.

All Korotva and I knew was the sound of the gun going off. Instinctively, we both ducked. The cable lashed down less than five feet from the port side of the Zodiac, like a massive sword cleaving the water. The gunner had waited for an instant when we had gone down in a trough between the waves and the whales had come up ahead, then fired almost directly over our heads.

Korotva needed no instructions to turn the Zodiac almost on its heel and speed us away out of the reach both of the cable as it came up, dripping, springing about like a guy line as it took the full, shuddering weight of the whale, and of the convulsions as the massive tail was now beating the water while fountains of blood and spray burst into the air.

If Leviathan screamed, we did not hear it over the roar of our engine.

All that came into my mind then was the recollection of a whale expert back in Vancouver warning us that the second most dangerous moment would come in the event that a whale was killed and we were close to the scene. The whalers usually shoot females first, he had told us, in order to enrage the bull who led the harem, so that he would turn to go to her rescue, leaving the rest of the pod leaderless. "If you're in the water when the bull whale attacks, he'll probably turn on you, because you're going to be the easiest targets," the whale expert had said.

But that particular whale expert, like most, had underestimated something in the whale.

The bull did turn.

I saw him off in the distance, breaching, then swiveling around like a mostly submerged locomotive and plunging through the water toward us.

"Let's get the hell outa here!" I screamed at Korotva. And we took off for the *Phyllis Cormack*, casting only a glance over our shoulders to make sure the whale wasn't right on top of us. It was only when we had reached the *Cormack*'s side that I realized the Zodiac bearing Easton and Watson had stayed behind. Korotva went flying back across the water to warn them out of the area, but even as he was racing across the water, we heard the sound of the harpoon blasting again.

The bull had attacked, not the Zodiac, but the gunner astride the bow of the *Vlastny*, as though knowing perfectly who his real enemy was. He had surged up out of the water, jaws snapping like a dog. He was not a large whale, no more than half the size of his ancestors, perhaps only the equivalent of a teenager thrust with the responsibility of leading the family of other whale-children through the darkest age their species had ever known. For all his still-considerable power, he could reach up no more than halfway to the man who towered over him on steel decks, calmly fixing another harpoon into his cannon. Once, twice, the whale— exhausted as he was—threw himself straight upward, snapping, crashing back into the water explosively. Then the harpoon fired. The whale went down in a boil of blood and water and came up in one last desperate, screaming lunge but could only get a third of his body out of the sea before sagging back, thrashing. The breath from his blowhole spouted out a pinkish mist. The Russians left him there to bleed to death while they bent to the task of lashing the dead female to the side of the boat. Easton and Watson moved to within twenty yards of where the bull was dying. Rolling over, he brought his head out of the water so that his eye fixed itself on them for a moment, and they stared back, silent, until he collapsed for good into the blood-stained sea.

Moments later, the *Vostok* and half a dozen harpoon boats pulled in close to where the *Cormack* and the *Vlastny* were bobbing in the water. We hauled the last Zodiac on deck.

While most of our attention had been concentrated on the Russians as they picked up the two dead whales, Melville Gregory had stayed on the upper deck, watching as the survivors of the pod we had tried to save kept moving steadily away, leaving only vapor traces in the rapidly cooling air. In all the confusion, the other harpoon boats seemed to ignore them. Gregory's last glimpse of them was when they were about three miles away, heading due north. As far as he could tell, there were eight of them, although there might have been ten. Those, at least, had escaped.

Ron Precious was standing on the deck in deep gloom.

"Did you get all dat on film?" Korotva asked.

"I dunno," growled Precious.

"You vant we should do it again?" Korotva joked.

Precious made an animal sound and clenched his fist. For a second I thought he was going to hit Korotva with his camera.

Easton was trembling. "I dunno if I got anything. I took a chance. I *think* I got it. Listen, I couldn't believe it. My battery was *dead*, man. Then, just like that, it came alive, like somebody'd thrown a switch, aye? I shot the sequence, then the battery died again, and it's dead now. I don't know how the hell it ever came alive for just that few seconds. I still can't believe it. But it did."

Cormack's only comment was:

"Close one, huh?"

The *Vostok* and her fleet closed into a convoy formation and began cruising southward. We followed. As darkness fell and the lights of the ships came on, it seemed we were in the midst of a floating city dominated at its center by a ten-story building. There was little excitement on board the *Phyllis Cormack*—we were too exhausted, too jangled. Also, the implications of having been fired upon were sinking in. We were faced with knowing that the next time we set out to block a harpoon shot, there could be no kidding ourselves that the Russians were unwilling to take a chance on killing us. We would have to crank ourselves up that much more.

In the morning, Rex Weyler awoke to see a huge sperm-whale cloud formation hovering over the rising sun. The sea was slightly choppier than the day before. Our position was about fifty miles offshore, slightly north of the Mendocino Ridge. At 6:00 A.M., we came alongside the *Vostok*. There was no visible response: the few people on deck ignored us completely. No whales were in tow, although there were still the remains of a few carcasses being processed on the flensing deck. By 9:30, four harpoon boats had pulled up to the factory ship, but none of them had yet killed any whales. When one of the boats—NK-2042—set out, presumably to begin hunting, we followed. The boat only went about a half a mile before coming about and retreating to the *Vostok*'s side. After a great deal of yelling back and forth between the two ships, the harpoon boat set out again, still dogged by the *Phyllis Cormack*. This time the factory ship itself changed course and plodded heavily along behind us. The three remaining harpoon boats fanned out in a search pattern.

By noon, no whales had been killed.

Following the same plan as the day before, those of us with wet suits were wearing them. The Zodiacs and engines were primed. We had every intention of going back in front of the harpoons. The difference was that instead of riding on jet sprays of adrenaline, we were all somber and thoughtful. So much emotion had been burned up the day before, we seemed to have little, if any, left. Most of us felt transformed, as though we had gone through an elaborate ritual that had had the effect of changing our status somehow. We would not know for certain until we got back to shore whether or not Fred Easton had gotten the footage of the harpoon incident, but one thing was definite—if he had, indeed, captured it, then as a media campaign the voyage was already a success. No network would be able to resist such footage, just as no wire service would be able to ignore the story. As a newsman, I knew we had achieved our immediate goal. Soon, images would be going out into hundreds of millions of minds around the world, a completely new set of basic images about whaling. Instead of small boats and giant whales, giant boats and small whales; instead of courage killing whales, courage saving whales; David had become Goliath, Goliath was now David; if the mythology of Moby Dick and Captain Ahab had dominated human consciousness about Leviathan for over a century, a whole new age was in the making. Nothing less than a historic turning point seemed to have occurred. From the purely strategic point of view of a media campaign aimed at changing human consciousness, there was little more that we could hope to achieve. So why go on?

The reason lay deeper than the immediate requirements of publicity or politics. However well the rationality of the Mechanic Gang was holding up, the *Phyllis Cormack*'s hard core of mystics were by this time utterly convinced that the gods had intervened so often, not only before our eyes but before our cameras, that there could be no further debate: we *were* blessed. We had entered a zone of profound mystery more fascinating in itself than even the wonder and travail of the whales. We had brought the ancient influences of both Tibet and China out into the water off the coast of America to confront a fleet from the Asian mainland, led by a ship, the *Dalniy Vostok*, whose name meant "Far East." Here, surely, was a yin-yang situation of global proportions, a true meeting of East and West, except that the East was coming from the West, and the West from the East, which suggested that some tremendous reversal, a shifting of the axis of the spirit of the world, had taken place. The West now was the East and the East itself had been turned into its opposite. What had just occurred in the waters off the Mendocino Ridge was a microcosm reflecting all of this, signaling, in our minds at any rate, a change whose outlines dwarfed any of our previous lesser fantasies. It was as though the world's collective-unconscious mind was making one of its periodic efforts to render itself conscious, and we were the immediate instruments of the transformation. Yet something remained to be done—some final ceremonial dance or ritual. No one knew what it was, but the mystics— myself included—were blissfully certain that the situation itself would soon instruct us.

Whatever the final scenario would be, we knew by 4:00 P.M. that it was not going to be a repetition of the previous day's events. The harpoon boats were now only specks on the horizon. The gray outline of the *Vostok* lay far to our port stern. The ships were steaming steadily, having long since ceased to make any attempt to hunt, which meant they had either decided to shake us off their tails by heading to another location, taking full advantage of their superior speed, or else, for some reason known only to the gods who were in charge of these things, no whales had appeared anywhere within the death ships' range.

At 4:30, we spotted a dead, marked whale in the water ahead. It was not much longer than the first one we'd seen, perhaps twenty-eight feet, but still under the legal minimum-size limit. Immediately, it became clear what remained to be done. Whereas, the day before, we had exerted a masculine aggressiveness, today we would adopt a feminine passivity.

Our preparations were detailed. Waterproofed information packages were put together, containing buttons, posters, tapes, T-shirts, and a detailed explanation of our position. Taking all this material, plus my passport, I would sit on the dead whale's body and refuse to budge when the Russians came to pick it up, forcing them to either dump me in the water or take me prisoner. Watson would stand by in one Zodiac and Moore in another to fish me out of the sea if necessary. The action would have to be swift, because there were at least three small blue sharks cruising about in the immediate vicinity of the body. What I really wanted to do was squat there, naked, in a full-lotus position—but the reality was that the air was too cold for my skinny frame, so a wet suit was in order.

The hours that followed seemed to stretch into days. Out on deck, Gregory played his flute to the dead whale. Korotva and Hewitt hung about the radio room, listening for messages passing between the now-invisible Russian ships. But only extremely terse bursts of voices came through, reeling off figures and numbers, and that was all.

"Not very conversational, are they?" Hewitt remarked.

Shortly after supper, I fell asleep and awoke to find that it was already dark. The Russians had still not come to claim their prize. From what Hewitt could tell by the RDF, the entire fleet was still proceeding southward. So, about ten o'clock, we too abandoned the dead whale. There had been some talk of sabotaging the marker beacon, but I found myself saying: "Render unto Caesar that which is Caesar's." We cranked up to full throttle and set out in the direction of the last sighting of the *Vostok*, which was proceeding far more slowly than the rest of the fleet.

By midnight, we could see the *Vostok*'s lights again. We spent the night following some five miles astern, moving at nine knots. Toward dawn, the factory ship increased speed to twelve knots and began to pull inexorably away from us. None of the other boats could be seen.

At 4:00 P.M., we found ourselves 135 miles southwest of San Francisco, with the *Vostok* by this time a tiny gray smudge on the horizon. There was no longer any question about what to do—we had fuel enough to make it to San Francisco, but not enough for even another day's full steaming in the direction the Russians were moving. We came about and headed toward the California coast. It was Sunday, June 29—sixty-four days since we'd left Jericho.

An odd feeling of inertia gripped us as we slipped toward the shore. All our thinking and planning had been directed toward the moment of confrontation with the whalers; beyond that moment, we had no plans at all. For myself, I wasn't sure whether I had expected to survive or not, but now that it was over and we were all still alive, it was as though we were entering a whole new lifetime. Everything from now on would be in the nature of a surprise.

The first surprise was the discovery that the days were over when I would have to write desperate "giant squid" stories just to get us a few column inches of coverage in the press. The American media had picked up on our encounter and was waiting for us en masse. The national TV networks were clamoring for film footage. The wire services were poised to grab our photographs. Reporters by the score were demanding interviews the moment we docked. Back up in Canada, television crews were being assembled to fly down to San Francisco to meet us. It appeared we were about to become instant electronic age celebrities. Although we had no inkling of it then, this meant we were now entering the most difficult and exhausting part of the voyage—and the most dangerous from the point of view of our sanity and ability to hold our collective trip together.

With the single act of filming ourselves in front of the harpoon, we had entered the mass consciousness of modern America—something that none of our previous expeditions had achieved. It was Walter Cronkite himself who introduced our footage to the mass TV audience, footage that was then run on every single television channel in the U.S. and Canada, spilling over into Europe and even Japan. Weyler's photographs went out to every country in the world that had a wire service.

After two months at sea, stopping in at no community larger than Prince Rupert, San Francisco seemed like the stage for a science-fiction movie, as though we had traveled farther through time than we had through space. We were all suffering from cabin fever to begin with. To be pitched from the sweaty confines of the *Phyllis Cormack* into television studios on the fortieth floor of a glass tower overlooking a great elongated pyramid, with the whole pastel expanse of the Bay Area reeling away over the hills until everything was lost in copper-pink mist, was to invite culture shock. To move directly from our mysterious natural world of rainbows and tides, whales and dolphins, our world of magic and routine miracles, into a universe of jackhammers and press conferences, vast silent radio-station studios with their arrays of sound equipment like the control panels of starships, to be toasted and cheered, wined and dined on some of the richest food in the world, after weeks of diesel-soaked waffles and coffee, was to invite complete disaster. Our existence on the boat had been so simple. We had only one goal: to find the whalers and save some whales. But now existence was brain-befuddlingly complex.

We had expected to be welcomed warmly by the other conservation groups who had been calling for a moratorium on whaling. There were some twenty or thirty antiwhaling groups based in San Francisco. Yet they had never gotten together in a single room before. Competition for funding was fierce. Group ego conflicts were rampant. Individual ego conflicts were equally rampant. Rather than being welcomed as brothers and sisters by our fellow conservationists, we were greeted with token smiles and congratulatory remarks that barely masked the underlying mood of resentment and suspicion. There were not a few California conservationists who felt silly because the Russian whalers had been discovered operating virtually under the nose of the most powerful antiwhaling lobby in the world. Joan McIntyre, the head of Project Jonah, one of the best-known antiwhaling groups, went so far as to accuse us of "machismo." Said she, "I want to see the whales saved, but not *that* way." It was very much as though we had usurped someone's turf.

There was a jagged air of craziness to everything that happened to us in San Francisco, a sense that we had come out of the pure salt air into a poisonous kind of nutrient tank. And it was not just the tangible pollution that you could smell, it was a psychological reality. Everything was either too rich or too rotten. The houses and apartments of the people we stayed with were for the most part opulent, marked with that certain San Francisco style that borders on decadence. Since these were mainly the homes of people who were involved in conservation, it was impossible to avoid noticing that the practice of environmentalism was the preserve of an elite. Down in the dangerous garbage-strewn streets, where human derelicts stalked and lurked, we might as well have been a million miles from the sea and its mysteries. There was an ecosystem here, too, but its laws were unknown to us. We were like babes in the woods. We had only been in San Francisco a day when a phone call reached us at the offices of the Oceanic Society at Fort Mason, overlooking the bay. It was from New York. It was an offer to make a multimillion-dollar movie about our adventure. The outfit involved, Artists Entertainment Complex, had apparently made such spectaculars as *Earthquake*, *Serpico*, and *The Godfather, Part II*. The lady in New York who was their agent was willing to fly immediately to San Francisco, along with a scriptwriter. Why not?

Twenty-four hours later, we were into the midst of our worst internal haggling yet. While it seemed to me only logical to let a multimillion-dollar movie about saving whales go ahead, the idea did not sit well with several of the crew members because there was a slight catch. First, it was doubtful that Greenpeace would see much money—if any—out of the deal. So far as I was concerned, that didn't matter, since the objective was to save whales, or at least raise "whale consciousness" around the world, not get rich. But second, each and every crew member would have to sign a release, giving the movie company the right to portray them as it saw fit. That was too much for Walrus Oakenbough and Paul Watson, who refused to sign. We tried to solve the problem by holding a "tribal council" meeting on the hill at Fort Mason, looking out over Alcatraz Island and the Golden Gate. The lady from New York, Amy Ephron—sister of the famous media guru, Nora Ephron—was allowed to speak, explaining all the reasons why we should go ahead and sign a contract. Unfortunately, Ephron's New York style had the effect of turning almost everyone off. In the end, despite the opposition, Patrick Moore and I went ahead and signed a one-page contract, giving Artists Entertainment Complex the rights to our story for one year. In my view, it was the only way to tie up the rights, otherwise anybody could have gone ahead and written whatever book they wanted or made whatever movie they wanted, drawing on the printed and televised accounts of our voyage. Screams of "sellout!" came from several of the crew members. A genuine feeling of bitterness toward both Moore and myself developed, and we, in turn, were furious with the others for failing to perceive that the signing of the contract was the best possible move for the moment, and in the best interests of the whales. The schism that developed around the movie contract was never to fully heal itself and was to lead to divisions that would plague us for years.

It took us nine days to get out of San Francisco. By that time we were a transformed group, bitterly divided, filled with an anticlimactic sense of having badly debilitated ourselves. At the last minute, Moore had gone around trying to get everyone to sign the releases for the movie contract. Rather than sign, Walrus had fled up the mast, screaming at Moore, and refused to come down until the boat had left the dock.

We spent the next two weeks tracking Russian voices steadily northward along the coast. Off the coast of Oregon, we got within radar range of the *Vostok* late one afternoon, but a heavy fog rolled in and our radar suddenly broke down. It was too clear a sign to ignore. It seemed obvious to me that we were not meant to meet the Russians again this year. The last leg of the voyage was otherwise punctured by only two unusual incidents. The first occurred only two days out from San Francisco, when I heard Ron Precious yelling: "Bob! Bob! Come quick!" Scrambling onto the deck, I was stopped in my tracks by the most awesome sight I had ever seen. It was a phenomenon known as a "fog rainbow." What it meant was that I could stand on the deck amidships and see a perfect rainbow curving up from the *Phyllis Cormack*'s bow, over the sail, and downward until it touched the stern. It was like a nimbus or a halo. Then, gently, it began to move off into the fog. So convinced was I that it was a miracle, I ripped off my clothes and plunged into the icy water and started swimming in a state of bliss toward it, my heart pounding, thinking that I was swimming into a divine embrace. Finally, the shock of the cold water got to me. I dived into the depths, going down, down, down as far as I could, and instead of being full of life, the ocean seemed an extension of the fog above. Lifeless. Empty. By the time I surfaced and got back aboard the ship, I was shuddering so hard I couldn't speak. A couple of nights later, I fell asleep with a cigarette in my mouth and awoke to find my bunk in flames. Korotva yanked me free while several of the others poured buckets of water on the fire. After that, nobody would speak to me for days. Causing a fire on a wooden boat at sea is the lowest of all possible crimes. By this time, I was so blown out I couldn't tell the difference between water and white wine. I was the *Phyllis Cormack*'s latest "bow-case."

Two days away from Vancouver, Weyler was up on the mast, doing daredevil stunts. He grabbed for a rope to swing out in the air, only to discover that the rope wasn't tied to anything. Down he crashed some twenty feet to the deck, snapping his anklebone. As he lay there on the deck, with his foot twisted almost backward, Cormack emerged from the galley with a rusty butcher knife and announced: "Guess we'll have ta amputate." For a second none of us knew whether he was kidding or not.

We stopped at Alert Bay, even though it meant going all the way around the northern tip of Vancouver Island, in order to bring our tattered *Greenpeace* flag to the Kwakiutl Indians, whose whale design we had borrowed. This time, instead of the Indians blessing us, as they had back in 1971, it was my task to give them the flag we had flown since embarking on the voyage to save the whales. They accepted it with due ceremony. There was a tremendous party in Alert Bay that night. The next morning, hung over, but with some of our group feeling of solidarity back, we set out on the last lap down the Inside Passage to Vancouver.

It was Sunday when we arrived in English Bay. There were some ten thousand people waiting for us at Jericho Beach. As we moved in close to shore, hundreds of them began to wade out into the water to greet us. It was a very Biblical scene. The Zodiacs roared forward, and we leapt out of them into the embrace of strangers, lovers, family, friends, splashing the last few yards through the water as though in a kind of ecstatic group baptism.

3

Foundation and Empire

Fear success.

—BEN METCALFE

For all our efforts, Greenpeace was now some forty thousand dollars in debt.

We had generated worldwide media attention on the issue of whaling, had exposed the presence of a Russian fleet off the coast of North America, had helped to stir up political pressure for a two-hundred-mile limit to replace the old twelve-mile territorial limit, and had physically saved at least eight sperm whales—for a day or so, at any rate.

All we had in our hands was a movie contract, which was probably worthless and which was the source of a great deal of internal acrimony. What we had gained was an expanded hard core of determined whale savers. The Quaker theory about "bearing witness"—that it changes the level of commitment on the part of the witness—had worked. No one who had viewed the disappearance of Leviathan into the giant robot anus of the *Vostok* could ever be quite the same.

I, for one, could not go back to writing newspaper columns. Scarcely a week after we returned to Vancouver, I quit my eighteen-thousand-dollar-a-year job as columnist to become full-time president of the Greenpeace Foundation, which meant I was faced with dozens of creditors pounding at the door, a whole new set of organizational and political problems that had not existed until the moment of our "success." The sense of magic—easy to keep alive when you are at sea among dolphins and rainbows and living completely outside the parameter of day-to-day, mundane realities—faded rapidly. We all started coming back down to earth. The ambition that had been inside me four years before on the way up to Amchitka—to lead a revolutionary environmental organization—was now a reality, except that the bills were stacked up knee-high; new Greenpeace groups were suddenly forming in our name, clamoring for a share of the prize of fame, or at least notoriety, which we had won at such a cost in terms of human energy; and within our group itself there was a division between a faction, led mostly by Paul Watson and Walrus Oakenbough, that liked to think of itself as the "grass roots," and the "executive branch," led by me, which was determined not only to pay off the bills but to put Greenpeace on the kind of financial footing that would make future expeditions possible on the scale we wanted. That meant learning the arcane arts of business. At that particular point in time, I didn't know the difference between net and gross, let alone have any idea what a "cash-flow chart" might be, what the difference was between an audit and a debit, and certainly had no idea what deficit financing might mean.

Our first action, after returning to Vancouver, was to throw a picket line up to prevent the loading of fresh vegetables onto a Russian supply vessel that we believed was destined to rendezvous with the whaling fleet. Vancouver and Prince Rupert were the only two ports on the entire West Coast where Russian ships were allowed to resupply. If we could close off Vancouver, it might hurt the whalers, might possibly force them to cut their hunting season short.

The problem was that our picket line was strictly illegal. Within half an hour, we were confronted by two squad cars, out of which grim-faced Harbors Board policemen were climbing, removing their sunglasses.

Longshoremen and dock workers stood around, glowering, and it was difficult to tell whether they were for us or against us. When the harbor police finally moved in, they took me aside in one of the warehouses and the officer in charge said, "Look, you're after publicity for your cause. All I'm after is removing you people without any trouble. How many do you want arrested?"

"Give me a minute to check with the treasurer, okay?" A few minutes later, I was able to come back and say, "Six of us. The rest will step back."

So six of us squatted on the roadway, singing *We Are Whales*, while the cops moved in and hauled us off one by one to the waiting squad cars. Only Paul Watson broke the rules of the game by stiffening and resisting, so that he had to be carried. A few minutes later, Patrick Moore, Melville Gregory, Watson, two women—Bree Drummond and Janet Cook—and myself were being booked at the Vancouver Public Safety building and led off to separate cells, charged with obstruction. Watson and Gregory and I had all been in jail before, for various reasons, but the experience was new for Patrick Moore and the girls. In Moore's cell, several skid-road junkies were fixing up with heroin, using dirty spoons. The girls found themselves locked up with half a dozen prostitutes. Within three hours, we were all released, to face trial later.

An odd thing happened the next day. The cop who had been in charge of the arrest came down to my boat to tell me that he was "fed up" with what was happening to the world, that he was afraid for his children, that he agreed with what we were doing. In fact, expecting us to be down at the dock again, he and several of his fellow constables had booked off sick rather than have to arrest us again. "Sometimes I want to tear off my badge and join in on what you're doing," he said. And he meant it.

Months later, when the case of the "Greenpeace Six" came to trial, the cop who had visited me—the chief witness against us—failed to appear. He had booked off sick. The magistrate became so irritated, he threw out the charges against us. Whether the cop got fired afterward or not, I don't know. But I do know that he booked off deliberately.

A classical raincoast autumn now fell upon Vancouver. George Korotva started up a small contracting company of his own, hiring Moore and several other unemployed Greenpeacers who could use a hammer and nails. They spent chill gray days clambering about on rain-slicked rooftops, trying to make enough money to keep alive. It was slightly ironic: no sooner had we decided to form an official board of directors, with Moore as vice-president in charge of policy and Korotva as vice-president in charge of operations, than the two of them were having to climb into dirty clothes and go out day after day in the rain, hammering shingles on leaky roofs, to survive. They had both taken tiny apartments in Vancouver's Kitsilano hippie ghetto not far from the office. Paul Spong had returned from Europe. He too was forced to go to work with Korotva and Moore in order to maintain himself and his family in Vancouver and carry on his whale campaign. Korotva frequently joked that he had the only small-time construction company in town that had two Ph.D.s as common laborers. None of us was getting any salary out of Greenpeace. All that was keeping me alive now that I had quit the newspaper was Bobbi MacDonald, who was still working as an electronic engineer when she wasn't down at the Greenpeace office. Melville Gregory tried working with Korotva & Company for a while, but he found the work too hard and quickly begged off.

Our third vice-president was Rod Marining, who was to be in charge of communications. Otherwise, the board included John Cormack, Bobbi MacDonald, Eileen Chivers, Moore's new lady, Michael Chechik, our film producer, Spong, Watson, and Gary Gallon, the executive director of the city's largest antipollution organization.

There had been a terrific burn-off rate during the summer's expedition. Katrina Halm, who had acted as treasurer, had ended up in the hospital suffering from half a dozen physical ailments, all of them brought on by nervous exhaustion. Blair Halse had taken her place, only to flame out in less than a month from the unaccustomed pressure. Elizabeth Dunn, who had helped a lot during the Mururoa campaign, had lost faith in the whale expedition about a week before we finally found the Russians and since then had not shown up. The crowd of familiar Greenpeace faces had gone through one of its periodic gestations, leaving us with new faces, new personalities. People who had played no role at all before the trip—like Korotva—were now part of the inner circle. People like Jim and Marie Bohlen, who had started the whole thing, were complete unknowns so far as most of the people working with us now were concerned. Ben Metcalfe? Bill Darnell—who had invented the word *Greenpeace*—Terry Simmons, Will Jones, Irving and Dorothy Stowe? It seemed we had gone through at least three complete "generations" in Greenpeacers in less than five years. By this time—autumn 1975—Moore, Marining, Watson, and I were the only remaining old-timers, and of the lot, at thirty-four, I was the oldest. None of the others were out of their twenties. Except for John Cormack, of course. And who would have thought in the fall of 1971, when we were setting out on what seemed to be a broken-down old halibut seiner, meeting for the first time the old grease-stained captain, that four short years later he would be a member of the board of directors of the Greenpeace Foundation, and that most of his crew, including all the founders, would be scattered to the winds?

Of David McTaggart, not much was seen because he was spending most of his time over in France, pressing his claims against the French navy.

Bobbi took on being treasurer.

And Rex Weyler introduced us to two people who were to change the very nature of our operation. One was Peter Speck, owner and publisher of a North Shore weekly newspaper. The other was his accountant, Bill Gannon, who also served as chief accountant for the biggest real-estate-development company in British Columbia. Working with Bobbi and myself, they proposed to give Greenpeace the one thing it had so far utterly lacked: financial savvy. They would show us the secrets of survival as an organization in the workaday world of marketplaces and banking exchanges. Speck was a tall, balding man with a full red beard, heavy-lidded eyes, and a beatific grin. He had built his newspaper up out of nothing and had struggled for something like seven years against creditors and banks to create what was rapidly becoming the biggest publishing empire on Vancouver's affluent North Shore. So far as he could see, it was all done out of a fertile kind of sorcery that had to do with budgeting, cash-flow projections, accountability, organization charts, projects areas, merchandising, lines of credit, assets, guarantors, demand notes, and so on. His guru in this department was Gannon, the accountant. A quiet curly-haired young man with a Che Guevara moustache, he had a perpetual air of innocence about him, a well-tailored look that completely disguised the fact that in his private life he was a mean electric-guitar player. And what he had that we needed so badly was the mysterious knowledge of how not only to balance books but to create on paper the necessary symbols that would magically spring money loose from the bank to carry on, despite the fact that we already owed forty grand. To Gannon, whose development company accounted for hundreds of millions of dollars a year, our Greenpeace debt was a breeze. In no time at all, he was showing us how to prepare budgets, work out our projections, and prepare a presentation to the bank quite unlike anything they had ever seen from a conservation group before. Thus, a revolution began.

With the assistance of Bill Gannon and Peter Speck, we would remake Greenpeace into an ongoing tool at the disposal of the environmental movement. We would begin to build a membership, set up a merchandising arm to sell T-shirts, bumper stickers, buttons, posters, and anything else we could think of. We would publish our own newspaper. We would reorganize the office so that human energy could be channeled instead of wasted, as so much of it obviously was. Peter Speck's dreams of organization dazzled Bobbi and me. We decided that here lay the solution to the long-term problem that faced us: how to cope with the rapid turnover rate, the merciless exhausting of individual energy, the clear-cut need to have more money at our disposal than we had so far been able to raise. It was certain that if we were to go back out after the whaling fleets again, we would need a bigger, faster ship, one that would keep up with the whalers. If we had gone forty thousand dollars in debt just putting the little *Phyllis Cormack* to sea, how could we hope to mount a greater expedition a year from now unless we learned new tricks? Here now were two men with the combined expertise to teach us how to organize and finance ourselves, making all things theoretically possible. But it was going to require an entirely new kind of discipline. Budgeting only works if everyone stays within their budget. Projections are only useful if they are realistic.

The immediate problem we ran into was with the Greenpeace hard core of roughly thirty people. The "grass roots" faction simply could not come to terms with such unheroic notions of cash flow and bookkeeping. These were the same people who had not been able to grasp the value of a movie contract, whether releases had to be signed or not. What did the "eco-revolution" have to do with contracting T-shirts out to a distributor? The situation stayed within the bounds of acceptable comedy, because one of the worst of the flipped-out mystics—namely myself—was now the chief advocate of organization, fiscal responsibility, and the budget system itself.

In late August, we got a chance to do some very "practical" whale saving of a slightly different kind. Instead of signs and portents being involved, it was pure politics. Six orcas were captured at a place called Pedder Bay, on Vancouver Island. Led by Rod Marining and Paul Spong, our fledgling little organization went straight to work to get them released. Half a dozen people took a ferry over to the island and drove to Pedder Bay to organize demonstrations and record the plaintive cries of the whales trapped behind a huge net, unable to rejoin the rest of their pod, whose members swam around anxiously in the open water, calling. We got a quick media campaign going by releasing the tapes to the radio stations and sending Spong to do interviews, pointing out that between 1962 and 1973, a total of 263 orcas had been caught off the coasts of Washington State and B.C. Of those, 48 had been kept and put into aquariums and oceanariums, where fifty-two percent of them had died within the first year of captivity.

While Spong was thus occupied, several volunteer lawyers went to work in the courts, challenging the legitimacy of the licences that were issued for "live-whale capture" on the grounds that orca populations along the West Coast were being depleted.

But the main thrust of the campaign took place behind closed doors in the office of the province's new premier, a Social Democrat named Dave Barrett, who had come to power at a time when antipollution and conservation sentiments seemed to be rising dramatically. Barrett himself was sympathetic to the whale situation. And his transport minister was so sympathetic that he refused to allow the British Columbia ferry fleet to carry the captured whales from Vancouver Island, forcing the "owners" to charter an expensive jet to get even one of the whales out of the province. The rest of them were released. And to the last minute, the whale kidnappers were hounded to the airport by a small fleet of honking cars and people chanting "Free the Whale!" The upshot of the action was an announcement by the B.C. government on September 12, of a total ban on the capture of orcas in the province, a move that was followed a year later by Washington State. Spong's debt to Skana—if that's what it was—had been paid off royally, and Greenpeace had scored a significant local victory.

During this time, Bobbi and I set off on a series of cross-country lectures, setting up Greenpeace chapters each place we touched down. The lecture circuit took us to almost every major Canadian city and university campus, with odd side trips to festivals and ecology conferences. We brought slides and projectors and sound systems with us. Except for the fact that we traveled by jet between stops, it was much like an old-time medicine show. I'd stand up on the stage, explaining what the slides were all about, describing the encounter with the Russians for the fiftieth time, and Bobbi would remain at a table at the back of the room, flogging T-shirts, buttons, and memberships. After each lecture, there would be at least a dozen people who were interested in setting up a group. We would take off afterward often to talk all night, leaving the next day with another clutch of names and addresses, more members, more chapters. As often as not, the hard core would prove to be people who had seen Paul Spong's whale show w hen it had passed through Canada on its way to Norway the winter before. By Christmas, there were a dozen Greenpeace branches functioning in Canada. Well, perhaps *functioning* is too strong a word. At least there were that many groups, with addresses and phone numbers, and a burning desire to find some way themselves to participate in the saving of the whales, the stopping of bombs, the prevention of nuclear proliferation.

During this period we had several flights down to San Francisco, where Will Jackson had arranged benefit concerts through his contacts in the underground music community. With the money raised from those concerts, we intended to open an office in San Francisco, the first of what we hoped would someday be hundreds. It was an autumn and early winter full of ambition and drive, as though the energy generated by our experiences that summer were a fire that refused to go out. At most of the campuses where we went, the main task was always to overcome the pessimism that existed. The students were already shockingly conservative, compared to their radical predecessors. It was not so much that they were apathetic as simply without hope. What we had to offer was a last shot of faith that the overwhelming ecological catastrophe could somehow be turned into a liberation. The line I used over and over again, that drew the most laughs and relieved grins, was: "Don't worry, we've already won. It's just that there's a two-hundred-year mop-up operation left to be done."

Back home, in Vancouver, there were a couple of dozen of us whose lives had become so intertwined that we might as well have been a family.

Top: *The core of the Don't Make a Wave Committee, from left: Jim Bohlen, Paul Cote and Irving Stowe. Vancouver, May 1, 1971.*

Above: *A toast as the* Phyllis Cormack *leaves Vancouver. Clockwise from left: Thurston, Fineberg (obscured), Metcalfe, Darnell (face partly obscured), Cummings, Hunter (face obscured), Bohlen, Moore and Simmons. Vancouver, September 15, 1971.*

Photos © Greenpeace/Robert Keziere

The crew of the *Phyllis Cormack*. Clockwise from top left: Hunter, Moore, Cummings, Metcalfe, Birmingham, Cormack, Darnell, Simmons, Bohlen, Thurston, Fineberg.

International waters, September 22, 1971.
© Greenpeace/Robert Keziere

Below left: *Bill Darnell (left) and Bob Cummings in the wheelhouse of the* Phyllis Cormack. *International waters, September 22, 1971.*

Below right: *Ship's doctor Lyle Thurston stands on deck for a cigarette. International waters, September 22, 1971.*

Bottom: *"The experience of arriving in Akutan Bay was a bit like landing on Mars." With the nuclear test delayed the* Phyllis Cormack *and her crew faced an indefinite wait. Akutan, September 26, 1971.*

Right: *"Now I was 30, an old man, heading home in defeat." Robert Hunter, hours after the decision to abandon the voyage to Amchitka. Sand Point, October 13, 1971.*

Bottom right: *"For the first time, we were heading away from Amchitka." From left: Thurston, Moore and Darnell, the day after the arrest of the* Phyllis Cormack. *International waters, October 1, 1971.*

Photos © Greenpeace/Robert Keziere

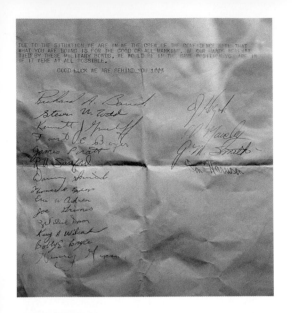

Left: *"My God, they're freaks!"* The crew of U.S. Coastguard vessel the Confidence *declared their support for the anti-nuclear voyage to Amchitka before arresting the* Phyllis Cormack. *Akutan, September 30, 1971.*

Top right: Meeting *of the* Phyllis Cormack *and the* Edgewater Fortune. *October 27, 1971.*

Bottom right: *"Suddenly we were bearing down on a dock jammed with several hundred people, including all our relatives and friends."* The Phyllis Cormack *is welcomed back to Vancouver. October 30, 1971.*

Photos © Greenpeace/Robert Keziere

"He had the air of an honest, but world-weary highway patrolman who had just pulled over a speeding car." Commander Floyd Hunter of the U.S. Coastguard on board the *Phyllis Cormack.*

Akutan, September 30, 1971.

Right: *"We felt like Pygmies beneath the overhanging Thunderbird beaks of the Totems."* The crew of the Phyllis Cormack *are made brothers of the Kwakiutl people. Cormorant Island, October 26, 1971.*
© Greenpeace/Robert Keziere

Below: *Greenpeace pioneers. September 1971.*
© Greenpeace/Robert Keziere

Bottom right: *"McTaggart and Ingram were grabbed and savagely beaten." French commandos board the* Vega *within the nuclear test zone. Polynesia, August 15, 1973.*
© Greenpeace/Ann-Marie Horne

Left: *Cetacean researcher and save-the-whales campaigner Paul Spong with orca Haida. Victoria, Canada, 1975.*
© *Gerry Deiter/courtesy Paul Spong*

Bottom left: *"Gently Watson leaned forward and pulled the eyelids shut." The* Phyllis Cormack *encounters a dead whale. North Pacific Ocean, June 1, 1975.*
© *Greenpeace/Rex Weyler*

Right: *Will Jackson and skipper John Cormack installing fuel tanks on the* Phyllis Cormack. *North Pacific Ocean, June 1, 1975.*
© *Rex Weyler*

Bottom right: *From left: Patrick Moore, Fred Easton and Ron Precious on board the refurbished* Phyllis Cormack. *North Pacific Ocean, June 1, 1975.*
© *Greenpeace/Rex Weyler*

Above: *"The little bullet-like boats buzzed like bees toward the immense floating fortress of death."*
Greenpeace encounters the Soviet whaling fleet for the first time.
North Pacific Ocean, June 1, 1975. © Greenpeace/Rex Weyler

Right, top to bottom:
The Phyllis Cormack, *repainted for the first whales campaign.*
North Pacific Ocean, June 1, 1975. © Greenpeace/Rex Weyler

"On the side of the James Bay *was the message:*
Save the Whales, Save the Earth.*"*
North Pacific Ocean, June 1, 1976. © Greenpeace/Rex Weyler

Paul Watson and Bobbi Hunter, next to a whaling ship.
North Pacific Ocean, June 1, 1976. © Greenpeace/Rex Weyler

"Even from 2500 feet, the crew could see the trails
of seal blood spattered beside and behind the ships."
Newfoundland, March 1, 1976. © Greenpeace/Patrick Moore

"Stop 'er Cap! Stop 'er! They still ain't a-movin!" Paul Watson (left) and Robert Hunter block the *Arctic Endeavour*.

Newfoundland, March 1, 1976.
© Greenpeace/Patrick Moore

Right, top to bottom:
Robert Hunter in front of the
Phyllis Cormack *during the first*
whales campaign.
North Pacific Ocean, June 1, 1975.
© *Greenpeace/Rex Weyler*

Kazumi Tanaka, a young
photographer from Tokyo, paints
Japanese text on the hull of the
James Bay.
Vancouver Island, May 1, 1977.
© *Greenpeace/Rex Weyler*

Film cameraman Ron Precious (left)
and stills photographer Rex Weyler
during the 1976 whales campaign.
© *Rex Weyler*

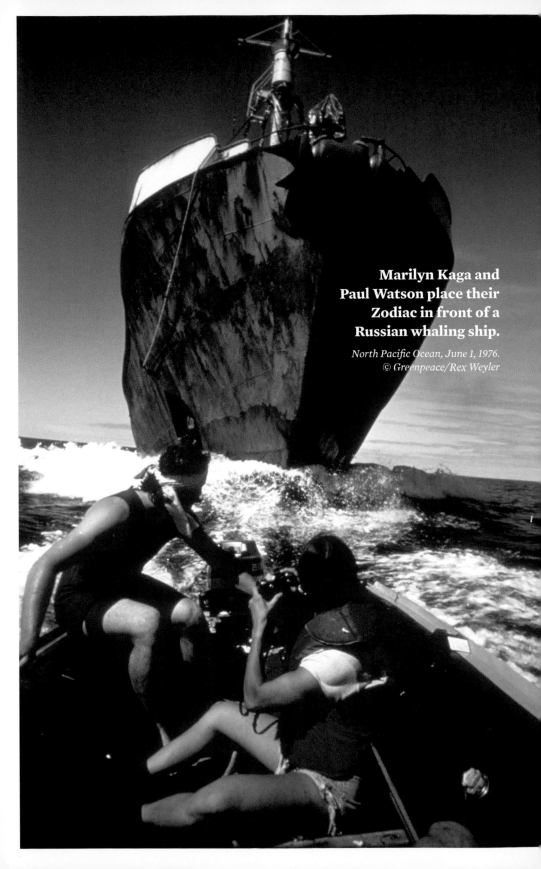

Marilyn Kaga and Paul Watson place their Zodiac in front of a Russian whaling ship.

North Pacific Ocean, June 1, 1976.
© Greenpeace/Rex Weyler

We shared each other's clothes, we shared our food, we pooled our resources for parties, and like an extended family network, we squabbled amongst ourselves, bitched about each other, counseled one another, and came to think of each other as brothers and sisters. The family included Melville Gregory and his lovely lady, Sybil, plus their two kids David and Nathaniel; Korotva and his equally lovely lady, Rita Outlaw; Moore and Eileen Chivers; Rod Marining and Bree Drummond; Walrus and Taeko Miwa, whose love had begun on board the *Phyllis Cormack*; Watson and his girlfriend, Marilyn Kaga; Paul and Linda Spong and their son, Yashi; Bobbi and I, and my two kids, Conan and Justine; then, of course, there were John and Phyllis Cormack, who presided over this little tribe like grandparents. Each couple had their own problems to deal with on top of the Greenpeace problems that confronted us, but the existence of the "tribe"—with its fantastic goal of saving a whole species and maybe even the planet—acted as a gyroscope, keeping our minds off our domestic tribulations. It seemed a general rule that the more engrossed you became in your own private problems, the more they consumed you. But if you could keep your attention focused on the great planetary crisis, your own problems somehow solved themselves without having to do so much as lift a finger.

Money problems, of course, were the source of most discontent. We were all living on the fringe of civilization—Bobbi and I in an old boat we owned, tied up to a rotting dock in a polluted lagoon in the center of Vancouver; the Spongs in a condemned house that was due to be torn down but whose destruction kept being postponed; the others in various crumbling old tenements and houses scattered through the hippie ghetto or the city's working-class East End. We were all what is known as "downwardly mobile," driving battered old vans and wearing clothes largely from the Salvation Army. It was a hand-to-mouth existence, with no end of worrying and arguing over payments that could not be made, but on balance, it was an existence of peculiar richness. At least we had all learned to get our cost of living down to a minimum, which meant that we then had to put in only a minimum amount of effort to make money to survive, and the rest of the time we were free to put our heart and soul into the one thing that mattered: the crusade. Few of us were willing to live communally—although Walrus Oakenbough had started one commune called "Greenpeace Harmony House." But what we did have was a kind of floating communal situation where you could never be sure who was staying at whose place or for how long. Bobbi and I found this arrangement particularly useful, since the first thing we would do upon visiting anyone was to head for the shower to wash off the diesel fumes from our boat.

While we were thus living what amounted to a tribal existence in the gray depths of Vancouver, a larger-than-life image was rapidly building in the outer world. Articles were being prepared—or had already been published—in *Playboy*, *National Geographic*, *The New York Times*, *Smithsonian* magazine, even *The Wall Street Journal*. Sitting in our office, looking out across Fourth Avenue, I found myself starting to get more calls from New York, San Francisco, Los Angeles, and other American cities than I was getting locally. At home, we were "old news," actually somewhat stale. British Columbians had been hearing about Greenpeace since 1971. It was part of the furniture. But in America it was something new. It had sex appeal.

The seed of the next campaign had been germinated on the *Phyllis Cormack* during the voyage back from San Francisco to Vancouver. It was the brainchild of Walrus Oakenbough and Paul Watson. One night in the galley they had been discussing the harp-seal hunt, which takes place every year off the eastern coast of Labrador and Newfoundland. The pups, scarcely two weeks old, are bashed over the head with a hakapik, and their snow-white pelts are stripped away on the spot. A man named Brian Davies, chief of the International Fund for Animals, had been taking newsmen out to the scene of the hunt for years, stirring up a formidable public-opinion campaign against the slaughter, but so far he had not succeeded in getting the Canadian government to call it quits.

Thinking about ways to interfere with the slaughter—just as we had interfered with the whale hunt—the warrior-brothers came up with one of those outlandish notions that instantly bears the stamp of brilliant originality. The harp-seal-pup pelts were valuable simply because of their pure-white color. Now if someone were to spray an indelible dye on the pups before the hunters got to them, then there would surely be no reason to kill them: no one would want to wear fur that had been messed up by the application of a dye.

It had not been long since we'd returned home from the whaling grounds before the two of them started pushing for an antisealing campaign. At first, it was out of the question: we owed too much money. But as September and October and November passed, the application of the new techniques suggested by Peter Speck and Bill Gannon were beginning to pay off. The cross-country lectures had also generated money. Oakenbough and Watson pulled together a group of hard-working seal lovers who were willing to stand in the rain, night after night, outside liquor stores and hotels, selling buttons, demonstrating their willingness and ability to fund themselves. It was only Paul Spong who was fiercely opposed to the idea—an irony no one could help noting. Only a year before, he had been the one having to buck the old guard in order for a new campaign to be attempted, moving us from antinuclearism to antiwhaling. Now he was in the position of trying to keep Greenpeace on track with the whales and beating back an attempt to move us on to yet another issue. There was another—more humorous—level to the debate that now dominated our board meetings: Spong's main source of inspiration had been the great toothed orcas, and it was to these creatures that he still devoted his deepest loyalty; seals happen to be part of the orcas regular diet. It was, indeed, an amazing exercise to sit seal lovers and whale lovers down in the same room, and try to plot a joint strategy that would allow us to save both types of creatures. Only in the ecology movement could you come up against such a problem: save the seals so the whales can eat them, save the whales so they can eat the seals. Or was someone willing to throw themselves between an orca and its meal?

No one could deny that Oakenbough and Watson were doing an excellent job of organizing. Also, now that our collective attention was drawn to the seal issue, the process of "selective perception" began to go to work. We started noticing small items in the newspapers, references in magazines, and found ourselves involved more and more often in discussions about the subject. Bit by bit, our awareness of the seals increased, and with the increased awareness came a growing sense of revulsion. The harp-seal hunt was the only one of its kind in the world, where nursing marine mammals were virtually pulled from their mothers' nipples and killed before their eyes. The idea of saving them by applying a dye to their pelts, crazy as it was, became too tempting to resist any longer. In late November, the board of directors gave the go-ahead.

We knew it was a totally "impractical" scheme, since to be effective it would involve flying a crew by helicopter to the ice floes off the coast of Labrador and having them spray over one hundred thousand seal pups before the icebreakers arrived, disgorging hakapik-armed hunters. But then, putting one Zodiac between one whale and one harpoon boat had been equally "impractical" in terms of saving *all* the whales. Overriding the factor of numbers was the symbolism of the gesture and its imaginative power to draw mass-media attention to what was taking place. The theory of mind bombing, which had worked so well at Amchitka and Mururoa and off the Mendocino Ridge, could not help but work on the ice floes of Labrador. From the theoretical point of view of changing mass consciousness, of expanding ecological awareness, of continuing our assault on the old industrial mind-set, the annual seal hunt was too tempting a target to ignore.

Most of the people who had come in with Oakenbough and Watson approached the issue far more passionately; it was an issue that brought out the worst forms of anthropomorphism and yet at the same time the highest forms of compassion. We knew we would have to walk a tightrope between a balanced "scientific" analysis that the hunt itself was simply bad for the ecology of the ocean—since it removed the seal entirely from the nutrient cycle upon which all life in the sea ultimately depended for food—and the depths of emotion that the killing of "babies" generated in the breasts of millions of urban people, who, otherwise, with their cars and swimming pools and electric gadgets, were the worst environmental destroyers of all. There was one other deep concern: it was going to be a bad situation if we had to go into Newfoundland and face-off against men who were at the bottom rung of the working class, because then the contrast between the "elitism" of the environmental movement and the agony of the workers could easily achieve such a high profile that more would be lost than gained.

All killing is inhumane and barbaric by definition, and all the slaughterhouses are Buchenwalds. What was going on off the coast of Labrador, however, was a massacre of a different order: it was a mass killing of pups so that their pelts might be turned into trinkets and trimming; the small amounts of meat on their bones was left behind to freeze. It was the life cycle of the ocean itself, finally, that was being wrecked. On this point—the integrity of the ecology of the seas—the whale issue and the seal issue became one. Both were marine mammals near the top of the food chain who were being eliminated, thereby punching a gaping hole in the chain of life. The oceans themselves had greater need of the whales and seals than the men who hunted them. It was not just that whales and seals ate fish: ultimately, it was the fish who ate the remains of the whales and seals, and without that nourishment the fish themselves could not survive.

When we announced our plan—to fly to the ice to spray the seals with a green dye to render their pelts worthless—it was greeted for the most part in the media by whoops of laughter.

Nevertheless, it got tremendous coverage in Canada—especially in Newfoundland, where the largest number of fishermen make extra money during the winter months by going "swiling," as they call it. There was no laughter in Newfoundland. They were furious. The seal hunt had a tradition that went back centuries. Tales had been handed down from grandfather to father to son. Hundreds of men had died on the ice, and others had struggled heroically to survive against wind and ice and sea. Newfoundland was an island that had only joined Canada in 1949, by a narrow electoral margin, and its people, inbred like all islanders, were a race apart. They had virtually their own language, their own traditions and culture, and swiling lay at the heart of their sense of identity, for it had always been their central myth that it was on the ice, in the darkness and cold, that manhood itself was earned. If there were not as many sealers as there once had been, that was because there were not nearly as many seals, but no one wanted to face the reality that the seals were gradually vanishing from the earth. Just as Americans cling to their identity as cowboys, so the Newfoundlanders cling to their identity as swilers. *Ecology* was a word that meant nothing. Absolutely nothing.

Within a day of announcing our plan, we picked up the Vancouver papers to read banner headlines that said:

NEWFIES FURIOUS OVER GREENPEACE "SPRAY" SCHEME

Newfoundland politicians were being quoted everywhere, warning that the first Greenpeacer to step on the island would be "mobbed." Threats from anonymous fishermen to "shoot the Greenpeace helicopters down" were widely printed. Now the story had substance in the media. The threat of a violent reaction from the Newfoundlanders changed the "green spray" story from a joke into a serious confrontation.

The task of organizing the details of the campaign fell mainly to Oakenbough and Watson. Patrick Moore was charged with the responsibility of finding an appropriate green dye that would be indelible but would not harm the seals themselves. As for the rest of us, our main concern was still the coming summer's antiwhaling expedition. We knew we'd have to do better than the previous year. It fell to George Korotva to start looking for the boat we needed, something that could travel at least as fast—preferably faster—than the whaling ships themselves, and that could stay at sea for prolonged periods. It was a tall order. Such ships are seldom available for less than a million dollars.

Korotva and Paul Spong would work together on that one—they were already close buddies. Until the arrival of the *Phyllis Cormack* back in Vancouver, Korotva had never met the diminutive scientist. When they did meet, it was on the beach at Jericho, where Spong, barefoot and dressed in tattered pants and T-shirt, had been handing out pamphlets and petitions. Until then Korotva had envisioned "The Spong" as a typical scientist and had been prepared to dislike him. Instead, he found a kindred wild spirit—"jost anudder goofed-up, crazy sonofabitch, aye?" The tall Czechoslovakian and the short New Zealander were soon the best of friends, and neither had any intention of being distracted from their main task by the upcoming seal campaign. "Okay," said Korotva, "you guys go save the furry little boggers, Spong and I, we'll get da boat ta save da fat blubbery guys. Simple like dat."

The winter went by like a blur.

From the time we'd staggered ashore at Jericho to the time the seal campaign was ready to go, at the end of February 1976, we were possessed by a feverish excitement, an obsession. We really did believe that some vast gate had opened at least a crack. A blinding light was pouring forth over the poisoned landscape. We had said it so often we were beginning to believe it ourselves: a small group of people, acting imaginatively and nonviolently, can affect the course of events in the "global village." On a planet where some four billion human beings live in gigantic ant heaps, each of them apparently powerless because of the sheer density of the masses packed around them, no more audacious idea could be conceived. Most of the thinkers on earth hold firmly to the belief that history is shaped by swarms, by the colossal momentum of numbers and bulk. The individual, even the rare genius who appears only once every few centuries, can do little more than deflect the course of events for a few years before the great current closes in and sweeps onward. How can any rational person convince himself that there is the ghost of a chance of having even the most minute impact on world events? And who but a megalomaniac can dream of actually changing the consciousness of humanity?

The answer is simply the existence of a planet-wide mass-communications system, something that had never existed before. Its development was the most radical change to have happened since the planet was created, for at its ultimate point it gives access to the collective mind of the species that now controls the planet's fate. One man can now command the attention of the world. One group—such as Greenpeace—could do the same. In my own mind, it seemed crystal clear that awareness itself is the cure. If a mass awareness developed about the seal problem, the problem would be solved. If crazy stunts were required in order to draw the focus of the cameras that led back into millions and millions of brains, then crazy stunts were what we would do. For in the moment of drawing the mass camera's fire, vital new perceptions would pass into the minds out there that we wanted to reach. Mass media is a way of making millions bear witness at a time.

If we use our brains, we can trigger a revolutionary change in the consciousness of humanity.

It was an exciting game to play, far more exciting than anything else that any of us had to do.

On March 2, 1976, the first Greenpeace save-the-seals expedition left Vancouver by train to travel across the country to Nova Scotia, from there by ferry to Newfoundland, and up the length of the island by van to the northeastern port of St. Anthony. There we would be joined by two chartered Bell Jet Ranger II helicopters that would take us out across the open North Atlantic waters to the ice floes where the seal hunt took place, a region known as the Front. At each stop along the way across the frozen expanse of Canada in late winter, the crew held press conferences, building up tension as they drew closer and closer to "Fortress Newfoundland," where the guns and the mobs were reported to be lying in wait.

The crew this time had been mostly selected by Walrus Oakenbough and Paul Watson. Watson had a bit of a militaristic streak in him and delighted in giving each crew member a specific title. He, for instance, was "expedition leader." Then there were "squad leaders" and "flight assistants" and even "squad quartermasters." He would have been happier if Patrick Moore and I had not insisted on coming along: there was some hostility between him and Moore stemming back to the first days with the whales at Long Beach, when Watson had forced his way on board against orders from Moore. But Moore was vice-president in charge of policy, and there were going to be a hell of a lot of policy decisions to make before this trip was over. As for myself, I was there simply because I was president and still the real leader of our crazy little band. Ron Precious would be going too.

The rest were going on a Greenpeace expedition for the first time. As to be expected, they were a diverse bunch. Paul "Pablo" Morse was a heavily built young fisherman, originally from the maritimes, who had once survived a sinking in the Atlantic where two of his buddies had drowned. Somehow, this had impressed the crew-selection committee. Morse was as strong as a half a dozen ordinary men, had a gargantuan laugh, and was willing to go anywhere and do anything. Yet, for all his physical presence, he was as mild a man as you could meet. If a dog bit him, you had the impression he would look down at it, smile, pat it on the head, and say: "Hi there, little feller."

Al "Jet" Johnson was in his early forties, a resident of Sausalito, California. His thing was wildlife. He loved bears, wolves, leopards, coyotes, and seals. He had a private plan to save the wolves: he would get someone to dangle him from a helicopter and fly ahead of the hunters, blocking a clear shot at a wolf. How many trees he might bounce off in the process was a detail he'd worry about later. Originally a Canadian, he had served in the Royal Canadian Air Force, managing eventually to crash a jet fighter—and yet walk away from the smoking mess. He also spoke Norwegian and had had arctic-survival training in the air force.

Carl Rising-Moore, a thirty-year-old veterinarian, was clearly our token hippie. His hair flowed over his shoulders. He had startlingly blue eyes, radiant with a kind of bliss. He, too, had been chosen because of his track record: he had been jailed in West Pakistan during the Pakistan–India war for trying to get through the battlefields to the jungles beyond in the hope that he might be able, somehow, to save the Bengal tigers from becoming the war's heaviest casualties. An American, he quickly displayed the habit of forgetting his equipment, and so came to be known as Carl "Falling-Down"-Moore. He is probably a saint.

The ladies in the crew included Paul Watson's Japanese-Canadian girlfriend, Marilyn Kaga, twenty-six, a steady, intelligent woman, whose dark eyes roamed every scene, alive to the details. Towering over her was Henrietta Neilson, a Norwegian woman in her early twenties whose face was truly that of a Viking chieftainess. She was powerful, intense, and brooding. When Marilyn and Henrietta stood together, it was enough to make a Warrior of the Rainbow weep with joy—for here, not in the myth but in reality, the East and the West had finally joined. Marilyn's ancestors had been shoguns in Hiroshima. Henrietta's had been Vikings. And here now were the daughters of shoguns and Vikings traveling together from the West Coast of Canada to the East Coast, to a town only half an hour's ride away from the place where the Vikings first set foot in what they called "Vinland." Every time I looked at the two of them, my sense of history and its amazing crosscurrents went spinning dizzyingly out of control. This *had* to be an age like none before.

The two other women were Patrick Moore's lady, Eileen Chivers, who, after a year of helping Patrick build their house at Winter Harbour, now listed her occupation as carpenter, and Bonnie MacLeod, a bank teller, who blushed but did not disagree when I suggested to her that "the reason you can relate so easily to beautiful baby seals is because so many guys have been after your own pelt so long." She could laugh, but more often than not there was a worried, pained expression on her face.

While the others traveled by train across the prairies, Moore flew ahead to Ottawa to do some politicking with other scientists he knew, on both sides of the seal issue. I left a few days after the train departed and joined the crew in Montreal. From there we went together by train to North Sydney, Nova Scotia, where we were joined by another American crew member, Dan Willens, from Indiana, a stocky young music teacher whose newly formed Greenpeace group had raised a lot of money for the expedition.

In North Sydney, we piled into a Hertz Rent-a-Truck van and drove to the ferry. The wind was bitterly cold, yet it was scarcely a taste of what lay ahead when we got to the northern tip of Newfoundland. All night, the ferry crashed and shuddered and roared through ice ridges and floes. Our first nighttime glimpse of the ice packs caught in the glare of the ferry's spotlight revealed a world that was far more hostile and alien than even our worst expectations.

At first light, the ferry slid into the dock at Port aux Basques on the southern tip of Newfoundland. Except for the huge pilings around the terminal, the town seemed to consist of wooden shacks scattered loosely about on outcroppings of rock that was glazed with ice. Even the snowdrifts had a rigid, permanent quality to them, as though they had stood, carved into their various shapes, without having moved or melted in ages. The only sound, once the ferry had stopped, was the distant, gigantic tinkling of the ice pans.

We were so overloaded with equipment that we couldn't all squeeze into the van, so some of us took a bus while the van went ahead to the town of Corner Brook, and we rented a second vehicle. Everybody had removed their GREENPEACE buttons to avoid attracting attention, but even by the time we reached Corner Brook, hourly radio reports were being broadcast across Newfoundland, giving our latest position. From Corner Brook, we had to travel along a road that steadily grew more narrow and slippery as we went. The wind began to pick up. By nightfall of the first day of driving, we could scarcely see twenty feet ahead. Long fingers of snow were piling up across the road. Moore, Morse, Willens, and Watson took turns at the wheel, while the rest of us huddled together in the backs of the freezing vans, passing a bottle of rum around to keep us warm. As the fury of the wind increased, the vans seemed to turn into small boats in a gale. Several times we rammed into snow banks almost as high as the windshield. "We're taking white ones over the bow!" Moore yelled. No less than five times, we plunged completely off the road and had to clamber out into the stinging blindness of the subpolar night and push against the screeching wind to get back on the road. Finally, we had to pull in at a motel. It was impossible to go any farther.

It was clear from the moment we entered the motel that everyone there knew who we were. People looked up from their pool tables and chairs, from the bar, and stared at us coldly. By the time we were able to get rooms and bundle the crew into them, most of us were shaken.

"Jesus, those boys don't like us do they?" Pablo Morse remarked cheerfully.

"Well, I guess we're in enemy territory, all right," said Oakenbough.

On television, we saw footage taken in St. Anthony that morning, where a mob of Newfies had stood out on the road in thirty-below-zero weather all day, blocking the entrance to town, waiting for us. They were as mean-looking a gang as any of us had ever seen. They did not look happy. Icicles hung from their moustaches. Many had scars all over their faces. The roadblock people pledged to be back into position first thing in the morning. One of them was quoted as saying: "Those Greenpeacers don't dare to show up here."

None of us slept very well that night. The wind howled around the motel, rattling the double-paned windows. Visions of the mob that was waiting for us played through all our minds. Yet no one suggested turning around. There was nothing to do but keep going and walk right into the arms of those angry, frozen men. According to the news, there had been two hundred of them. I had never been badly beaten before. I wondered what it would feel like.

Yet as we staggered out into the wind the next morning to push the vans back onto the road, everyone appeared calm. Ron Precious had a serious but basically untroubled look to him as he cradled his camera into his arms, surveying the sky as though automatically doing a light-meter scan. "Looks good," he said. That was somehow comforting. If we were going to get our heads bashed in, at least the film footage wouldn't be overexposed. Precious had orders to hang back and keep shooting, no matter what happened. But for our inner tension, that last part of the drive would have been a magnificent experience. To our left lay the Gulf of St. Lawrence, with ice ridges driven up on top of each other, tilted icebergs stabbing at the sky like broken swords, and great scoured pans from which the light flashed blindingly. A blue haze hovered over the sheets and floes, unfolding like a lens all the way to a horizon where the line between the sky and ice was impossible to fix. To our right, ahead and behind, were patches of small gnarled trees and long shallow valleys where frozen rivers could be detected only by the contours of the land around them. Once, we glimpsed a moose, crashing with seeming clumsiness through ice in a swamp, plunging into water up to its belly.

"Oh, let's go back!" cried Henrietta. "It might drown!"

"Oh, for Christ's sake," muttered Eileen. "The moose'll take care of itself."

Snow was still breaking like spray over the ridges of the banks along the road, blinding us. But finally the road began to level out with the land, and we found ourselves emerging on a wind-blasted ice plain that we suspected stretched all the way to St. Anthony. It was just after 10:00 A.M. The hourly newscast had reported that the mob of Newfies had been blocking the road ahead since 7:00 A.M. "Hope they're frozen stiff," grunted Watson.

After an intolerable stretch of straight road, there was a curve. Beyond the curve was the mob. There had to be *at least* two hundred of them, and any thought of just slowing the vans down and butting our way through was immediately lost, for there were thirty-five to forty cars, trucks, pickups, and Ski-doos parked across the roadway behind them, plugging it for at least half a mile.

"The laws of impenetrability are against us," someone remarked.

Pablo Morse and Jet Johnson were in the van behind us. Moore was driving our van. Watson was in the front seat beside him. And I was right behind Watson. The plan was for the three of us to step out and for the rest to stay inside until we knew what was happening.

We slowed to a stop about forty feet from the solid line of Newfies, some of whom were waving picket signs that said things like: GREENPEACE OUT! MAINLANDERS GO HOME! GO SAVE B.C. DRUG ADDICTS! There were about a dozen tough young guys in the front row who were swinging ropes with nooses on them. The moment we stopped, they surged forward, almost pushing each other over in the rush to get at us. Within seconds, they had crashed against the van, slapping their picket signs across the windows so that we couldn't even see out, and began with incredible energy to rock the van in an attempt to push it over. Inside, it was all we could do to cling to our seats. Moore remarked as he turned off the engine with a resigned kind of motion, "We seem to be here." Then he put out his cigarette, and with a deep sigh, both he and Watson climbed slowly out into the press of shouting, fist-waving men. I was out right behind Watson, shaking badly, trying to remember the best method of protecting my head while being stomped. There was a brief urge to clown—to come out of the van with my hands up in the air like a bandit who had just been caught by a posse. But they might take that as mockery. ...

A lot of men in the crowd were wearing little yellow buttons on their caps and parka hoods, which indicated they were sealers. Quite a few of the boys had obviously been into the booze, because there were empty bottles underfoot—a bad sign. Moore and Watson were surrounded, jostled, jabbed at, shoved, howled at, and threatened with the butt ends of pickets waved in their faces. It was impossible to answer any one man, because the moment Moore and Watson opened their mouths to reply to one attack, another would be coming, and another. Both Greenpeacers had on orange arctic-survival suits, which made them stand out like flames in the crowd of men dressed mainly in black. It polarized the sides, turned our guys into unmistakable alien targets, heightening all the divisions that already existed. The clothes that the Newfies wore were mostly old and dirty and tattered, and their faces were covered with stubble. I happened to be wearing an old navy jacket, which meant that within seconds of stepping out of the van, I was in the midst of the mob, but not easily identifiable. The Newfies ignored me almost completely, concentrating their wrath on Moore and Watson, allowing me to slip to the edge of the road and up onto a snowbank to gain a vantage point.

Then Eileen Chivers was out of the van, also highly visible as a Greenpeacer in her orange mustang suit, her dark hair loose over her shoulders, crowding in close to Moore and Watson. The violence in the air, which had been building rapidly to a breakpoint, suddenly slackened. The power of sisterhood had arrived, and the mob became immediately confused, sensing that if there were to be a riot, it would be impossible in the rage of it to distinguish between the two men they wanted to stomp, and the woman who was standing there with them. Now a space opened between the three Greenpeacers in front of their van and the restless, still-howling mob that circled about them, thrusting forth a new man to yell and wave his fists every few minutes, while Moore and Watson continued to try to respond and Eileen merely kept smiling at the hecklers, destroying the machismo madness as effortlessly as someone snuffing out a candle.

From the top of the snowbank, shoulder-to-shoulder with some of the less-animated Newfies, I could survey the scene, looking for an opening. There was a Mounty car parked down the road; it was obvious the two officers inside were not eager to have to move into the midst of the mob. The answer—the way through—therefore lay with the mob itself. Within a few seconds, I had spotted what I was looking for. The crowd was not quite an amoeba: it had a leader, a stocky, double-chinned man in his mid-forties, wearing a black Ski-doo outfit, but with a white shirt and green tie underneath. Okay, I thought, we're okay, a group with a center is no longer a mob. A posse, perhaps, but not a true mob.

Immediately, I skidded down the snowbank, still making use of my low profile, to wiggle my way through the crowd to the man's side. He stared at me blankly for a moment, then made the connection that I, too, was one of those damned Greenpeacers. I offered him my hand, which he took reluctantly. I yelled over all the yelling that was going on: "Can we arrange a meeting?"

The man, whose name was Roy Pilgrim, coordinator of the Concerned Citizens Committee of St. Anthony Against Greenpeace, had already figured this one out. A meeting had been prearranged for 9:00 P.M. at the St. Anthony Elementary School auditorium.

The noisy debate between Moore, Watson, Eileen Chivers, and the seal hunters raged on for another twenty minutes, getting nowhere, but giving every person in the crowd a chance to vent their anger on a real, live target. Then Roy Pilgrim took over firmly and announced that there would be a meeting at the school that night. The Greenpeacers would get to present their case. The people of St. Anthony would have their say. "And then we'll see what happens."

"Ain't gonna make no difference, b'ye," snarled one old man, who had been jumping up and down with excitement. "Meetin' or no meetin', these b'yes are gonna be on their way back to the mainland by midnight." There was a great cheer.

We were led into the town convoy-fashion. Ten trucks and cars were ahead. At least thirty were behind, driving, for some ritualistic reason, at less than five miles per hour, as though we were something the men of the community had hunted down and caught and now they were proudly parading their prize homeward. Two guys in the back of the pickup truck directly ahead of us amused themselves by taking turns slinging a rope around each other's neck while the victim imitated a man being hung by the neck until dead.

We finally arrived in front of Decker's Boardinghouse, a white and lime-green two-story old house at the edge of a frozen bay around which the town seemed to be built. Two months before, we had booked reservations, and had half expected the owner, a Newfie grandmother named Emily Decker, to cancel us out. But even with the whole town against us, Emily Decker's code was that a deal was a deal, and she had never welshed on a deal in her life. As we arrived at the boardinghouse, amid honking horns, jeers, catcalls, and the occasional whiskey bottle flying through the air, there was one last struggle to get the crew safely out of the vans and through the door into the house. Most of the St. Anthony posse had closed in again. Teenagers on Ski-doos had come racing across the ice to see the action. Again, the focus was on Moore and Watson, but just as one of the young seal hunters was working himself up into a frenzy, ready to throw a punch at Watson, Walrus Oakenbough, warrior-brother of the Oglala Sioux, suddenly made his presence known. Until now, we had been conciliatory, desperately reasonable. Not any longer for Walrus. Dropping his pack on the ice, he suddenly strode up to the seal hunter, jabbed his finger furiously in the man's chest, and screamed: *"Listen, I haven't heard one of you guys say one word yet about Mother Earth! And that's what this is all about! We're here to protect the seals, the whales, the birds, everything! It's all part of Mother Earth! And you're not gonna stop us, because it's Mother Earth's will! Those seals are my seals too! So just get out of our way, get out of our way, that's all!"*

Within minutes the crowd dissolved. No one knew how to talk to *that* guy. It was like the old Mexican tradition of never touching a naked, crazy man.

But we still had a few problems ahead.

No sooner had we made it into Decker's Boardinghouse than I was on the phone to Rod Marining in Vancouver to fill him in on what had happened, only to learn that the Canadian government had just passed an order-in-council making it *illegal to spray seals*. An order-in-council is a law or statute that is passed by the federal cabinet behind locked doors, is not subject to the will of Parliament, and does not need any further sanction. With the flick of a finger, new laws may be invented to suit just about any purpose. We had perhaps been naive to think that the so-called

Liberal government of Canada, under Pierre Trudeau, would not dare to tamper casually with the laws of the land. We had underestimated the totalitarian streak that existed in Ottawa. Whatever the case, we now faced not only the wrath of the people of St. Anthony, but a whole new legal situation. The new law meant that the moment a Greenpeacer went out and sprayed a single seal—assuming we could physically get the choppers off the ground—he or she would be immediately arrested. And so would the next one. And the next. Until we were all in jail. It seemed wise at this point to change tactics—rapidly.

From several of the Newfies I had listened to in the welcoming party, I had gathered that what they were really upset about was not so much that we might hassle the large icebreaker fleets that broke into the pack ice on the nursing grounds. Rather, their anger was beamed at the "green spray" plan, which, in their minds, might mean that as they climbed out over the ice floes themselves to go kill a seal or two, the seals they found might be marked, and therefore useless.

We knew that there was a tremendous disparity in Newfoundland between the company-owned ships that go out to the Front ice, and the "Landsmen" themselves. The Landsmen, numbering some three thousand, were the men who had gone out from their tiny outports with their fathers to kill a few seals here, a few there, close to the shore. And as they had gone out with their fathers, so they went out with their sons. It was as normal for them in the spring to go swiling as it was in the summer to go fishing, or in the fall to dig up potatoes. Less than two percent of the Landsmen made more than a thousand dollars a year out of sealing. The majority made less than two hundred dollars. These were not the men who had been seriously depleting the seal populations. In fact, the Landsmen and the seals might have lived in a kind of ecological balance with each other for eternity but for the introduction of great steel icebreakers into the very heart of the breeding grounds, far out to sea. The icebreakers were divided between a Norwegian fleet and a "Canadian" fleet, but much of the Canadian fleet was in fact Norwegian-owned. Most of the kill occurred on the Front ice, and it was carried out by a relative handful of men—a couple of hundred—operating from the ships. Only a fraction of the profit went to the men themselves. Most went to the shipowners. There was resentment over this—not as much resentment, by far, as there was against mainlanders coming in and messing with local custom, but nevertheless a deep division existed between the men we were now confronting and those out on the icebreakers who were the ones we really wanted to confront. As any general knows, if you drive a wedge between your opponents, you are that much closer to winning the battle.

There might just be a way. ...

While everybody was still unpacking and coming down from the excitement of our arrival, I got hold of the local telephone directory. There was only one Roy Pilgrim in the book. I called him immediately.

"Mr. Pilgrim, I think we should talk privately. There's certain things you don't want to see happen and there's certain things I don't want to see happen, and I think together we can probably do each other a lot of good."

We met a half an hour later. I had to slip out of the boardinghouse and wander a block away to meet him, then I climbed into his car, and we sat there, out of everybody's earshot and made a deal.

Pilgrim knew, as well as I did, that several planeloads of reporters and cameramen were on their way to St. Anthony. The world media was only a day away from arriving en masse. Already, television crews were disembarking from the airport. The four major hotels in town were booked solid by the press. He knew that footage and photographs of peaceful eco-protesters getting stomped to a pulp would hardly do his people's cause any good. We'd look like martyrs. They'd look like an arctic version of the red-necks in the Deep South.

I put it to him straight: We wanted to get our choppers out to the ice where our intention was to focus our attack on the big commercial fleets. That way, we'd be able to make a case for a two-hundred-mile fishing limit. In Pilgrim's mind, that would mean cutting the Norwegians out of the hunt, leaving more for the Newfies. In my mind, it meant a major reduction in the kill. In my mind, it also meant playing on the general Newfie resentment against foreign fleets and the specific Newfie Landsmen resentment against all the icebreakers. The deal was this: We would give up our spray plan and hand over our cannisters and hoses and dye in exchange for a promise not to interfere with our helicopters and a promise to allow me to try to recruit Newfies to join us in a protest against the commercial fleets, particularly Norwegian. From a purely political point of view, it was the only way I could see to reverse the tide that was moving against us. The Canadian government was obviously content to leave Greenpeace and the Newfies to slug it out, standing, as it were, above the fray. There was nothing from our point of view to be gained from getting caught in a trap that pitted us against the workers. If, by the simple expedient of handing over the dye, which was now illegal to use, we could manage to get a whole lot of local people on our side, then the government would be faced with the specter of having to deal not just with the eco-freaks from the mainland but outraged Newfies, who were angry with the way the present hunt was being conducted. The hope was to outflank Ottawa and pull off a worker–conservationist alliance that would dumbfound everybody. I knew that a lot of our supporters back in the big cities would not be able to fathom the political subtlety of it all, and that we would have a few bad days in the media when it appeared we had "given up." But with luck, so long as we could get the choppers out to the ice, we would recover, and, if everything went perfectly, we might yet emerge with a formula for bringing the environmentalists together with the grass-roots people in Newfoundland against the government and the handful of fur barons who had so successfully exploited both seals and sealers for centuries.

Pilgrim and I shook hands. The deal was struck.

Walking back to Decker's, bent against the wind, I was just in time to see the helicopters, like fantastic violent insects, coming down on the ice beside the house. Several snowmobiles were converging rapidly on the spot, and our own people were pouring outside. There were plenty of local people gathering in the dusk. But I knew that the choppers were safe now and that we would get out to the ice.

On board the helicopters were three more crew members: Marvin Storrow, a top-notch lawyer from Vancouver, who had volunteered to pass up his usual Hawaii vacation to act as our guardian angel; his wife, Colette, whose maturity and good sense was going to be badly needed; and Michael Chechik, who had not yet finished our whale documentary but was prepared to start making a seal documentary.

The meeting that night in the St. Anthony Elementary School auditorium went off like the piece of theater it was intended to be. The whole town had turned out, and the auditorium was packed, standing room only. There were about six hundred people present. Much of the advance wave of press had arrived: a man from Germany's *Der Stern*, two cameramen from France's Gamma News Agency, CBC, CTV, the *Toronto Star*. "Christ," Patrick Moore whispered, his eyes slightly glazed from awe, "I was just on the phone to New York. The *Washington Post* is sending a guy up here. NBC already has a crew in Gander. NBC, man!"

The crowd had a pre-hockey game feel to it. There was no doubt we were the visiting team. As we entered the auditorium there was a great

"*BOOOOOOOO!*" Surprisingly, though, there was also an undercurrent of approval. I actually heard some cheers.

Six of us climbed onto the stage with Roy Pilgrim, who took the podium and began his speech by praising the good sense and fantastic courage of the people of Newfoundland. The whole audience agreed with that and responded with cheers and ovations. It was a good moment for Roy Pilgrim. He ended up emphatically stating that there was no *way* the Newfoundland people were going to allow these people from the other side of the country come and tell them what to do, just *no* way. To my relief, he gave no indication whatsoever that the two of us had been sitting privately in his car hours before. Watson and Moore, who had been filled in on the plan and had agreed, now took their turns at the microphone. It was rough going. They were heckled and booed. Then came my turn, and knowing I had a card to play, I spun it out. Only a handful of people in the room were aware that there was now in existence a law that made it a crime to "spray or mark any living seal," and that the crime carried with it a penalty of one thousand dollars or up to a year in jail.

I talked to them about Greenpeace, right back to Amchitka. About the courage of men like McTaggart and Cormack, about high seas adventures with Russians and Frenchmen and Americans, about noble ships and people who were willing to die for their beliefs. "Greenpeacers are not frivolous. They are rugged, independent people, too."

Then I made the big concession. "Out of respect for the serious economic hardships experienced by the people of Newfoundland, the Greenpeace Foundation will drop its plans to spray the seal pups with green dye—"

Astonished cheers. Total audience participation.

"—and, instead, Greenpeacers will go out to the Front, to the icebreakers operating in international waters, focusing primarily on the Norwegians, and we will throw our bodies between those seal hunters and the seals."

More astonished cheers. Aha, they like that idea of concentrating primarily on the Norwegians. "Tomorrow we'll hand over our spray cylinders and dye." The stories went out on the wire services that night:

GREENPEACE SURRENDERS! THE DYE IS CAST! GREENPEACERS "CONVERTED" TO SEALERS' SIDE!

Back home, there was shock, dismay, confusion. By the next day the switchboard in our Vancouver office was lighting up with people demanding their donations back. In Toronto, a newly formed Greenpeace

270

group crushed all their Greenpeace buttons under their heels and mailed them in a brown bag to me personally in St. Anthony. Decker's telephone wouldn't stop ringing. People were calling up from all over, demanding to know why we'd sold out. There was one remark I made during my speech at the school auditorium that stuck out and was widely reprinted, to the enragement of many a seal lover. I had said: "Everybody's a cute little pup at one time." We had lost a lot of support at some levels. It now remained to be seen whether we could pull the Landsmen in on our side in a combined assault against the big ships and the fur barons, thereby forging a political alliance that would endure long after the headlines had faded.

It was one of those events that seemed to want to happen. While we were in the auditorium in St. Anthony, trying to make overtures to the Newfie working class, Carl Rising-Moore found himself in the town of Corner Brook, several hundred miles south, sitting at a table with a group of union leaders. Rising-Moore had been left behind in North Sydney, Nova Scotia, because he'd lost his sleeping bag and survival equipment on the train from Montreal. He'd had to wait around a day before the stuff could catch up with him. By that time, the rest of us were closing in on St. Anthony and there was nothing for the long-haired veterinarian to do but start hitchhiking. Slouching along the road in his U.S. Army–surplus coat, with a thick pack slung over his shoulder, he was anything but unnoticeable. Not often in winter do mainland hippie hitchhikers pass through the Province of Newfoundland, heading *north*!

At Corner Brook, frozen stiff and windburned, he slipped into a hotel beer parlor to rest. He was on time to catch the six-o'clock news. There on the screen were the scenes of the mob greeting our van that morning and the news that a meeting was scheduled between the townspeople and the invaders from the mainland. Rising-Moore quickly became aware that a group of men at the next table were grumbling and swearing about those "damned Greenpeace assholes." Almost automatically, Rising-Moore trudged over to their table and introduced himself by name, hinting that he was an observer sent by the United Nations. He had a blue UN emblem patched to his parka. He was so soft-spoken, so confident yet unassuming, that none of them challenged him.

The group proved to be dominated by a slick politician who was filling in time between elections running the Newfoundland Fishermen, Food, and Allied Workers Union, a nine-thousand-member group that included nearly four thousand registered sealers. Richard Cashin looked a bit like Archie Bunker, the famous TV red-neck, but he was an elegant professional speaker, silver-haired, very intelligent. He had served as a member of the Legislative Assembly for a while, but had been tossed out by the voters. Right now, he was cursing the "Greenpeace b'yes," and trying desperately to dream up a way to take advantage of the tremendous concentration of media that was coming together up in St. Anthony.

In the conversation that followed, it was Carl Rising-Moore, the diplomat, who suggested that there might be areas of agreement between the unionists and the eco-freaks. For instance, Greenpeace was in favor of a two-hundred-mile limit and would probably be more than willing to join forces in a push for its implementation.

By dawn the next morning, Richard Cashin and the executive council of the NFFAWU were down at the airport boarding a chartered plane for St. Anthony. Even though they knew every hotel room in town was taken up now by media people, the union men had lots of members' homes where they could stay. Carl Rising-Moore was left behind to push back outside into the Glacier Age, a lonely radiant figure plodding along the highway between the drifts, icicles hanging from his beard, a faint but seemingly permanent smile playing on his frozen lips.

A watch had been posted over the helicopters in case any of the local boys got it in their heads to go to work with sledgehammers. A bulldozer had cleared the snow out of the parking lot at Decker's Boardinghouse. The helicopters sat there, shrouded like horses, electrical umbilical cords running across the parking lot to supply heat to keep the engines from freezing. Our real concern wasn't so much the spectacle of a sledgehammer attack, it was the fear that somebody might sneak up during the night and simply unplug the electricity. That would sabotage us quite effectively, since the temperature dropped to thirty-five below that night.

We stared at the helicopters with pride and excitement. They were sleek red-and-white Bell Jet Ranger IIs. Their presence meant that even if we were still surrounded, we had the power to leap over the heads of the encircling brigades and reach out across the ice floes toward the seal herds and the steadily approaching death ships.

Our senior pilot was a crisp but immediately lovable fifty-year-old veteran of the Royal Canadian Air Force, who had risen to the rank of captain before leaving the service. His name was Jack Wallace. He wore an impeccable dark-blue uniform with a silver helicopter-shaped badge pinned on his left chest. He was completely bald.

The second pilot was a NATO-trained veteran of the modern West German air force, Bernd Firnung, a younger man who, in contrast to his impeccable senior, quickly took to loafing around the boardinghouse in crumpled slacks and checkered shirt, his red face glowing with good health and amusement.

In the days and weeks to come, Wallace and Firnung would fly in the first ecological air battle of the North Atlantic, skimming the edges of whiteouts and blizzards, dropping down on crackling sheets of ice, dodging around eight-hundred-foot cliffs to escape from hurricane-force winds. They would be flying against a superior force of federal-fisheries-department choppers, chartered fixed-wing aircraft, and roaring jet-engined Canadian Armed Forces Argus surveillance aircraft. At one point, in fact, there were to be no less that eleven choppers airborne, three fixed-winged spotter planes, and one Argus swooping and diving over a fleet of seven icebreakers and one squad of Greenpeacers on the ice. In the course of the battle, the fisheries department was destined to lose two helicopters. One would be grounded by mechanical failure. The other would be blown by the worst storm in a decade off the deck of a sealing boat onto the ice.

All through the night, blue-faced watchpersons staggered in at the end of their hourly shifts guarding the helicopters, while the wind whipped around them—chill factor about sixty below—and the sky blazed with enough light from stars alone to cause the snow-impacted landscape to gleam. With the approach of the full moon as well, the polar world around Decker's was totally visible—the starkly outlined sleeping helicopters, parked trucks, and Ski-doos, the little wooden houses hunkered in the snowbanks, the great frozen pan of the bay in the center of town, crossed and crisscrossed with a seemingly infinite number of snowmobile tracks. To walk out on the bay under the moon was to step upon a tremendous stage. It was an invitation to profound private monologues with only the cosmos as a witness. The lights of the little houses along the shore were almost as distant as the stars themselves. The wind whistled in from out on the open North Atlantic where it had run for so long without interference. The helicopters shuddered in the savage gusts, wobbling on their rubber pontoons, almost as though alive. Each had a long sensor probe thrust out like a proboscis from its nose, and these quivered in the night wind like antennae. Immense folded mosquito wings, the propellers jiggled, clicking restlessly. Moonlight gleamed in blinding slashes along the unhooded parts of the hulls. Ecology had an air force now, and it was a thrilling thought.

Our second day in St. Anthony would be another occasion for political guerrilla theater. No sooner had we finished a tremendous homemade breakfast than the press guys began to arrive, followed by Roy Pilgrim and his Concerned Citizens. It was time for us to go outside in front of the cameras and hand over the half dozen converted fire extinguishers that were to have served as our spray cans, along with several packages of powdered organic green dye.

The ceremony took place in sparkling sunlight in the parking lot in front of the helicopters. The TV cameras seemed exceptionally noisy in the frigid air. The clicks of the cameras were magnified. A couple of dozen seal hunters had accompanied Pilgrim, who was having his finest hour as the "victor" accepting the laid-down weapons of a vanquished band of invaders. Wire-service photographs show Watson, Moore, and me looking very grim, while Pilgrim looks triumphant, and the swilers in the background look plainly surly and suspicious. The converted fire extinguishers sit in open neutral ground between us. Hooded helicopters wait behind the scene of the ceremony.

Immediately after the ceremony, we retreated into Decker's for a strategy session. In this exercise we were joined by Marvin Storrow, who had gotten us out of jail back in September when we'd tried to blockade the loading of fresh fruit onto a Russian supply ship. Storrow was a partner in the biggest and most-established legal firm in western Canada. Formerly, he had been an Ottawa lawyer, prominent and upwardly mobile in political and legal circles, but something had gone sour for him in the corridors of Canadian power. He had come to Newfoundland to put his skills and brainpower entirely at our disposal. Thus, when we gathered around the charts in Decker's living room to begin our strategy session, we were joined by one of the finest legal brains in the country. Colette Storrow, an attractive, quiet woman in her early thirties, was the only mother included in the crew, and she exerted a subtle but stabilizing influence on the boys and girls who had assembled in Decker's. Again it was a family.

Storrow's great ability, at the onset, was simply to throw practical questions in our faces every time we came up with a new scheme, keeping us in touch with the real world. A brainstorming session involving him, Moore, Watson, and the two pilots was bound to come up with something—and it swiftly did. We knew by this time that we had been invited to meet with Richard Cashin and the fishermen's union bosses—well and good. But we didn't know if anything productive was going to come out of the meeting, and we didn't know if the moment the meeting was over, our local opponents and the unionists might not take some action to prevent us from getting out to the ice. Whether that meant wrecking the helicopters or just sitting on top of them en masse or wrecking us was a moot point. We had to make sure we were going to get clear.

The plan was to get everything ready to move an expeditionary force out by helicopter to an island roughly thirty miles north of St. Anthony. There, a base camp could be set up, which would be that much closer to the hunt. A fuel depot could be built up, giving the helicopters a greater range, and guaranteeing that no one would be able to control our freedom of movement. Belle Isle was uninhabited, except for a lighthouse keeper in the summer. In winter, the lighthouse was abandoned. Not a single tree grew on the island. Surrounded by stretches of open water and patches of jagged "slob ice," the island was unapproachable now except by helicopter. The idea was that if we could get a crew and a fuel depot established out there, no one could possibly take any action to prevent our helicopters from getting airborne, otherwise they'd be condemning our Belle Isle advance squad to death by freezing and starvation.

Sure, we would go to a meeting in town with the fishermen's union, but while that was going on, and everyone's attention was focused on the meeting, the choppers would lift off for Belle Isle to start setting up the base camp, leaving people out there right from the very beginning, as kind of reverse hostages.

With each flight out to the isle, the helicopters would carry one less person than normal, but would have several thirty-six-gallon fuel drums slung underneath. Within a matter of only a few flights, we would have a forward fuel depot with the capability of filling up both choppers for complete flights. A hand pump would have to be used in subarctic conditions, but that was no matter.

We were halfway through the strategy session—feeling very pleased with ourselves for our generalship—when Pablo Morse came charging through the door with the news that the local Imperial Oil dealer was refusing to sell us fuel. He had been threatened by officers from the Canadian Department of Fisheries with imprisonment and heavy fines if he sold any of the precious black stuff to Greenpeace. And that was all there was to it: he wasn't selling.

We'd been grounded.

But no. Marvin Storrow leapt nimbly to his copy of the Criminal Code of Canada, a section of which makes it an offense under the law for anybody to threaten or intimidate anyone else with any kind of penalty if no offense has yet been committed. While the rest of us were still staring at each other with doom-filled eyes, Storrow was already on the phone to the local RCMP office. In tones that dripped with chilly reason and calm, he advised the Mounties that they had best send an officer down to Decker's Boardinghouse, because the Greenpeace Foundation was about to file a formal complaint of attempted extortion against employees of the federal department of fisheries.

Within fifteen minutes, two Mounties had arrived, slightly breathless. Still oozing with the great dignity and certainty of a jurist, Storrow explained the situation—and the relevant sections of the Criminal Code. The Mounties did not argue much—they immediately promised an investigation, and disappeared in their black squad car in the direction of the Imperial Oil station.

Less than an hour later, Pablo Morse was able to return to the station to find that the oil pumps were suddenly flowing freely. The boys from the fisheries department had already been visited by the Mounties, and had quickly called the oil dealer to advise him that there had been a misunderstanding, and that of course there was nothing wrong with selling fuel to Greenpeacers, nothing at all.

That night, one of the national Canadian television networks ran a half-hour special on Greenpeace. The interviews had all been done a month before, along with the filming and editing. It was powerful stuff— footage going back to the voyage of the *Greenpeace* and the *Greenpeace Too*, McTaggart's adventures, and a lot of stuff from the whale voyage. A TV crew had even captured scenes from the meeting in the St. Anthony Elementary School auditorium the night before, bringing the story right up to date. In winter, at any rate, the Newfoundlanders are a television people. Almost every man, woman, and child on the island saw the program, and it was especially well watched in St. Anthony. By the end of the evening the mood of the town had shifted considerably. It was now more or less safe to go into the beer parlors. We might be idiots about the seal hunt, but otherwise, there was a grudging acknowledgment that we did have other things to our credit. There was a certain macho element to it, which many of the men had to admit they liked. The bouncer at the St. Anthony Hotel was the first to advise me: "If anybody bugs you or your b'yes, let me know."

In the morning, Moore and Watson and I dutifully went off to meet with Richard Cashin and several dozen members of the NFFAWU. While we were gone, preparations went ahead at the boardinghouse to get all the equipment loaded on the helicopters.

The meeting lasted for an hour. There was a long narrow table at the head of the union hall, with Cashin and three of his executives seated on one side, the three Greenpeacers sitting opposite them, and a room full of grim, scar-faced seal hunters and fishermen behind. The purpose of the meeting was very simple: the unionists wanted to get us on their side in favor of the two-hundred-mile fishing limit, something we were eager to do. The only real exercise was to see if we could "agree to disagree" on the issue of the seal hunt, because the union wasn't willing to concede an inch in any of its arguments in favor of the hunt, and—by dropping the spray plan—we had already conceded as much as we could. In the end, we emerged to face the press with a joint Greenpeace–NFFAWU statement, and a whole new project to consider:

> On the matter of the overall need for conservation and resource management, both organizations urge the Canadian government to make a formal commitment to unilaterally declare a two-hundred-mile fishing management zone. Both organizations are pledged to take further action if the government of Canada refuses this plea. We are jointly planning a blockade of the ports of St. John's and Vancouver to all foreign trawlers and draggers by June 1, 1976, if Canada has not yet taken the action we have recommended.

The press guys made a rush for the nearby Vinland Hotel to file their stories and then get back to the beer parlors. Most of them therefore missed the lift-off of the Jet Rangers. The propellers whipped the snow into a mini tornado as they howled their way heavily into the sky. Looking down through the windows and forward bubbles of the choppers as they hovered briefly over Decker's Boardinghouse, the members of the first Greenpeace expedition to Belle Isle flashed peace signs at the rest of us. The expedition carried Watson, Moore, Jet Johnson, and Walrus Oakenbough. The first thing they did was swing over to the Imperial

Oil station, where Pablo Morse and Dan Willens bent their backs and muscles, hoisting the thirty-six-gallon fuel drums—loaded—onto strap slings under the choppers' bellies. Then, like insects carrying almost more food than they could handle, the machines turned and began to move northward, across the bay, over the low rocky hills and up the coastline toward a body of water called the Strait of Belle Isle.

When, finally, they left the sight of land behind, it was to find an infinitely delicate network of small ice floes beneath them, a pattern that looked like a great frosted mirror shattered by thousands of stones. The horizon all around—tremendously distant—was lost in a gauze of mist, even though a hard, fierce sun was blazing above.

The weather report was bad. The wind was picking up. A storm was approaching. By nightfall, there were expected to be gusts up to forty knots.

Belle Isle was perched about thirty miles north of St. Anthony, just seventeen miles due east from the coast of Labrador. Its cliffs rose eight hundred feet out of the encircling crushed jigsaw puzzles of ice. The isle was composed of igneous rock, a few protrusions of quartzite, with thin scabs of lichen clinging to its surface. The only plant life in the summer was heather, berries, and low scrub. Nine miles by four, Belle Isle's jurisdiction itself was in question. Nobody could say for sure whether it was the northernmost point of Newfoundland or the southeastern most point of Labrador.

The Jet Ranger IIs came down on Belle Isle to a silence that went all the way out to the northern horizon. Moore found that the silence, once the chopper engines had stopped, was so tremendous he could hear the icebergs grinding miles out to sea. Watson was the first to put his big astronaut boot down on the surface. It made no impression whatsoever on the diamond-hard ice. It was roughly thirty-five degrees below zero.

By late afternoon, sweating profusely in their survival suits, the forward squad had established a three-tent base on the ice, and the helicopters had shuffled four fuel drums out to the depot, a quarter of a mile from the tents. The quarter of a mile would prove, later on, when the crew was stranded, hungry, and freezing, to be a long, hard, frozen road standing between life and death. The moment the tents were up, Watson drove a steel rod down through the impacted tooth-like snow to the rock itself and ran up a big blue United Nations flag. All the crew members were carrying Planetary Citizen passports with the UN symbol on the cover.

A few press photographers and network cameramen were shuttled out to the island during the afternoon. For the color-camera boys, it was a feast: fantastic, blazing available light, bright yellow and blue and red double-skin arctic tents, blue UN flag, Greenpeacers in phosphorescent spacesuits, a landscape of dune-drifted diamonds and razor-blue rock, a giant sun blasting through purple-orange skies, and the Jet Rangers sitting there looking like time machines that had taken their crew back to the Ice Ages.

At the base of the northern cliffs, near the campsite, mysterious temple-like ice caverns were discovered coiling some 120 feet inward. Mastodons might lurk there. And of course there was always the possibility of a polar bear wandering in off the ice fields to sniff at the juicy mammals inside their tissue-thin tents. Just before dark, the two pilots, "Smilin' Jack" and the "Red Baron," leapt back into the sky and streaked over the blue-veined mosaics of ice down to the parking lot at Decker's. That night, newscasts carried the report that Norwegian and Canadian icebreakers were converging somewhere off the coast of Labrador, drifting southward among the ice pans and the seal herds, bearing steadily toward the northernmost tip of Newfoundland, waiting for the official opening of the annual hunt: within forty-eight hours.

The next day, Friday, March 12, was spent ferrying fuel drums and press photographers to Belle Isle, a task made more complex by the fact that there were now dozens of newsmen who wanted to get out there but a limited supply of helicopters and much equipment to be moved. We had to charge the newsmen just to cover the costs of getting them airborne, and while most of them accepted this as only fair, a few took violent exception. Since it was my job, back in St. Anthony, to "coordinate the media," that meant making dozens of agonizing decisions over whom to give a berth to on the choppers and whom to reject. No one liked being rejected. A crew from NBC that had just flown in from New York was particularly indignant. Led by a ex-football player who was the producer, they finally refused to pay any money, even though it would have cost us an entire return helicopter flight—several hundred dollars—to get them to the island and back. "That would compromise our objectivity," the chief thundered. Then he jabbed his finger at me, like a New York cop who had just fingered a hippie, and added: "You're trying to pull a fast one, kid!" With that, they marched swiftly out of the boardinghouse and back to their hotel; the producer's parting shot was: "You're dead in New York, kid."

To make life just slightly more complex, David McTaggart arrived from Paris along with a crew from a big French television network and a photographer from Gamma, the French photo-distribution agency. The seal hunt was big news in France. Good coverage for Greenpeace might mean extra support within France itself for McTaggart's ongoing case against the French navy. So far as he was concerned, Greenpeace owed it to him to get these French guys out to the ice. Years before, I had insisted that the photographs of McTaggart's beating at Mururoa be released immediately, regardless of whether McTaggart himself had any plans for their sale, so I could hardly refuse to help him out now.

"Okay, David, we'll get these French guys out there somehow." Howls of outrage promptly were torn from the chests of Michael Chechik and Ron Precious. As the official film crew, they were the ones who by rights should have had priority on any helicopter flights for documentation purposes. While both of them "appreciated" the politics of McTaggart's situation in France and Greenpeace's debt to the ex–badminton champion—our only real martyr thus far—they could still not quite accept the idea that I was tossing them aside in order to get a handful of commercial French media hustlers through. ... Ah, global politics. Here I was, having to let a French film crew go first to the site of the Labrador seal hunt in order to help maintain the pressure against nuclear military fortresses in the middle of the ocean on the other side of the planet, and, in the meantime, we'd lost New York. In that atmosphere, it was easy to get the feeling that one had lost one's mind, as well.

The photographer that McTaggart had brought along with him from Paris was, in fact, an impressive figure. His name was Jean-Claude Francolon, thirty-two, a strangely delicate man, like a true European aristocrat. He had flown from Paris direct from Beirut, where he had been covering the Moslem–Christian war. For him to be switched from Beirut to St. Anthony meant that someone up there in the control tower of a major worldwide photo agency had decided that the seal hunt was bigger news than the holy war in the Middle East. He was dressed in the best arctic-survival gear that money could buy—complete with sealskin boots, a fur-lined parka, and a cap made of the downiest brown deerskin imaginable. Henrietta Neilson and Bonnie MacLeod stared at him in horror and were flabbergasted when they finally learned that I was, indeed, going to allow him to go out to Belle Isle.

"But he's a *murderer*," seethed Bonnie.

In St. Anthony, press guys swarmed from hotel to hotel, mixing with a small army of fisheries department officials. Suddenly, more long-distance phone calls were being placed out of remote little St. Anthony than were coming out of the provincial capital, St. John's, itself. The entire Newfoundland communications grid had shifted its axis northward. Reporters from Quebec argued into the night with reporters from Toronto and Ottawa. The big-time American network boys all hung around in one corner, dominated by the New York NBC producer who'd told me we were "dead." Everybody in town was now suddenly a potential TV star. Dozens of them had already seen their images flickering back from the tube after they'd talked to reporters. It was heady stuff. There were a few of the boys that still wanted to wring somebody's neck, but the mood had shifted in town, and it was suddenly uncool to stomp an eco-freak. It was now possible for the "recruiting team," meaning Carl Rising-Moore—who had finally succeeded in hitchhiking all the way into town—Dan Willens, and myself, to walk openly into beer parlors, sit down, order a drink, and wait to see who would show up. We made no bones about it: we were looking for Newfies to volunteer to join us against the big commercial fleets.

Within an hour of our first visit to the Vinland Hotel, we had at least a dozen local people—and a couple from out of town—with their chairs pulled up, jammed around our table. The talk wasn't always sweetness and light. Some of them were still mad. But they could see our point: the big ships going into the heart of the nursing grounds were ruining it for the little guys along the coast. Like fishermen in British Columbia, these men were perfectly aware that the oceans were dying before their eyes. "Usta be, b'ye, ders seals come inta thu bay. Jes go out thu door an pick 'em up. Now there's naught one t'be seen come spring in this 'er bay. It's troot, b'ye! Dey's gun!"

I recognized one young man who had approached us back at the blockade, where he'd carried a sign that said: GO SAVE B.C. DRUG ADDICTS! He had showed up later at the auditorium, and a photograph of him flashing a peace sign, still holding his picket, had been carried on the wire services and published right across Canada. Now he was sitting down with us, grinning. His name was Doug Pilgrim. He was the younger brother of Roy, with whom I had negotiated the hand-over of the dye.

"Hey man," he said, "I dig what yer doin'."

Before we were to leave, Doug Pilgrim would become the leader of the St. Anthony branch of the Greenpeace Foundation. Brother was now pitted against brother over the seal hunt in Newfoundland.

In any event, the St. Anthony branch was not fated to last for long—not much more than a couple of hours after we eventually left town. And Doug Pilgrim, who was relatively safe so long as the media remained, found the day after all the strangers had pulled out that he was no longer welcome at the beer parlors. The first time he ventured into the Vinland Hotel, he was elbowed, butted, punched, and generally treated as a traitor. His brother, Roy, disowned him, absolutely refusing to speak to him. His wife couldn't stand the social disgrace of being married to the town pariah and left him, taking their small child with her. Things got so bad, in fact, that Doug finally had to leave town. A year passed before things cooled down enough for him to return. Even then, he was still viewed with suspicion by the majority of the townspeople.

Less than thirty miles away, locked in the ice like a broken chess piece, Belle Isle had been turned into a giant wind tunnel. Before we arrived, Watson and his crew had had to pack up the whole camp and move it three-quarters of a mile because a wind change had suddenly left them exposed to fifty- and sixty-mile-an-hour winds blasting straight down from the north. Walrus's tape recorder had frozen. The two walkie-talkies had frozen. Only two of the three camp stoves were working properly, and the spare didn't work at all. In the process of moving camp, several items got caught by the wind and swept away over the nearby cliffs before anybody could catch up with them. A dozen cans of soup had been dropped and gone skidding like pucks down an ice-glassed hill out of sight. The tents had metal poles with tiny screws that could only be handled with gloves and mitts removed, but if one's hands remained unprotected for more than thirty seconds, they began to stiffen. The crew members with moustaches all wished they were dead. Marilyn Kaga had to treat everyone for minor problems like chapped lips, windburn, frozen cheeks, and burned fingers and thumbs. Everyone was coughing and spitting up phlegm. It was discovered that if we broke off bits of ice and sucked them to try to ease our thirst, we came down with stomach cramps within half an hour. By the first night, everyone was suffering from indigestion and nobody could manage a good bowel movement.

The walls of the tents smacked and crackled ceaselessly, always seeming to be on the verge of snapping apart. To conserve fuel, the stoves were only used for meals and for melting ice into drinking water. Everybody stayed in their sleeping bags with the strings around the openings pulled tight, leaving only tiny air holes at the top. Also to conserve fuel, it was decided that the Coleman lamps would not be used. There were three flashlights, but their batteries were frozen. So it was pitch black, and inside the sleeping bags, tight as cocoons, claustrophobia was setting in.

On Monday, March 15, Moore, Walrus, and Eileen woke up simultaneously just before dawn and found themselves in the midst of a great stillness. "After all that wind and the sounds the tents were making, suddenly everything was absolutely quiet, like on the moon," said Eileen later. "We all felt very high and tingly."

At 6:45 A.M., the Jet Rangers arrived from St. Anthony, bringing along a tough older Newfie named Art Elliott, who had volunteered to act as the expedition's guide over the ice pans. Walrus, Jet, Watson, and Moore climbed stiffly aboard, still slightly dazed from the pounding they'd taken during the storm. The choppers were quickly refueled, and without a pause, they banked and swirled away. Very little could be said over the noise of the engines. But Captain Jack and Bernd Firnung had already been handed all the information that could possibly be obtained in St. Anthony. Some of it was slightly contradictory. We had built up a pretty good idea of the position of the fleets, but none of our sources had been in total agreement. To complicate matters, Brian Davies of the IFAW had his own pipelines, and they, too, provided slightly different answers. The main problem was that the two-day storm had scrambled the situation, since the ships had done nothing for the last forty-eight hours except drift.

The helicopters moved out on a sweep that was to take the crew northeast 110 miles over the ice. Approaching the limits of their fuel capacity, they began to circle, but there was nothing below except tributaries of water, splinters, and pie chunks of ice. No seal-hunting ships. They retreated to Belle Isle for more fuel and huddled in the tents with navigation charts spread out over the burned, food-stained floormat. Captain Jack's twenty years of military service and Bernd Firnung's harsh, precise training under NATO brought forth tactical thinking of the highest caliber. Watson perched tiredly, clumsily, on his knees, thinking back to the days when he was a deckhand, trying to pull together the equivalent of a street hustler's sense of where the ships might be. Jet Johnson's stiff white finger jabbed repeatedly at the charts. Next door, in what had come to be known as the "spiritual tent," Walrus was busy throwing the *I Ching*. He got a reading that counseled a movement directly toward the east. Back in the military tent, the heavy-duty strategists were reaching exactly the same conclusion. Eileen Chivers found herself scanning the eastern horizon, her attention drawn there in the same way that for the last two days she had automatically been looking northward.

"A feeling," she said.

The Jet Rangers were refueled, and without pausing to eat more than a handful of nuts and figs and a few sips of cold tea, Watson and his crew were back in the sky, climbing to three thousand feet.

Within twenty minutes two black specks appeared on the horizon. The helicopters dived toward them, coming in close enough to be able to make out the little yellow buttons pinned to the caps of the sullen-looking men on the decks. The ships were the *Martin Karlsen* and the *Theron*, Canadian registered but Norwegian owned. They were traveling at four to five knots, easily breaking through the slob ice. In radio contact with each other, Smilin' Jack and Bernd Firnung scanned their instruments, established the exact bearing of the ships, and then leapt forward and upward—heading in the direction the seal hunters were headed. The hunt had to lie somewhere ahead, over the horizon.

The ice thickened below. There were fewer and fewer tributaries. The pans expanded and interlocked. The helicopters were nearing the Front, the famous nursery of the harp seals.

Black specks ahead. Two ... three ... six ... eight!

"Well boys," said Captain Jack, "we've got them for you. It's up to you now."

There were long strange marks on the ice, blood red, as though the pans themselves had been wounded. Even from twenty-five hundred feet, the crew could see the trails of seal blood splattered beside and behind the ships, like claw prints of a massive, invisible monster. The ships were moving languidly. It was not, after all, a *hunt* that was taking place—more like men walking through a field of ripe berries, squashing them with motorcycle boots.

Seen through film clips, slides, and black-and-white glossy prints, the first Greenpeace assault on the seal-hunting fleet looked like a ballet on a tremendous stage. At the last minute, finding himself leaping onto the ice, with the seal ships three miles away across the surreal landscape of ice and pups and mother seals and a torch-bright sky, Paul Watson shed the last fragment of his guerrilla thinking, the last vestige of his sense of being a lieutenant in a disciplined army, and reverted to the identity he had acquired in a sweat-lodge ceremony during the battle of Wounded Knee. He was back to being what the Oglala Sioux had named him: Gray

Wolf Clear Water. And he *charged* across the ice. Tough old Art Elliott, the Newfie who was supposed to be the guide, had trouble keeping up with him. At his side, loping effortlessly along, unrecognizable but for his walrus-tusk moustache—now firmly sheathed in ice—ran Watson's warrior-brother, no longer David Garrick, archaeologist, no longer even Walrus Oakenbough, now an Indian himself whose name was Two Deer Lone Eagle. Keeping abreast the whole distance across the ice was Jet Johnson, moving despite his forty-one years with the vigor of a trained athlete; he wasn't even gasping. Clutching his cameras, because he was also doubling as photographer, came Kwakiutl-brother Patrick Moore, furious because this had changed into an Apache-like attack instead of a disciplined advance. But he drove his powerful frame forward, easily matching the pace of the others.

Soon he could see the ships and their names clearly. He could see men on the decks. He could see skinned carcasses of baby seals. He paused once to get a picture of a crazed, shocked mother sniffing at the tiny-ribbed remains of her pup in a puddle of brownish-ruby goo. In death, the eyes of the pup remained open, but they had lost their quality of depth and sparkle. They were frosted a pale blue-gray from within, like stones. With the fur and the blubber stripped away, the dead pup looked like a thin Pakistani child of famine, except, like a victim of Thalidomide, it had no arms.

The seal hunters had probably been expecting a bunch of cringing city slickers to approach them, wailing at the sight of all the blood and guts. They were not quite prepared for what happened.

Two sealers were making for a pup that was wiggling desperately across the ice. The pup tried to bury its head in a drift of snow. The sealers both had their hakapiks raised, but Jet Johnson was suddenly skirmishing behind them and leaping ahead, throwing himself abruptly on his knees, his head ducked instinctively, hands pinioning the seal so that it was fully shielded from the hakapiks—the first seal in history to be so protected by the loving flesh of a human being. The hakapiks froze in the air. The absolute power of the sealers on the ice had finally been matched and checked; they could not chop their way through the body of a human to get at the animal beneath him. There is one photo of Jet on his hands and knees over the pup, which has its head thrust straight up in panic, that is reminiscent of the statues of Romulus and Remus feeding at the nipples of their wolf-mother. A seal had been saved!

Elsewhere a furious seal hunter was trying to close in on two pups that were close together. An orange-clad mainlander with a moustache like a walrus was pacing him steadily, butting him shoulder-to-shoulder every time he tried to get in close to the seals. He jammed Walrus's shoulder in a rage, but instead of Two Deer Lone Eagle giving ground, he butted back, and for a mainland sissy creep, he felt as if he were made of iron. Finally, the sealer gave up and retreated toward his ship, plainly confused. Another seal had been saved!

Gray Wolf Clear Water had meanwhile been planting himself like an unmoving wedge between seal after seal and their approaching killers. Several seal hunters waved hakapiks at him, gesturing with their knives. Two of the younger ones were so livid they wanted to make straight for him, but were held in check by an older, seasoned fellow. Finally, frustrated because he could not get between all the seals and their hunters at once, Gray Wolf Clear Water's brain began to approach the problem more methodically. He perceived that the ships were in constant motion, crashing forward ten to fifteen feet, then backing up thirty or forty feet, and making another rush at the ice, chopping through like an axe. Often, he saw, there were live pups in the way. He saw pups being crushed and mangled and drowned.

Abruptly, he made up his mind. Brushing aside two federal fisheries inspectors who were trying to head him off, he galloped toward the bow of the nearest ship, the *Melshorn*. The ship had just finished backing up for another run at the ice pack. Its turbines whined and the whole ship rattled as it gained speed. A seal pup was lying directly ahead, looking blankly up at the oncoming ship, not understanding that the eight feet of ice now separating it from the huge machine would soon be reduced to nothing. Gray Wolf Clear Water did not know if he had enough time to sprint across the ice to reach the seal before the massive iron bow came down on its body, but he did know, once he was in motion, that he could not stop. The seal had seen him. Tens of thousands of pounds of steel were on their way down on top of its life. Gray Wolf Clear Water did not know, either, how far ahead of the bow the ice began to fracture and turn over like pieces of asteroid. All he knew was that the ship was towering over him like a wall as he grabbed the pup, stumbling and falling to his knees, quite unprepared for the weight of the creature. The ice was crunching apart, exactly like animated scenes of earthquakes. Gray Wolf Clear Water found himself on a piece no bigger than an average living room, and it was beginning to tilt. The *Melskorn*'s bow was less than ten feet away, breaking through the ice like Thor's magic mallet. In its terror, the pup writhed and thrashed in Gray Wolf Clear Water's powerful arms, and scratched his cheek with its teeth. But then, somehow realizing that it was being held and protected, not crushed to death, it snuggled against him limply. He bore the pup at least sixty feet across crumbling ice pans before he could stop, heart crashing, to look back and watch the *Melshorn* crunch through. A seal hunter approached. But Gray Wolf Clear Water was blazing now. The sealer retreated.

Two Deer Lone Eagle and Moore had to yell at him repeatedly to get him to leave the seal's side. The signal had come from Captain Jack. Winds were coming up. The margin of remaining light was down to its minimum. Time to pull out. The shore was still a long way away. Gray Wolf Clear Water would have rather stayed there forever until by sheer exercise of his will he could cause the icebreakers and the hakapiks and the men in bloodstained clothing to melt away, but as Paul Watson he remembered that the battle had just begun.

"It got kinda personal at that point," he said later. "It wasn't just *any* seal any longer. I kinda knew that seal by then, and he kinda knew me. It was like we were buddies." He tore himself away from the pup, unable to look back for fear that he would lose his power of vision.

The confrontation squad got back to Belle Isle slightly crazy. They all wanted to have a bath in town, as though to wash away even the memory of what they had seen. It was late in the day, but there was still barely enough time to make a quick run to Decker's, let the boys clean themselves, and still get to Belle Isle and back to Decker's once more. Captain Jack had in fact left a wide safety margin. Quickly, Watson, Walrus, Moore, and Jet piled into the choppers, leaving Marvin Storrow, Art Elliott, and myself at the base camp, along with the regular crew: Marilyn, Eileen, Ron Precious, and Jean-Claude Francolon.

Decker's became a madhouse of sweating males thundering in and out of the bathtub, telephone blitzkriegs, cameramen, reporters, fisheries officers, RCMP constables, Newfie friends, the Decker family. Not many things were getting done efficiently. In the heat of the moment, the support team got sloppy and did not get into town right away to purchase badly needed supplies of food and fuel for the camp. Bonnie and Henrietta, who were supposed to be in charge of that, were miffed at not having been taken out to the ice and were not functioning as they should have. Unaware of the breakdown, several males were leaping back into their survival suits. Pablo Morse took Moore's place, leaving the ecologist free to stay in St. Anthony and help with public statements, press releases, interviews, and ongoing negotiations with friends and foes alike. Pablo went charging like a centaur out to the Red Baron's screeching metal bird, and the chopper was airborne.

It was going to be a tight run because the sun was sinking fast and there was a wind building out on the Gulf of St. Lawrence. The Red Baron flew furiously, but somewhat in vain. Less than a minute after his chopper had lifted, the van that had been dispatched at the last minute to pick up the supplies came skidding into the parking lot and Dan Willens leapt out, screaming: "The food! The food! They forgot the food!" They had, in fact, also forgotten the fuel for the camp stoves.

Bernd Firnung got back to St. Anthony at precisely the moment the sun went down. His Jet Ranger was already rocking and hopping in the sky, her movements jerky and uncertain. The winds were gusting to 30 knots. Anything more than 45 knots would have put the Red Baron out for good. Coming back across the slob ice separating Belle Isle from Newfoundland, he had had to fly roughly six feet above the ice. Any higher and the wind would start to cause sideways rolling tilts. The propellers screamed their outrage, and everyone in the chopper—including Storrow, Elliott, and myself—had groaned as though hit. The Jet Ranger, so arch and powerful on the ground, seemed then to feel extremely light, no more potent, finally, than a Christmas toy, a wind-up tin dragonfly. Approaching the tall cliffs of Newfoundland, still skimming along only six feet above the ice, traveling at 85 knots, with the needle at 120 knots—indicating a headwind loss factor of 35 knots, less than 10 knots from the knockout level—the Red Baron surprised everybody by speeding directly toward the cliffs as though bent on suicide. At the last minute, he banked, and the Jet Ranger was suddenly flowing in a current of air along the cliff faces, out of reach of the winds from Labrador.

"Veil, how's *dat*?" asked the pilot, his usual apple cheeks gone rather pale, but his voice steady. "Ve'll be jost fine now."

The discussion that night at the boardinghouse concerned the situation now facing the crew on Belle Isle.

At last report, they had been down to one pint of white gas for the three camp stoves, two cans of soup, and a bag of figs and peanuts and crunchy granola. There were now eight of them on the island. Storm warnings were up everywhere. The airport at St. Anthony was rapidly closing down as advancing winds threw up drifts that finally stopped even the snowplows. Gander itself, the nerve center of air transportation on the entire East Coast, had already closed. A blizzard was on its way, coming in a heaving movement like a tidal wave. The puny thirty-five-knot gusts we had felt in the helicopter were just the first delicate tendrils of the real wind to come. The main storm front was still over Labrador. In weather like this, it could quite easily be a *week* before aircraft of any kind could get off the ground again. There was no icebreaker around that could get through the ice to Belle Isle, and it was far too broken for walking. The base camp crew was going to have to tighten its collective belt rather drastically.

The first wave of the storm all but buried northern Newfoundland. A school bus with twenty-three kids got trapped overnight in drifting snow between two nearby towns, beyond the reach of any snowplows. A snowmobile search party was ready to go out after them, but the RCMP had to forbid it. Conditions were approaching the whiteout stage, where nobody goes anywhere. There was not the slightest possibility of getting our helicopters airborne until the storm was over.

On Belle Isle the next day, there was surprise at first, then confusion. While the winds were howling over St. Anthony and all points to the south, on the island itself it was a crisp, clear day. Where were the choppers with the food and fuel? All day long, there was at least one person outside, waiting impatiently and with a growing feeling of alarm, for the Jet Rangers to show. The temperature averaged around ten degrees below Fahrenheit. Overnight, the wind had gusted to forty-five to fifty knots.

Late in the afternoon, the real winds began to set in. Pablo Morse tried to do something outside without wearing his gloves, and suddenly his hands were rigid. He couldn't move the fingers. He stared at them in shock for long precious seconds before stumbling desperately toward the tent, hands held out in front of him as though made of wood. It was Jet Johnson who saved everyone from freezing. He devised a drainage chute out of cardboard, covered it with spit and water, and put it outside to freeze. Then he took it over to the fuel depot, where he and Watson and Pablo wrestled one of the thirty-six-gallon fuel drums on top of two others to gain some leverage. With one man tilting the drum from behind, another could unscrew the cap on the drum and let the oil flow down the chute, with so much spillage that Jet's survival suit became soaked in oil, making him a potential human torch. Painfully, one at a time, Jet would fill up the pint-sized empty containers of camp-stove fuel, then carry them against the wind back across the ice to the tents. For some reason that no one could fathom, Jet also turned out to be the only person who could succeed in getting the camp stoves to work. He did it by tearing up pieces of San Francisco underground comic books, including a *Freak Brothers* comic he dearly loved, to make a little fire in a folded tin pie plate. Then he would place the entire camp stove, filled with jet-engine fuel, on top of the fire. There was a whoosh like a flamethrower, leaping as high as the tent roof. After that, the stove would simmer down and work just fine for perhaps two hours before another journey to the fuel dump had to be undertaken. To Jet fell the sorcerer's task of making fire. His tent, which he shared with Pablo and Ron Precious, became known as the "black lung" tent. Its interior was quickly coated and recoated with black soot that thickened into scabs. The flapping of the tent in the wind guaranteed that the soot stayed in motion, filling the tent with dark fumes and black pollen. After a while, things got so intolerable that Jet had to stagger outside to start the fires, leaving the blazing torch out on the ice while he crawled back into the tent, zipped the doors closed, and built another fire out of comic books to vaporize the air so that when he brought the stove back in, it wouldn't immediately turn into a flamethrower again.

Three times the fire almost got out of control. Once, Jet's fuel-stained survival suit began to burn. It was quickly and urgently smothered in sleeping bags. A second time, they lost half the floor of the tent before they could control it. A third time, Walrus grabbed what he thought was a jam can filled with water to throw it over the blaze, only to discover that it was the crunchy-granola can. The granola did succeed in smothering the fire.

As the wind mounted, Jet went around poring boiling water over the tent pegs and at the base of the UN flag so that the tents and flag had less chance of being torn out of the ice.

The French photographer, Jean-Claude Francolon, somehow managed to remain aloof from all this messy survival business. Even in the tent, when the food was all gone and the camp stove was empty, Jean-Claude remained supremely indifferent to the situation. Eileen had to boil the ice down into water. It took half an hour to get even a third of a soup can filled with drinkable water. Then she would pass it to him. He made no effort at any point to participate in the struggle. He was there to be fed or be ignored, to suffer or enjoy, to freeze to death or be lifted back into the sky, and it did not seem to matter much to him which part of the planet he was on. He stayed in his sleeping bag, keeping his cameras warm against his body and ceaselessly, meticulously, making fine adjustments to the equipment, dabbing at the various lenses—he had no less than five cameras—so that when the moment came to capture the image he was seeking, he *would* capture it with fantastic perfection, showing every nuance within reach of the machinery and his eye.

By the time the crew had been trapped on the island for two days, we began to get desperate about their condition. The weather reports predicted several more days of worsening winds. Fisheries officers and RCMP shrugged. There was nothing they could do—they were grounded too. Fortunately, we had a secret source of information about the weather that was superior to anything the airports had. It was "Pa" Decker himself, who sat in the kitchen at the boardinghouse, eyeing his barometer and making predictions that we quickly came to learn were awesomely accurate, sometimes to within five minutes.

It was he who saw a small trough coming, a lull in the storm that would only last a matter of hours before the main body of the winds from Labrador arrived.

"B'ye, she's gonna let up at first light. If y'go in a beeline straight t'Belle Isle and straight back, you'll make it, b'ye. But no foolin' around, mind," said Pa.

Sure enough, as the first light came stealing across the snowbanks, there was a ringing silence. Smilin' Jack and the Red Baron were airborne before the sun peeked over the horizon. The trough in the storm held just long enough. The helicopters were back with the camp crew with a fifteen-minute margin of safety before the full winds hit. The tent crew staggered into Decker's, faces blackened like Al Jolson, all of them spitting and coughing up black phlegm, and no one saying much at first about how crazy they'd all gotten once tent fever had begun to set in. "What amazed me," reported Eileen, "wasn't that we were getting just a little weird, it was how *fast* it happened that we got weird. People were really starting to freak out."

The storm that followed was described in press stories as the worst of the winter, possibly the worst in a decade. Even the telephone system broke down. It was dangerous just to try to walk a couple of blocks through drifts up to ten and twelve feet deep. Out at Belle Isle, the winds got up to close to one hundred miles per hour. The sealing fleet itself was reported stalled, unable to put any men out on the ice.

Walrus's eyes gleamed. "The winds have come up at Mother Earth's bidding," he said. When the storm finally abated, on the morning of Friday, March 19, ours were the first helicopters up in the air on a reconnaissance run to Belle Isle.

Pablo came back, reporting: "Sorry, folks, the camp's gone. Poof! Vamooshed! There's no camp anymore." Everyone thought he was putting us on, but it turned out to be true. All that was eventually salvaged were a few tatters of tent and a third of a bottle of Scotch whiskey, still unfrozen. Everything else had been torn apart and blown away over the cliffs.

It wasn't until the storm passed and the Warriors of the Rainbow were able to get back out to the ice that I got my own first glimpse of the seal-people on their vast birthing ground. By noon, we were down on the ice, some thirty miles south of the position where the first encounter had taken place. We had to land some two miles from the sealing ships because of a Canadian government regulation—called the Seal Protection Act—which forbids anyone, other than a sealer, to land anywhere within a half a mile of a seal, or to fly less than two thousand feet over the seal herds. The government claims the regulations are designed to prevent the herd from being "disturbed." That is, you may not disturb a seal unless you are definitely going to kill it. Canada's Seal Protection Act protects no seals at all—but it does protect the sealers, and George Orwell himself could not have invented a nicer title for a piece of legislation aimed at the destruction of an animal. The legislation had, in fact, been passed to keep troublemakers such as ourselves away from the slaughter.

Stepping gingerly from the Neoprene-rubber landing floats of the helicopter, its Allison engine still screaming, I had the feeling we had landed on another planet. The moment the whirling guillotines of the propellers stopped, the silence that descended was thunderous. It was only after the buzzing in my ears had ended that I began to hear the distant cries of the seal pups and their mothers. There, not too far away, was a seal.

There were tiny ice crystals clinging to its exterior like thousands of bracelets. The pup was covered, as though God loved it more than almost any other creature, in a pelt that evoked awe. The animal looked to be as helpless as a cushion. It was almost comical—Charlie Chaplin would have loved to try to imitate one.

The eyes.

The seal had eyes like the lens of the costliest, most heavy-duty camera ever invented. Black, tending to blue, and bottomless. Getting down on my knees and leaning forward to look into those eyes, I got a perfect fisheye reflection of the world as the pup must have seen it.

It was crying. It was afraid. Of course, we stood over it like giants, our helicopters lurking in the background like cyborg insect monsters with quivering proboscises. And out of these things had come bizarre entities in phosphorescent space suits, Ski-doo wraparound goggles, carrying an array of Pentaxes, motor-driven Nikons, French-made Eclair movie cameras, Bolexes, and Nagra sound-recording systems, each of which looked like a death-ray machine being carried by fire-bright orange demons.

The infant cried for its mother. But Mother was herself so freaked out that she had just leapt down a blowhole into the safety of the water below. She was not one of the ones who would stand their ground against all odds to defend their living children, helpless as eggs before the onslaught of tanks.

The landscape around us was spectacle of the highest order. We were eighty miles east of the coast of Labrador. There were fractures between the ice pans, and between the fractures, sometimes three feet wide, you could see the dark blue tides of the Atlantic rushing. We had arrived on a giant wrecked marble table in an atmosphere as severe as an operating room. Above, a blinding painful light. Below, a white more perfect than hospital sheets. It was, I suddenly realized, a sacred place. A nursery. A horizon-to-horizon bed. There were frozen birth sacs everywhere. And from these had emerged creatures astonishingly like the "shmoos" in the Li'l Abner comic strips.

The pup before me was obviously utterly useless in terms of taking care of itself. It could neither swim nor run, nor even waddle. All it could do was wiggle at roughly the speed of a turtle, like an animated pillowcase. Yet it faced me, and perhaps as its one and only defense, it cried to *me* for help, me, the advancing nightmare monster, as though death itself could be trusted in the end. There were, of course, no fathers around, for the adult male harp-seal hides out in the Arctic Ocean, leaving the mothers to come south to the nursing temple to fend for themselves against the destroyers who burst every year through the doors to begin hakapiking and skinning the babies before the mothers' eyes.

The second Jet Ranger came down on the ice. Paul Watson leapt out, followed by the rest of his squad. There were now eight of us on the ice: Ron Precious and Michael Chechik, Patrick Moore, Eileen Chivers, Jean-Claude Francolon, Watson, myself, and Doug Pilgrim, our new Newfoundland guide, the only one not wearing a Mustang survival suit. He had his own blue winter garb and he knew the ice just well enough so that he might be able to keep us from falling through the cracks and being crushed in the cold Atlantic tides pulsating beneath us.

"Bad conditions," Pilgrim yelled. Watson and Moore huddled with the pilots and synchronized their watches. Painted bright red with white trim, the sealing boats lay on the horizon like blood-caked ax heads. The pans between us and the ships were broken, with their dangerous edges often hidden in snowdrifts.

The reading we'd gotten from the *I Ching* before leaving Decker's that morning had warned that we were "Treading upon the tail of the tiger." I had thought it referred to the power of the seal hunters and the government combined, but now it was easy to see that the tiger was nature itself, ready to snuff us out in a blink if we made a single false move.

Doug Pilgrim took the lead. Yes, we were being led by a Pilgrim. "Okay, single file, and keep at least eight feet apart," yelled Watson. "Some of these chunks are gonna be small, and we don't want more than one person on them at a time, 'cause, like, I don't wanna see any of you guys goofin' off on the job 'cause ya wanna take a little dip, if ya know what I mean."

I was hanging back, still on my knees, with my face as close as possible to the seal. I saw in the lenses of its eyes that another orange space monster was coming up behind me. It was Eileen, who said: "Come on, Uncle Bob, you can pray later."

But the truth was that I hadn't stopped praying since the night before. I had been praying nonstop during the entire helicopter ride, and I had no intention of stopping now that we were on the ice. I had held the small brown pouch around my neck in my fingers and prayed to the Gyalwa Karmapa XVI, to the Buddha, to God and Christ and St. Francis, I had tried to center myself by visualizing the faces of every friend and loved one I had ever known, had tried slow Yoga breathing exercises, had remembered the ceremony in the Kwakiutl longhouse, the fog rainbow that had appeared like a halo over the *Phyllis Cormack*, and, although in the months between the whale expedition and this particular trip I had tried to be rational and businesslike and had left the mystical stage behind, it was time again to fill myself up with a sense of the holiness and sacredness of all life.

Maybe I was praying too much. We had scarcely gotten going across the ice pans when my left leg sank through a patch of snow and kept going, down into the Atlantic. I managed to throw myself forward onto the pan ahead and was pulled out by several strong arms. From the thigh on down, my left leg was soaked, and my boot was full of water. Furiously,

I emptied the boot, wrung the sock out, and put everything back on. Within minutes, we were moving across the ice again, more warily than ever.

It took close to an hour before the ice around us was stained with blood and the seal hunters themselves were close enough so that we could make out their faces. Many of them had dabbed blood around their eyes to reduce the glare of the sunlight off the ice. Heavy snotty icicles hung from their moustaches.

It was the power of sisterhood in the temporal form of Eileen Chivers that led the attack. While the rest of the group had become bogged down for a moment among all the carnage—talking to sealers and fisheries officers—Eileen had kept her cool. Around her everywhere were lifeless skinned babies and crazed-out mothers, purple wads of gut, the pink steaming inner stuff spilling out of twitching harp-seal pups split in half on their backs.

She was the first of her sex to come to this place where, for centuries, Newfoundland males had entered their manhood by steeling themselves to kill the most beautiful infant creature they had ever seen. Eileen had kept herself centered by focusing on the ships and the *men* with their knives and hakapiks, their harsh stubbly faces, their flowing physical power. Then, ahead, she saw a seal pup wiggling frantically. A man was crunching across the ice in its wake, not even having to run, yet coming up on it swiftly, and the pup's mother was bobbing up and down in a nearby blowhole, powerless to answer the desperate wailing of the infant that was not long ago safe inside her. Eileen moved at first almost in a dream state. Then she caught her first full look at the pup.

It had turned around and was snapping at its destroyer, even though it could not hope to reach with its small teeth any higher than the man's boots. In that instant, it was no longer a seal, but a small defiant boy with fantastic courage. Eileen began to run. She slipped on the ice and failed on that first try. Not five feet away, the hakapik came down and the blood splashed several feet across the ice and the seal-boy began his series of throbbing death contractions. The next seal hunter found himself blocked by a furious young woman, breathing harshly and gasping at him: "No! No! No!" He elbowed her aside and her light body could not resist the rocky muscles of his arm. But she was still at his side, crowding him, too close for him to get a clear swing. It was only when she slipped again that he got his chance, and the hakapik came down again. A dozen times Eileen fell or was brushed aside, and a dozen times pups died almost within her reach. But a dozen times she got back to her feet, scrambling to block the way again. Not far away, Jean-Claude Francolon was there to capture the moment he had been waiting for, when a woman from the twentieth century would rise between a man armed with primitive killing tools and an animal that died in its infancy to service equally primitive female vanity, a new age in collision with darkest antiquity.

Meanwhile, in full view of our camera crew, Watson and I were about to take an existential leap. Chechik positioned himself on one knee with his Sennheiser-815 shotgun microphone thrusting forward to catch the last words of the martyrs. Ron Precious was crouching directly ahead, his 12-pound single mag Eclair fitted to his eye. Patrick Moore had three Pentaxes ready.

We had found a vessel called the *Arctic Endeavour* and had planted ourselves directly in front of its bow, standing with our eyes fixed straight ahead, our arms across our chests in very firm positions. This was not to be a mere game of chicken. We had turned our backs on the enemy, and if he wanted to split us in half, it would be entirely up to him. There would be no fight.

On the starboard side of the vessel, a crewman was hurriedly securing lines to a bundle of blood-stained pelts. He yelled at us: "Ya better move, b'yes, the ole man ain't one ta tink twice bout runnin' ya into the ice."

I yelled back: "Tell the old bastard to do what he wants, we're not moving."

The ship backed up. Then she reversed engines and plunged forward. Everybody heard her coming. Everybody felt her coming. The vibrations of the diesel engines ruptured the air and tingled the soles of our feet, even through the thickness of arctic boots. The ice trembled, cracked, crunched, shattered. Blocks of chunky ice tumbled upward, stabbing like thick blunt broadswords. Dark lightning bolts of fracturing ice leapt outward from the steel bow as it reared over us. We stood there with our heads slightly bowed, as though before a guillotine. From far above on the deck, we heard someone yell: "Stop 'er, Cap! Stop 'er! The stupid asses ain't a-movin'!"

The engines were cut and reversed frantically. The bow of the *Arctic Endeavour* ground to a halt less than ten feet from our backs. Watson had writhed a bit during the exercise but had stood his ground. My body had trembled vigorously, but I had kept my mind centered as much as possible on the Clear Light, a Tibetan Buddhist meditation technique. It was a bit like meditating while squatting on a train track in front of an onrushing locomotive, not the most serene environment, but it fended off most of the fear of dying. And at that moment, with the vast panorama of ice and carcasses, hakapiks rising and falling, the shrieks of the seals—the whole destruction of the natural world crystallized and outlined in blood before our eyes—it seemed we had arrived in the midst of a tremendous battlefield. What we did now had a weary sense of inevitability to it. As the ship ground to a halt, Michael Chechik leapt forward with his shotgun microphone.

"Could you tell us what you're doing, Bob?"

"I'm standing in front of a boat, Michael. We'll talk about it later."

The skipper of the *Arctic Endeavour* was by this time in a very bad mood. His crewmen stood on the deck, screaming obscenities at us. Moore shook his fist back at them and howled righteously: "You'd kill a man, you fuckers! You'd kill a man, wouldn't you?"

The ship began to draw back, this time a full forty feet, like a bull retreating to the other side of the ring to gain momentum.

"She's gonna come fast this time," Watson said. "It's a crummy way to die."

"*Just don't look back, Paul.*"

Everyone present who was able to see it agreed later that the *Arctic Endeavour*'s last charge was magnificent. She made a splendid full-engined roaring, vibrating lunge through the trench she'd already broken, and the ice heaved and began to buckle under our feet, and I could tell from the faces of our friends in front of us, from the way their eyes widened, that they thought we were dead, yet, ironically, I had never felt stronger or more certain in my life. That ship could *not* get through us. It was possible to *will* it to stop, and just as a chunk of ice the size of an automobile burst upward not five feet to my left, cutting off any avenue of escape, the cry from the deck came: "Stop 'er, Cap! *Stop 'er*! They *still* ain't a-movin'!"

There were howls of fury from the wheelhouse and confused, urgent shouts from the engine room, but, shudderingly, as the last glaciers of the Ice Age must have stopped when they finally surrendered to the heat and light, the sealing ship wobbled to a dazed halt, with the margin of room left between its bow and our backs down to roughly a yard.

All of us entered a new psychological realm at that point. The icebreaker had stopped as though it had impacted against an iron wall. Of course, the ship had only stopped because the captain had made her stop, but the captain had made her stop because *we* had made him do it.

Any way you looked at it, we had succeeded in halting a sealing ship in its tracks. It was the first time they had ever been stopped by anything other than storms and mechanical failures and ice.

Everywhere on the horizon the other ships were still moving, almost as though grazing. We had not slowed the overall operation by much more than a fraction, but it was a start, it was a momentary victory, it had been good guerrilla theater, and it was bound to create some great media images. That was the practical view of it; but there was also a shivery, excited feeling that we had tapped some fairly cosmic sources of energy in the process, that angels had guarded our backs. A particularly overwhelming case of paranoid grandiosity seized us then. It seemed our bodies had grown gigantic and the ships had shrunk to the size of toys that could be pushed away at will. It was a power rush of considerable proportions. In the final few seconds before the ship had stopped, and the sound of the ice had been like a mountain-sized molar breaking apart in the middle of my head, I had felt a blaze of anger like a blowtorch. It was righteous wrath. It brought with it absolute conviction and an ecstatic, exultant feeling of strength.

But we could keep the ship pinned down no more than half an hour. By then, Bernd Firnung was warning that the winds were picking up and if we didn't head back now it was going to be too late.

We made an orderly Boy Scout–like trek back across the ice pans to the helicopters, so distant that they looked like nothing more than insects, and there was always the danger, of course, that cracks might open up in the ice, separating us from the machines. Captain Jack had stayed behind to make sure there would be at least one pilot to come to our rescue, but the tension of knowing that you are on ice that is only a thin scab over the great Atlantic itself, and that we had seventy miles to fly yet in our little air machines against winds that could blow us around like bubbles, meant that no one's adrenaline level would come down until long after we were safely back at Decker's.

The Jet Rangers bumped and wobbled along three thousand feet up in brilliant golden light. We seemed to be inching our way with painful slowness across horizon-to-horizon ice that was turning dark blue as the sun sank ahead of us. No one said much. Our minds were still filled with vividly etched images of seal pups and whole highways of blood where piles of pelts had been winched across the ice, and even though we were in new danger, our minds couldn't stop replaying the scenes that had occurred back on the ice.

298

None of us had ever seen such profound contrasts of beauty and savagery. "So that's what they mean by a *mass* slaughter," commented Moore, his voice exhausted.

By the time we got back to Decker's, the sun was a red ball half sunk on the horizon, the lights were already coming on in town, and Smilin' Jack had only seven gallons of fuel remaining.

"Well boys, that's the closest I've ever cut it," he said, not smiling. But we had stopped them—for however brief a moment. We had stopped them.

Back out at the Front, the ships came to a halt as darkness settled over the field of battle. The men filed into the galleys sullenly. An American reporter who was on one of the Norwegian ships told us later:

"They kept telling themselves that they'd won, how they'd scared the whole pack of you off, and that you were the stupidest bunch of sons of bitches they'd ever seen. You were just publicity hounds. But they certainly weren't having a victory party. No one would say it, but they knew they'd lost *something*, the initiative, maybe. They were feeling it. There was a lot of drinking and a few fights. It was a sad scene, actually."

The next day, March 20, was the coldest day yet, but the helicopters were up in the air by 8:00 A.M., heading directly for the fleets. This time the seals were spread out over a wider area, and this time three fisheries-department helicopters were following right behind, obviously waiting to try to catch the Greenpeace choppers coming down too close to a seal. The choppers circled for twenty-five minutes before the pilots could pick out a spot where they were certain there were no seals nearby.

But no sooner had the crew set foot on the ice than the fisheries choppers came pouncing down, and a fisheries officer came straight over to advise Smilin' Jack and the Red Baron that they had flown lower than two thousand feet over the seal herd and that they had landed only a quarter of a mile from a seal, and had thereby violated the Seal Protection Act. Both Firnung and the captain argued furiously, and when they demanded to see the seal in question, the reply from the officers was that the seal that had been on the ice when they landed had now gone back into the sea.

With Marilyn Kaga beside him, Watson led the rest of his crew across the ice toward the distant ships, leaving the pilots to debate with the fisheries authorities.

This time the crew found itself approaching the fleets from behind, passing through an area where the hunters had already been. The ice was pitted where hot blood had melted the surface before coagulating and freezing over. Only the heads of the animals were recognizable once they had been gutted and stripped of their hides; the eyes were dull and fogged. Only the whiskers around the nose and a tuft of fur under the chin remained of the normal exterior. Snow had begun to drift around the carcasses. Here and there, whimpering and trying to hide their heads in the snow, were a few scattered survivors and dozens of mothers still sniffing at the small skinned bodies, nudging with their noses, making mewling sounds.

Then with permission from the fisheries officers, the pilots leapt over the intervening broken ice and came down beside the small band picking its way toward the ships.

"There's a whiteout coming!" Captain Jack yelled. "Come on, everybody on board. We've got to get out of here!"

They flew low, not daring to climb into the twisting currents of wind. It seemed like a long time before the cliffs of Newfoundland were visible ahead.

Late that afternoon, the RCMP arrived at Decker's to formally seize the helicopters. Jack Wallace and Bernd Firnung were charged with two crimes: violation of the Seal Protection Act for "unlawfully landing a helicopter less than one half nautical mile from a seal that is on the ice," and "unlawfully operating a helicopter over the seals on the ice at an altitude of less than 2,000 feet, contrary to Section 12 (5) (b)" of the same regulations.

Two Mounties set up a rope fence around the choppers, and for the next two days they rotated a two-man guard over the machines, day and night. Mostly they sat huddled in their truck beside the machines, avoiding the wind, sipping from a thermos of hot coffee. We were advised that we would have to post bonds for a total of twenty thousand dollars before the machines would be released.

It took us two days to arrange for the bonds, mainly by signing forms and giving personal guarantees, even though none of us had any money.

By the time we were able to get back up in the air three days later, the fleets had moved northward, just slightly out of range. We took one more run out to find them, but finally had to turn back, still not having spotted a single ship.

We learned that the captain of the *Arctic Endeavour* had had a minor heart attack two days after the confrontation during which we'd stood in front of his boat. He had been lifted out by helicopter and flown to hospital in St. Anthony. "Bad karma," someone muttered.

It remained that there was nothing more we could do. During the time we had been grounded by the Mounties, we had talked to hundreds of people. Carl Rising-Moore, Dan Willens, and myself had driven to several nearby villages to explain our position. In one small outport called Raleigh, no less than seventy of the villagers signed a paper indicating they were willing to come out with us to oppose the large commercial hunt.

Most of the fisheries-department people had left St. Anthony, along with almost all the media. Richard Cashin and the other union officials had gone back to St. John's. The excitement was pretty much over, and none of the wire services were particularly interested in a story to the effect that Newfoundlanders themselves were willing to join with eco-freaks in opposition to the great mass slaughter on the ice.

On the day of our last flight out over the ocean, support from the local people had expressed itself in the form of Emily "Ma" Decker, the pixie-sized grandmother who had been industriously feeding us since the day of our arrival, and who now volunteered to actually climb into a helicopter to express her own personal protest against the hunt. Ma had never been in a chopper in her life, and she was more than a little nervous, but she was willing to go right out there. Another first: Newfoundland womanhood against the slaughter. No soft, spoiled child of the cities, either, but a tough and unsentimental lady of a tough and unsentimental land. If even Ma Decker was against the hunt, then there were currents in Newfoundland itself that might yet spell the end of an era.

If we had, in fact, saved any seals, it would only have been a handful. But we had made political inroads inside Newfoundland, acquired allies, divided the opposition, and brought the great glass eyes of the mass-communication system to bear with more intensity than ever before.

There were a couple of surprises waiting for us at home. First, Bobbi MacDonald had found us a new office, which gave us four or five times as much space as we'd had before.

The second surprise was that George Korotva had tracked down exactly the ship we needed to pursue the whalers again.

After having searched through boat brokers in places as far away as Holland and Japan, Korotva had finally stumbled across a lead that took him to Seattle, only a couple of hundred miles south of Vancouver. Looming high over the dock was a ship that was the exact replica of the vessel that had rammed David McTaggart at Mururoa. It was a minesweeper, the *James Bay*, 150 feet long, one of the decommissioned "Bay"-class ships that had been built by the Canadian government only to be let go before they ever got a chance to be used in a war. While most of them had been handed over gratis to the government of France, at least two remained on the West Coast. One of them we were familiar with: the *Edgewater Fortune*, which had attempted to get to Amchitka. The *James Bay* had been purchased by a shrewd businessman, who had stripped away most of the high-powered electronic equipment, then sold what remained to a retired American executive who had fantasies of fixing the boat up on weekends, with a little help from his family, and chartering it the rest of the time. The new owner, Charles Davis, had not really bargained for the amount of work he had on his hands. So the *James Bay* was sitting in Seattle, perceptibly beginning to rot. Yet her wooden hull was still in excellent shape and her engines were basically sound. And she could easily travel faster than the whaling fleet—provided one could afford the fuel.

There was not much chance to rest after getting back from St. Anthony. We had only two months in which to put the *James Bay*, which we would rename *Greenpeace VII*, into shape to take her out into the North Pacific. Since, as usual, we didn't have the money we needed to simply charter the boat, we had to work out a complex agreement wherein we undertook to put fifty thousand dollars worth of work into the craft ourselves, using Greenpeace volunteers, rebuilding the interior almost completely, covering all the insurance, fixing up the engines, and replacing all the missing electronic gear.

At the same time, we had to deal with what then seemed to be a tremendous political opportunity. The United Nations was to stage the largest conference it had ever organized, called "Habitat," to try to come to terms with the problems of man and his environment. It would be held in Vancouver in early June. The site would be the abandoned air-force hangars at Jericho, which had been chosen because of the obvious success of the launching of the first Greenpeace antiwhaling expedition. The Jericho hangars were now being transformed at a cost of one million dollars, and the man in charge of the work was Al Clapp, the organizer who had put together our launching the previous summer. It was on the basis of this success that he had been hired by Ottawa to prepare a conference site that would impress the delegates from 120 countries. The great echoing hangars would be changed into theaters and forums; tapestries would cover the walls on the inside, and gigantic Indian symbols would be painted on the outside. One hangar would even include the world's longest bar. Almost everything was being constructed from recycled materials to demonstrate a basic conservation point. This would be the first major UN conference since the one in Stockholm in 1972 that had passed a motion calling for a ten-year moratorium on whaling and that had also opposed French nuclear testing in the South Pacific.

It was as though the mountain had come to Muhammad. Greenpeace had always flown a UN flag and it could not have been more appropriate, from our point of view, for the UN to come to Vancouver to address itself to the question of man's survival. It took a fair amount of negotiating and submitting of briefs, but in the end we managed to get the United Nations to agree to extend the conference two days—until Sunday, June 12, so that the launching of the *Greenpeace VII* on her mission to save the whales would be the grand finale of the whole exercise. We would also have a chance to try to get a further UN resolution approved reiterating its stand in favor of an end to whaling, and if we could get that, we would, in effect, be setting out this time with the formal blessing of the UN.

Representatives of almost every Indian tribe in North America would be present. The elders of the Hopi tribe would be conducting special ceremonies on the day of our launching—an "earth-healing" ceremony. It was as though not just one wheel was turning, but many, and large ones at that. In exchange for these special privileges at the conference, all we had to agree to do, other than launch the boat, was to construct a traditional Northwest Coast Indian barbecue pit—something I agreed to without thinking twice, only to discover that a traditional "pit" was actually a structure the size of a small barn, and that it would require forklifts and at least twenty workers to build it. Of course it had to be completed in time for the conference. What with all the work that had to be done on the *James Bay*, we were taxing our volunteers to the limit— and then some. In addition to practically rebuilding an entire warship, constructing a great wooden temple out at Jericho, organizing a large office of our own, preparing displays and events for the UN conference, publishing our own newspaper, and rounding up Zodiacs, outboard engines, diving equipment, and food supplies for the voyage, the spring of 1976 found us also involved in legal actions on three fronts, and in political actions on more fronts than anyone could count. In the legal department, we were fighting the charges against our helicopter pilots through a lawyer in Newfoundland, and few experiences are more frustrating than a long-distance court battle, especially if you don't have any money to spare. Six of us were still then facing charges arising out of the previous autumn's attempt to blockade the Russian supply boat in Vancouver. But the worst legal hassles were those surrounding getting insurance for the *James Bay*, without which the owner wouldn't let us use the boat. None of us could blame him. But even Lloyd's of London was nervous about covering an old warship going out to interfere with a fleet of ships from the Soviet Union. In the same general realm, there was the question of registration. Which country would want to have anything to do with the *James Bay*? Certainly not Canada, and the U.S. was wary. Maritime regulations concerning large ships are more complex now than they have ever been in history. With passage of various bits of environmental legislation by both Congress and Parliament, travel by motorized vessel along the North American coast is a nightmare of red tape. It takes at least one full-time lawyer just to hack a path through to the dock, let alone get the boat out again without being slapped by at least a dozen different fines and penalties.

The main problem finally came down to the fact that if we registered the boat in either Canada or the U.S., we would have to pass Steamboat Inspection, a very tough seagoing test of all systems to make sure that everything on the boat was working perfectly. It was a test we knew we would fail. Yet without registration, there was no possibility of insurance, and without insurance, the trip was off so far as the owner was concerned.

The only way out was for us to borrow a trick from the oil companies and apply for Panamanian registry. It is the very trick all the leaky old oil tankers use, Panamanian standards being so low. Telexes began to fly between Vancouver and Panama City. The long distance phone bills mounted with awesome speed. Deposits were required. Guarantees. There were fees and more fees. At one point there were nine lawyers at work.

As vice-president in charge of operations, it was George Korotva's job to pull the boat together. That meant countless parts for the engines and electronic gear, countless trips through marine-supply shops and warehouses. It meant getting engineers, and no professional engineer comes free. It meant hiring welders and shipwrights and marine-radio specialists whose jobs are so intricate that they travel in teams. It meant bookkeeping and begging friends for help, borrowing money, haggling over prices, looking for secondhand parts because new ones were too expensive. It meant bullying and cajoling to get tired people to work harder or to get parts to arrive on time or to fend off one bill for another couple of days in order to pay an angry creditor who was coming to reclaim the stuff he'd sold a month before on the strength of a handshake and a promise. It meant roaring over the phone at any one of the dozens of flinty-eyed local businessmen whose line was: "Look, I think what you're doing is great, but I'm sorry, cash in advance or no deal."

The only sources of income we had were T-shirts, posters, bumper stickers, buttons, our newspaper, membership cards, a computerized mailing list, and our access to the local media, particularly the radio stations where we were permitted to beg for money, volunteers, equipment. At that point our prestige at home had never been higher. And there could be no doubt about it: membership was booming, up from a handful in late 1975 to close to ten thousand by early 1976. From one branch—with tentative connections in England, New Zealand, and Australia—in 1972, we had grown to twenty-eight scattered across North America by May 1976.

By spring of 1976, a Space Age community had sprung up around Greenpeace. Its members were linked by telephone and the mails, by car, by train, by jet. There was Margaret Tilbury in Portland, John Bennett in Toronto, Peter King in Victoria, Bob Thomas in Ottawa, Ottar Ottoson in Reykjavík, Carol Koury in Boston, Thierry Garby-Lacrouts in Paris, Lowry Toombes in Inuvik, Michael Manolson in Montreal. We were a long-distance subculture, complete with gods and demons, heroes and villains, factions within factions, an ecosystem within itself. We had all seen the same slide shows, shared Xerox copies of news clippings, read many of the same books. Our communication bills were already one of the heaviest expense areas, far out of proportion to any normal corporation. But it seemed worth it: with the emergence of each new office, it was another room being added to a house that already sprawled crazily all over the world map we had up on the wall, with green pins showing the offices and contact points. We were concentrated most heavily, of course, along the West Coast of North America.

Our projected budget for the year was three hundred thousand dollars.

By this time, I had been forced to acquire a gray three-piece business suit. Most of the male board members—Korotva, Moore, Spong, Watson, and myself—had trimmed our hair to look more conservative and had taken to cleaning up our language in public. But there were stubborn rebels, like Walrus Oakenbough, Rod Marining, and Rex Weyler. "That's going too far!" Weyler howled, refusing to cut his ponytail.

Acting on the advice of our financial gurus, Peter Speck and Bill Gannon, we had hired a full-time office manager, a mature older lady named Betty Rippy, who had once performed that job for the biggest timber company in the province. She quickly became known as the "Betty Monster." The hippies who had gotten in the habit of dropping in from Fourth Avenue to hang around and talk quickly found themselves being given work to do. Any who refused were firmly led to the door. The Betty Monster almost never stopped smiling. Yet she methodically swept the office of loiterers, organized the volunteer work force, and put the fear of God into anyone caught making "unauthorized" long-distance phone calls. There was no doubt she increased our efficiency—but she was not without her critics. Many a bruised Greenpeace ego developed. There was constant pressure to fire her, grumblings and mutterings from the troops, but she was an integral part of the package. If we were going to hope to generate the kind of money we needed, we had no choice but to adopt a successful corporate model. If there were to be budgets, someone with a professional background was going to have to implement them and ride herd on everyone involved. And in order to do that she was going to have to have authority. The board handed it to her, uneasily, reluctantly, with a feeling of guilt because we knew it was our own friends who would feel her whip. Yet it had to be done. An element of chilling efficiency now entered the atmosphere of what we had come to call the "Greenpeace Factory." But if it had less soul, it was neater, less wasteful, and more people got more work done.

We made an unexpected and quite spectacular connection that was to catapult us into a whole new league. It came to us through high-level involvement with the United Nations conference. It involved an offer of one ounce of pure plutonium.

The "X" connection came to us from America. Only a few weeks remained before the start of the United Nations conference in Vancouver, and the town was rapidly filling up with advance men, organizers, assistants, and security people. The city's usual Sleepy Hollow atmosphere was being affected. There were fears that terrorists might strike. There were going to be a lot of famous political people. In fact, there was considerable nervousness in the little Greenpeace Factory over the discovery that our display would be right next to the Israeli display. The Canadian government was bringing two navy destroyers into town just for the occasion, the army was on alert, and the border crossing points were being screened with twice or three times their normal intensity.

It was against this background that the offer of one ounce of pure plutonium came through an intermediary who had in turn been approached by an underground group in the eastern U.S. somewhere. They had plutonium—it had to be stolen plutonium. And they wanted it delivered to the top spokesman for the United Nations in the midst of the Vancouver conference. The point of the exercise was to demonstrate that the material with which to build nuclear weapons had already passed beyond the control of governments, that, security around nuclear plants was sloppy, and that the days of nuclear terrorism were only a step away. The group involved wanted Greenpeace to be the organization to hand the stuff over to the UN, partially because we were already insiders at the conference, but also because we had a high profile as nonviolent antinuclear crusaders. We would be believed when we said we had no intention of hurting anybody.

Or would we?

Initially, Patrick Moore and I were the only ones to know of the offer. We didn't dare talk about it even in my office. Instead, we walked around outside, still speaking in whispers and looking nervously over our shoulders. There were a lot of risks involved. First, we'd have to become involved in a complex plot to smuggle the stuff across the border at a time when border checks were frequent. Then we would have to "hold" it until the right moment, get it out to Jericho and into the hands of the UN without word leaking out that we had it, which could result in either a premature jump by the Mounties or maybe even the army—leaving us in the situation of being caught with stolen plutonium before we could make our point—or, worse, having the stuff grabbed from us en route by a real terrorist group. There were some fairly serious penalties possible in all of this, the least of which would be public mischief, the worst—well, who knew?

There followed several days of intense agonizing.

In the meantime, the *James Bay* was ordered removed from the dock where we had tied her up in downtown False Creek. She was too long. Her stern thrust out into the harbor, cutting off maneuvering space for the smaller fish boats. Frantic phone calls and lobbying among local politicians to put pressure on the harbor master finally led to our being

307

given a temporary berth in Burrard Inlet, just around the corner from where the Canadian navy had tied up her destroyers. This worked enormously to our advantage, since the bored, restless young sailors and technicians had nothing much to do except hang around our boat, helping out, frequently "borrowing" equipment from the navy and slipping it aboard the *James Bay*. George Korotva, Al Hewitt, and Patrick Moore labored like Trojans, leading their team of renegade sailors and local volunteers in an assault on the old minesweeper that saw her practically disassembled and put back together again in record time.

And while the chain saws shrieked and the welding arcs threw strange shadows against the walls of the engine room, a whole other area of complexity entered our lives. It came in the form of a decision to turn the *James Bay* send-off into a major benefit concert, using the United Nations facilities at the Jericho hangars. The musicians who agreed to involve themselves included Country Joe McDonald, Susan Jacks, Ronee Blakely, Paul Horn, Bruce Miller, Valdy, Goose Creek Symphony, Pied Pumpkin, Danny O'Keefe, the Paul Winter Consort, and our own favorite Vancouver group, the Cement City Cowboys, who were also doubling as security around the Jericho site, riding back and forth on their horses. No benefit concert of any size just "happens." It involves dealing with agents, the entertainers themselves, promoters, lawyers; it involves arranging transportation and accommodation; it involves printing of tickets, security arrangements, police, customs officials, and that final juggernaut, the weather. The great advantage of using the Jericho hangars was that if it rained, we'd move the whole thing indoors.

Between rebuilding the *James Bay*, mounting displays and arranging for lobbying at the UN conference, building the barbecue pit, arranging for insurance and Panamanian registry, fighting our other court cases, trying to keep the financing for all this on track, picking a crew from the three hundred applicants from all over the world, maintaining our office, mail-outs, and merchandising, and setting up a major benefit concert, the last thing it seemed at the time that we needed was to become embroiled in a plutonium-smuggling caper.

As it was, we sighed with relief when word came through that we would not have to join the Newfoundland Fishermen, Food, and Allied Workers Union in a simultaneous blockade of both Vancouver and St. John's harbors against foreign fishing vessels. The blockade, which we had agreed to join back in March in St. Anthony, had been scheduled to take place by June 1, right about the time the *James Bay* was to be launched. Fortunately, the Canadian government bowed to the pressure from fishermen—and ourselves—and announced that a two-hundred-mile fishing limit would be implemented by January 1977. There was no need to back the demand up by a blockade.

The ordeal over what to do with the plutonium offer finally came down to one of those decisions where one didn't know—and knew that one never would know—whether one had turned back from a great gateway or a bottomless abyss. The timing, however ideal it might look in terms of influencing opinion against nuclear proliferation, could definitely not have been worse from the point of view of a fledgling ecology group with its resources overstrained on every front.

In the end, I opted to issue an affidavit—timed for release in the midst of the conference's debate on nuclear power—stating that I had been "personally contacted through a series of intermediaries whose names have been screened from me on a need-to-know basis with an offer to provide me with one ounce of pure plutonium." I explained at a press conference the purpose of the operation: to prove that plutonium is no longer in the hands of governments alone. And ended saying: "After serious consideration I decided that the Greenpeace Foundation could not undertake the responsibility of possessing the plutonium because of the risks involved of it falling into the wrong hands."

The story went out on the wire services. The RCMP quickly arrived to grill Moore and myself, and it looked for a while like they might send us to jail because we wouldn't divulge the name of our contact. The following week, the Canadian government changed the locks on all its nuclear installations. The justice minister of Canada expressed "concern" over the suggestion that there was black-market plutonium floating around. And whether it was just another coincidence or not, there did appear to be slight policy changes in both Canada and the United States in the months to come concerning the "nuclear terrorism" aspect of atomic power. Slight changes—that was all. But even slight changes were welcome relief.

The last weekend before the launching of the *James Bay* was like a lifetime in itself, jammed with events and situations and scenes and characters enough to keep a novelist busy with material for years.

The day of the benefit concert, the wind howled in from the west, sliding a dark hatch cover of cloud across an otherwise brilliant blue sky. By the time Country Joe McDonald climbed on the stage, half the sky was covered with swirling storm clouds and the other half held out all the hope of summer. A few flecks of rain splattered on the stage. Then he started singing a whale song he had written, and the rain stopped. The clouds paused. Then started gradually to back away. By late afternoon, delicious open sky prevailed from horizon to horizon, and possibly the best mass party ever held in Vancouver was fully underway. Had Country Joe—one of the stars of Woodstock and heroes of the antiwar movement—really turned the clouds away? I doubt that out of that crowd of fifteen thousand or whatever it was that day, that you could have found a dozen people who did not believe it completely. All around us rose the wooden hangars, transformed into massive longhouses, totem poles everywhere, tremendous silken tapestries covering whole walls, depicting every kind of Indian design. In the hills ringing the site were displays of every kind of energy-saving, nonpolluting, solar-powered, wind-powered device yet invented, scattered among tepees and sweat lodges, and the feeling was that the tribes were gathering from all over the world to undertake a sacred duty to turn the tide of industrialism. To some of us—those who had been involved a year before when we had launched the *Phyllis Cormack* from that same site, back when the hangars had been scarcely more than ruins—the arrival of all these people from around the world, all these Indians from all over the continent, the emergence of the symbols on the walls, and, lastly, the appearance of the *James Bay* out in the water directly in front of the hangars, its bow blazing with a brilliantly painted rainbow, made it as though our fantasy of a previous summer had been miraculously converted into a reality, and that somehow, accidentally, we had tripped a cosmic lever. Here were the results.

Then came the finest moment.

The *James Bay* was ready to go. The *Phyllis Cormack* was tied up at the foot of a wooden dock built especially for the occasion, ready to shuttle crew members out to the minesweeper. The flags of every nation snapped crisply in the wind above the rails along the edge of the dock. The blue UN flag topped the highest pole and likewise fluttered from both our vessels. We were sailing out this time with the official endorsement of the United Nations conference—we *were* the world community. A stage had been mounted at the top of the dock for final speeches. The last speaker was an Indian: Fred Mosquito, medicine man for the Cree people whose legends had foretold the coming of the Warriors of the Rainbow.

"You *are* the Warriors of the Rainbow," he told us.

There is a photograph taken then that shows the *Cormack* in the background, with an orca painted on her sail, a dozen national flags flying from the posts, and a cluster of mostly long-haired people around

Fred Mosquito as he speaks into the microphone, with the North Shore mountains in the background. Everyone is wearing whale buttons with cloth-made rainbows attached, and almost everyone has their head bowed: everyone looks very serious, very still, listening very hard. It was a benediction. But it was something more: what had been an idea to us was now being proclaimed by possibly the only person on earth who could proclaim it as a truth. We *were* an ancient Indian prophecy come true. Or, at least for that moment we were.

No one had planned for his arrival. No one had even known he was coming. He had just ... appeared.

Within an hour, the *James Bay*'s engines had roared to full-throated life, the vessel had shuddered, and with thirty-two men and women on board representing seven different countries, smoke exploding from the midship stack, backed up by a medicine man's blessing and an endorsement from the closest thing to a world government, the Warriors of the Rainbow threw back six-foot bow waves and once again set out from the pastel towers of Vancouver.

Impressive as our brightly painted minesweeper might have looked to the people back on the shore at Jericho, she was, in fact, a long way from being ready to go deep sea. There were a thousand mechanical problems that hadn't been solved. The electronic gear needed testing. Most of the Zodiacs themselves had not even been blown up. There simply had not been enough time. Money had been too scarce. There had been too many other jobs to do. So we only traveled as far as Vancouver Island, and nudged ourselves carefully toward a dock at the town of Sidney.

Back at home, Bill Gannon was counting the money that had come in from the concert. He knew we needed a minimum of twenty-seven thousand dollars just to pay off enough of our outstanding debts to allow the boat to go any farther. If the concert didn't generate that kind of money we would not be able to fuel up or buy the extra parts that we still needed to make the boat seaworthy. After working all night with Lorene Vickberg, our sympathetic loans officer from the Royal Bank of Canada, Gannon had been able to grab a couple of hours of peaceful sleep, knowing that when the switchboard started lighting up at 9:00 A.M., he'd be able to announce that everything was fine, everyone would be paid. A Loomis armored truck was on its way from the Jericho hangars to the bank with precisely twenty-seven thousand dollars. That's how much we needed, that's how much we got. Bill Gannon's eyes twinkled. "Cosmic accounting," he called it.

The crew this time included quite a few veterans from previous expeditions: Korotva, our captain, Moore, Hewitt, Watson, Oakenbough, Gary Zimmerman, Melville Gregory, Rex Weyler, Fred Easton, Ron Precious, Rod Marining, Matt Herron, Taeko Miwa, Eileen Chivers, and Marilyn Kaga. The new recruits included some people from branch offices, like Bob Thomas and Alan Wade, a folksinger, from Ottawa, Michael Manolson from Montreal and his girlfriend, Mary Lee. There was also Kazumi Tanaka, a young photographer from Tokyo; Mike Bailey, a young sheet-metal worker with enough energy for ten men; David Wise from Woodstock, New York; Chris Aikenhead, a cheerful, shaggy sound technician; Bree Drummond; Lance Cowan, our helmsman, an intense professional seaman who wore earrings and looked like a pirate; Don Webb, our doctor, whom we quickly came to name "Doc Spider"; Susi Leger, a magnificently structured young blonde lady from Newfoundland; most of them in their early twenties. Two engineers had been hired: an old sea dog named Ted Haggarty who had once served as chief engineer for not only the *James Bay* but the whole West Coast minesweeper fleet, and an aloof technician named Bruce Kerr. Finally, we had been joined by two other people who had done more than any of the others to initiate the first antiwhaling expedition, but had not traveled on the boat: Paul Spong and my bride of one month, Bobbi.

There was a shipboard joke that referred to the current voyage as "Captain Korotva's Honeymoon Cruise," which summed up part of the style of traveling. We had opted to create a communal family-like atmosphere on board, which meant that each woman on board had to have a mate with her so that everyone would know where they stood. Of the couples on board, only Bobbi and I were actually married, but the rest were already living with each other. For each couple, we'd built double bunks. Patrick Moore and Eileen Chivers shared one. Paul Watson and Marilyn Kaga shared another. Rod Marining and Bree Drummond. Al Hewitt and Susi Leger. Michael Manolson and Mary Lee. Walrus Oakenbough and Taeko Miwa. We felt rather pleased with ourselves for having come up with such an excellent solution to an ancient problem. The bachelor males all had their single bunks to go to.

It was a good system. And it almost worked. Only three things went wrong. Within a month, two of the women discovered they were pregnant.

And one of the couples broke up, the lady switching from her boyfriend to the captain, and all hell broke loose ... but that was much later.

Captain Korotva's Honeymoon Cruise had something else going for it besides double bunks, rainbows painted on the side, and a doctor who fortunately turned out to be a gynecologist. It also had "Black Ron" Papadropolis, a professional chef who normally worked at a luxury hotel on Vancouver Island, the same hotel in front of which the gray whales cavorted every summer. Black Ron was so named because of his temper, which was so violent that it actually frightened some of the more gentle young souls. Everyone had to admit that the food was deluxe. Not expensive, but brilliantly prepared. It was too bad about Black Ron's temper. It wasn't long before the rest of the galley crew—David Wise, Bobbi, Eileen—were in open rebellion against him, and he had to be asked to leave.

There was one other character on board whose temper became legend: George Korotva himself, "Captain Cruel." The Czechoslovakian had rarely displayed his temper the year before, but now that the responsibility for the whole ship rested on his shoulders, his personality underwent a rapid change. It was not just that he was nervous about handling the boat—he was—it was also that most of his crew were, as he put it, "fokking hippies," and they frankly drove him to distraction. Everybody on board was a rebel or iconoclast of some kind. About the only way to get them to follow orders was to terrorize them. It was a lesson John Cormack had learned long before. In the absence of a formal military hierarchy, it was only by force of personality—raw dominance—that a working crew could be fashioned out of the human material that existed, men and women alike. There was one more problem: when he became excited or angry, Korotva's accent would thicken, and when he bellowed furiously from the bridge at his deckhands, half the time they couldn't understand him. We quickly learned to judge the quality of our various attempts to dock the ship according to how many times we'd hear the captain scream. A "seven-scream" landing was a bad one, although not the worst. A "two-scream" landing was considered a miracle. A dock would be hurling toward us, Captain Cruel would be on the flying bridge, his face reddish-purple, roaring at the top of his lungs, and there would be as many as a dozen deckhands standing with ropes in their hands, staring at him blankly with a sense of rising panic as disaster drew closer. Somehow or other, at the last minute, we'd make it—with a couple of exceptions where we did crash into docks.

Even by the time we arrived in Sidney—less than a day's sailing from Vancouver—a strong tribal feeling had developed. The trouble with that, as McLuhan has pointed out, is that tribalization guarantees maximum disagreement on every point. It is the very opposite of lining people up in rows so that they are not endlessly, restlessly engaging one another at dozens of levels. Relations all became very in depth, very complex. To make matters even livelier, we were joined almost immediately in Sidney by the *Phyllis Cormack*, with a boatload of American whale savers called the Mendocino Whale War Committee.

At the last minute, old John Cormack had not been able to resist the urge to go out and do battle again. He had been under heavy pressure from the fish-packing company who held the mortgage on his boat to stay in B.C. waters this year and earn some money fishing. But he was getting addicted to these crazy Greenpeace voyages. Much more interesting than hauling nets around in the water. We had not been able to offer him any charter money, because we were already beyond the limits of our resources, but it so happened that the Mendocino Whale War people wanted to take his boat out to join us.

We'd first connected in California some months before while Bobbi and I had gone down to put the San Francisco office together. We'd been invited to the village of Mendocino, perched on cliffs overlooking the very place where we'd encountered the Russians the year before. Led by a huge craggy-faced sculptor named Byrd Baker—wearing a seaman's cap and jacket, looking like a character out of a Jack London novel—the Mendocino art community had banded together into the Whale War Committee, promptly published a newspaper, collected a bunch of memberships and money and announced that they were going out to "fight the Russians." Byrd Baker was a natural media personality. A deep, rasping voice, steel-gray eyes, jutting jaw, hawk nose. He was an evangelist. It was not just that he wanted to save whales, he wanted to save *God*'s whales. If any of us thought that our own Greenpeace approach tended at times to possess mystical overtones, we were rationalists compared to "The Byrd." Whenever a camera or a microphone came near him, he'd roar: "Save God's whales!" For a while, the Mendocino Whale War Committee had limited itself to having a woman broadcast appeals in Russian to the ships out beyond the horizon, urging Russian sailors to throw down their harpoons and come ashore where they'd be guaranteed sanctuary. But soon Byrd Baker grew more ambitious and decided he wanted to do as we had done: get in front of a harpoon. Unable to find a boat to take him and his fellow Mendocino warriors out to the rescue, he'd contacted us. We, in turn, had put him in touch with John Cormack. And so by June 14, 1976, the *James Bay* and the *Phyllis Cormack* were sitting on opposite sides of a dock in Sidney, on the edge of Vancouver Island, the *Cormack* looking as small beside the *James Bay* as *Vega* had looked the previous summer beside the old halibut seiner herself. It was Cormack's turn to be skipper of the smaller boat, but if he had any feelings of inferiority about it, he covered them up admirably by jeering at Korotva's docking style. Between the young and the old skipper there was a rivalry that was probably inevitable.

The bows of both vessels blazed with rainbows. On the side of the *James Bay* printed in both English and Japanese was the message: SAVE THE WHALES, SAVE THE EARTH, with a sperm whale and her calf painted on the hull above. We had borrowed the design of the rainbows from the U.S. Coast Guard, which meant that when we entered American waters we were often taken at a distance to *be* Coast Guard, something we didn't mind, since the Coast Guard's motto was: "Save people from the ocean, save the ocean from the people." The Coast Guard would prove to be very sympathetic to what we were doing anyway. For an organization that had once been busted by that same Coast Guard, it was an odd experience to have them on our side.

On the hull of the *Phyllis Cormack*, behind the rainbow, Byrd Baker laboriously painted the words: SAVE GOD'S WHALES. Cormack was furious. "That Byrd Baker fella, he's a regular dingbat." But there were worse problems to worry about than the Mendocino whale warrior's religious notions. There was the bad news that the fish-packing company was sending a sheriff to seize Cormack's boat, drag him back to Vancouver, and force him to go out fishing. "Hell with it," said Cormack, squinting. "Once I'm out of Canadian waters there's not a damn thing they can do." Promptly, he solved the problem by taking off at dawn just a couple of days after we had arrived and heading straight down toward the Mendocino Ridge to mount a patrol, just in case there were any Russians sneaking around. Byrd Baker and his crew waved good-bye. Two hours later, the sheriff showed up. By then, our crew had turned wearily to the task of finishing off the work on the engines and Zodiacs, still not getting very much help from their bow-case of a leader: I was so far gone by that time that I was refusing to eat on the grounds that "I don't need it." Dr. Webb was beginning to worry about my health, and Bobbi was becoming frantic. "Some honeymoon," she muttered.

Paul Spong reported that the Russian whaling fleets were not operating according to pattern. Normally, by this time of year, they had swept past the Hawaiian Islands and were deep into the eastern North Pacific en route to the edge of the continental shelf. This summer they seemed to be holding back. From our point of view, that was good news. It meant we were being given the time we needed to finish the work on the ship. But it also meant we were doomed either way to spend more time in port, since just running about at sea, burning fuel, was too ghastly an expense to contemplate.

We were, in fact, face-to-face with one of the great limitations of naval warfare until the arrival of nuclear power: fuel consumption.

In theory, the *James Bay* was capable of getting us all the way to Japan, although we would be foolhardy in the extreme to venture into the middle of the ocean in what was basically a coastal vessel beyond late summer, when the typhoons began. There were many observers who figured we were crazy as it was to try to take the ship so much farther out to sea than the Canadian navy had ever tried. But our plans were totally open-ended: we would go wherever we had to go, or could go, in order to find whalers, whether Japanese or Russian, although the chances of finding the Japanese were slim indeed since they used modern telex-like communications systems which couldn't be picked up by our RDF in the way that the Russians could.

We did not get out of Sidney until June 19, five days after we'd arrived, and then only with the help of the Sidney Volunteer Fire Department, who agreed to fill up our water tanks, despite a severe local drought. As usual, the moment we were at sea, life became more tolerable. Port stops had a way of bringing out negative energy. So long as there were regular shifts, regular eating times, regular watches, regular duties, plus the sobering thought that if any of us neglected our work everyone was liable to drown, we were all right. But tied up to a dock, with nothing much to do except focus all of our incredible energy on one another, we came apart like a jigsaw puzzle dropped from a great height. Much of the blame for the chaos and tension rested with myself for failing to be able to keep a firm grasp on the situation, but mostly the blame rested with the situation itself: everyone on board was to one degree or another frightened. Could Korotva *really* run this boat? The captain shared the doubt himself. That was why he tended to be completely uncritical about the fact that I had turned into a bow-case, even though we were supposed to be the basic operating team who would keep the ship on track and everybody safe. Korotva just wanted me along as a good-luck charm and a drinking buddy. In that sense, so far as I was concerned he was the least-demanding person on board. Repeated attempts were made to get me to leave the ship. It was always for "the good of the expedition," for the good of Greenpeace, for the good of the whales, for the good of the planet. At any one of a dozen points I would gladly have left, but Korotva would look me in the eye and say: "If you go, I go." In order to keep the captain on board, I had to stay. In my completely blown-out state of mind, that was about the only responsibility I could still appreciate. The rest all seemed meaningless.

Therefore guided by a blustering Czechoslovakian, left leaderless by a bow-case, carrying Canadians, Americans, and Japanese, male and female, with seven couples and seven Zodiacs on board, flying the United Nations and Panamanian flags, the seventh seagoing Greenpeace expedition passed through the Juan de Fuca Strait and—briefly—into the open North Pacific, burning fuel at a rate that in itself was enough to make you cry. In the galley a wooden model of TV's starship *Enterprise* dangled on a string from the ceiling, and depending on how much it swung back and forth, the crew could determine the weather. They didn't do it in nautical terms, such as Gale Force 6. Rather, they would relate to it as Warp Factor 6, taking their cue from Captain Kirk rather than Captain Korotva. The ship also got a new job description. Instead of being a minesweeper, the *James Bay* would be known as a *mind*sweeper. The movie camera crew was braced. The photographers—from East and West—were ready. The kamikazes were all in their places. The engineers and their team of lady oilers had the engines throbbing and purring down below. Meals were exquisite: a setting for vegetarians, who were the majority at the beginning of the trip and a minority at the end, and another setting for the meat-eaters.

There was one other passenger on board the ship who was, at some levels, the most important entity of all. She was a three-foot long iguana known as Fido, Melville Gregory's favorite pet. Fido's skin changed color according to mood and the vibes around her. There was much evidence that Fido did not like being at sea at all. So what were a bunch of animal lovers doing taking a reluctant lady iguana out into the North Pacific? The answer was simply that Gregory had insisted. It was the Oriental Year of the Dragon. Dragons and whales were always linked. The *I Ching* spoke highly of dragons. And there was still a chance that the *James Bay* might find its way into Tokyo Harbor that summer. The idea of entering Japan in the Year of the Dragon on a whale-saving ship with its own little dragon on board was too rich to reject. To make Fido as comfortable as possible, I built a nest in the skylight hatch above the "executive-suite," where Eileen Chivers, Patrick Moore, Bobbi, and I lived. The nest was composed of branches and boards, complete with a shelf where I could place the iguana's food: lettuce and tomatoes and cabbage and chopped-up celery. It let in lots of light and gave Fido the security of being *above* everyone else, something she seemed to appreciate. The only problem was that every four or five days, Fido would void on the floor below, leaving our running shoes, shirts, and pants covered with pale green piss. I tried to remedy the situation by taping a transparent plastic sheet under the nest, which succeeded in capturing the waste, but, when the boat lurched from one side to the other, also succeeded in splashing the green piss on all the walls, where our other clothes were hanging. As latrine officer, it became my duty to clean the "executive suite" regularly, as well as the heads where all the noniguanas left their messes.

From time to time, Fido would get tired of clinging to the branches in the nest and would decide to go exploring, only to discover that the rest of the environment—with its roaring engines, power cables, slamming hatches and heavy-footed human beings was no place for any iguana that wanted to survive for long. Outside was a world beyond iguana comprehension, or at least too hostile a realm to consider visiting. There were periods when Fido became lost for days at a time, but somehow, Gregory would always find her again. At other times Fido would seem to long for human companionship, content to climb on someone's head and hang there, long scaly claws with longer fingernails clinging to her carrier's hair. The more hair you had, the easier it was for Fido to get a grip. She therefore loved to cling to the hair of women and hippies best. One night, Fido coiled down out of the nest and went scuttling across Bobbi's naked breasts, causing my bride to smash her head on the aluminum beams above our bunk as she lifted straight into the air, screaming. Several times, Fido got loose in the galley, leaving a trail of people bumping into each other as they tried to catch the swift-moving little lizard. So far as any of us knows, Fido was the first iguana to reach the middle of the Pacific Ocean and travel in the midst of a pod of whales. And she did it during the Year of the Dragon!

Most of us wore small golden emblems in the shape of a dragon in addition to our Kwakiutl-design T-shirts. It was a source of tremendous pride to everyone involved that funky as we were, we were nonetheless driving an old warship that might fall apart around us at any moment out into the ocean to put our lives on the line to save the largest creatures that had ever lived. There might be bow-cases and egomaniacs and bad-tempered captains aboard, but the basic feeling that held us all together, whether crazed or sober, was that we were Warriors of the Rainbow, and at heart we all respected each other for being willing to take the risk of dying together in a boat whose deep-sea capability remained to be tested. The *James Bay* seemed like a wondrous starship to most of us, but she had also only been designed for offshore patrols. She had not been intended as a deep-sea traveler, and certainly the Canadian navy had never intended to take her as far as we were now considering: to Hawaii or maybe even Japan. She had a planing hull that tapered off like an ironing board only six feet below the surface. It seemed too much to hope that a vessel with ninety percent of her body above the water line could be stable in any kind of heavy sea. And we all knew we had a captain who had never commanded such a craft in his life. He was going to have to learn on the job with our lives as the price if he made a mistake.

Captain Korotva's Honeymoon Cruise did not promise to be a chance to get away from it all. Rather the opposite.

But it also held out the promise of magic. Just as we were about halfway out through the mouth of the Juan de Fuca Strait, Bobbi, Eileen, and I were sprawled on the back deck inside one of the Zodiacs, talking casually. I was almost back to being rational. But then I said, "Well, here we go. Expect whales to rise up any minute now." Eileen laughed. Bobbi paid no attention.

A couple of minutes later, a large black-and-white orca surfaced just off the port stern.

The Russians had still not pushed east of Hawaii so we decided to hole up for a few days at a place called Bamfield, on the southwest tip of Vancouver Island. It was there that the Canadian government tried to scuttle the voyage. With only an hour to go—after three days at anchor—before we planned to depart, a small Canadian Coast Guard cutter with two Mounties on board came roaring out to the *James Bay*. The sergeant in charge grasped the rope ladder we'd thrown down to him and announced: "I am seizing this vessel."

Technically, the excuse was that we had overlooked some minor bit of paperwork before leaving Vancouver. But had that been the real purpose of the arrest, it would have taken place back in Sidney. In fact, someone with a keen sense of timing had decided to wait until the last minute before we left Canadian waters to hit us. The sergeant wasted no time going straight to the locker where our bonded stock was kept. Ships leaving for foreign ports can buy bonded stock but must keep the stuff locked up, with the lock sealed, until they reach international waters. Whoever it was who'd sent the Mounties after us at the last minute had obviously calculated that it had been ten days since we'd left Vancouver and they doubted that a gang as undisciplined as ourselves could have refrained from breaking into the booze by then. Captain Korotva had indeed been hounded to open the lock, but had stubbornly refused. Now it was worth the dry spell just to see the look of dismay on the sergeant's face when he saw that the lock was still sealed.

All they could do at that stage was proceed to charge us with the minor technical infraction of having failed to fill out some customs clearance forms. It took seven hours to straighten out all the paperwork, pay a four-hundred-dollar fine, put out press releases charging the Canadian government with harassment, and finally head out through Satellite Pass into the ocean. The moment we reached the twelve-mile limit, we broke the seal on the bonded stock and celebrated our "escape" from Canada with war whoops, Bob Dylan tapes played at full blast over amplifiers mounted on the gundeck behind the smokestack, and dancing on the forward deck.

But still, so far as we knew, the Russians were nowhere near the coast.

The fuel limitations were as real as ever, so we decided to head up the Columbia River into Portland, where the antiwhaling movement was well-organized. The reception we got could not have been more sympathetic. City council immediately voted to waive the normal $160-a-day moorage fees we would otherwise have had to pay. The fire department provided us with water. An ex-governor came down to the boat to present us with an Oregon flag to fly. And thousands of citizens trooped down to the dock to visit the ship. Several local co-ops donated food. Local musicians immediately organized small benefits. And a group called Oregonians Cooperating to Save Whales moved down in force to help us raise money. A pilot, Don Ayres, who owned his own flying club—"Ayre's Flyers"—offered to fly reconnaissance flights out over the ocean for us.

We were joined in Portland by Peter Fruchtman, a sandy-haired former Berkeley student who had been acting as our Washington liaison officer and who had done much of the organizing for the Jericho benefit

320

concert. He wore glasses, had a trim beard, and looked very scholarly. He was at all times hyperactive in the nonstop buzz-saw American manner that left slower-moving Canucks with their heads spinning. As liaison man and benefit organizer, Fruchtman was great—he had helped to open enormous doors. But as someone to be trapped on a boat with for any length of time, he was not quite your ideal laid-back shipmate.

It was therefore with some perverse delight that we took him up in the air in one of Don Ayre's twin-engined Apaches to go scout for Russians. We knew he was prone to motion sickness.

We flew out as far as fifty miles, then turned south, zigzagging our way down the coast. Every time we'd see a ship, Don Ayres would take us into a wild roller-coaster ride of a dive-bombing attack. I'd scan for the names and numbers of the ships, shouting them out as we swept past only a few yards above water level, Bobbi would frantically scribble them down in a logbook, and Peter Fruchtman would gag and gasp and convulse with nausea. Then, when we were still about thirty-five miles from shore, the starboard engine began to cough and splutter. A tense half hour followed while pilot Ayres flipped switch after switch, pushed buttons, twirled dials, and cautiously edged us eastward, the plane wobbling unsteadily. By this time none of us were laughing, the pilot's forehead was beaded with sweat, Bobbi was almost as pale as Fruchtman, and I was clutching my pouch with the scrap of cloth from the Karmapa. The shore still seemed a long way away when the troubled engine suddenly resumed its normal throbbing hum. Ayres shook his head.

"Well, I dunno, but she seems okay now. Might as well keep going."

Fruchtman groaned. Bobbi and I nodded numbly.

We flew down as far as San Francisco. By the time we were zipping past the Golden Gate Bridge, we had dive-bombed fifty-three different ships, leaving their captains and crews scratching their heads. Except for a few tankers, freighters, and tugs, the rest of the ships were all trawlers and draggers, mostly Russian and Polish, although there were a few Korean ships. By this time it was well understood that the United States would soon be following Canada's lead in imposing a two-hundred-mile limit. The phenomenon we had witnessed—dozens of huge foreign draggers plowing leisurely along the continental shelf, unimpeded as they scooped up everything their nets could snag—was about to pass into history. This would be the last summer that the foreign fleets would be able to function freely in these waters. It seemed clear they were trying to grab everything they could in the time left.

In San Francisco, we learned that the mayor, George Moscone, had proclaimed Save the Whales Week in recognition of the work of groups like Greenpeace and Project Jonah. Our new office was crowded with volunteers; rainbows had been painted on the walls; T-shirts were being printed. It only remained for us to arrange docking for the *James Bay*, then we were back in the air and heading for Portland. A day later, Captain Korotva gave the order for the old minesweeper to nose out into the Columbia and begin the long, winding journey to the open sea.

There were still no whalers along the West Coast. They were reported north of Hawaii and on the move, slowly, eastward, perhaps getting ready to push toward us, but maybe not.

Outside of the mouth of the Columbia, we were greeted by squalls and seven-foot swells. The *James Bay* rolled far more than Korotva liked. "God damn dis bucket of bolts," he muttered nervously. Five crew members immediately crashed into their bunks. Down below, Taeko Miwa and Bree Drummond were feeling queasy too, but for different reasons. It was the onset of morning sickness.

In the wheelhouse, there was a mood of depression. If the boat handled this poorly in easy seas like this, how would she perform a thousand miles or so from shore, say in a typhoon? Korotva's agitation increased by the hour. "God damn, God damn, God damn," he growled. To add to the uneasiness, we had installed an automatic pilot. All the helmsman had to do in open water was perch on a stool behind the machine and watch the lights flashing and the dial swinging. It was like driving a bicycle with no hands, except that the mass and weight of the ship gave it the feel of a drunken elephant wobbling along on its own. It seemed hesitant, casting its nose around as though groping in the dark. It was far from a pleasant sensation. Korotva had a stash of vodka, which we broke into early to steady our nerves. And then, to demonstrate to each other how confident we felt, we'd take turns at the bow, riding skyward, only to be thrown downward so suddenly that our feet would almost lift from the deck and the sea would be exploding all around. Loud war whoops had to accompany the downward ride. The edgier we became, the more we cavorted. At night, in gloom broken only by a small red light bulb, those on watch would cheer each time the spray broke across the front windows and the boat shuddered and thumped. On would come the tape of Bob Dylan singing "Hurricane." Back and forth would go the bottle of vodka. *Ping! Ping!* went the automatic pilot. Down below, people were thrown against walls and dumped from their seats. The model of the starship

Enterprise seemed to hit at least Warp Factor 5. Fido glared down at us from her nest, clinging rigidly to the branches, refusing to eat or to have anything to do with humans.

By 6:00 A.M., July 1, the weather had calmed and we had reached the point near Mendocino where we'd encountered the *Vostok* the year before. Korotva throttled the engines down. It was here we were to rendezvous with the *Phyllis Cormack*, which had maintained a patrol on the water while we'd remained at Vancouver Island. Once we were in motion, John Cormack had taken his Mendocino Whale Warriors into San Francisco to resupply and was now on his way back out to join us. The rendezvous was expected to take place about noon.

At 10:00 A.M. some twenty Risso's dolphins appeared.

Some of them were big enough to be orcas. Within moments, everyone was on deck, including the seasickness cases. And within only a few more minutes, divers were going over the side and swimming directly toward the oncoming dolphins. Young Mike Bailey was the first in the water, followed by Gary Zimmerman and myself. "Watch dem! Day big guys!" Korotva warned.

It seemed significant to us that these creatures should have appeared at the exact spot where our confrontation with the Russians had occurred. Here in water that a year ago had been stained with blood, the quiet shattered by the roar of engines and explosions, there was now just the splashing of people and dolphins in the sea together, music coming from the deck of the *James Bay*, ecstatic shouts, and the clicking of cameras—a celebration of life in the midst of what had once been a battlefield between cetacea and man. One by one, crew members were flinging themselves from the safety of the minesweeper into water already alive with huge toothed creatures, any one of which could have swallowed us in a gulp. It was a frolic, a mass demonstration of trust. Instead of ignoring us, the dolphins broke ranks and nosed forward, coming as close as three or four feet, circling us with magnificent ease, surfacing, blowing, wheeling around the *James Bay*. In the midst of the pod was an albino baby, perhaps eight feet long, flanked on either side by adults who escorted it close to the people in the water, very much reminding us of parents guiding their kid through a zoo: "Here, look at *this* one! Look at *that* one!" A pure white Risso's dolphin surely had to be at least as rare as Moby Dick in his time. As I floated there just a couple of feet beneath the surface, face-to-face with that white child-dolphin, already larger than me, with his two twinkling-eyed guardians, none of us having the least intention of hurting the other, merely mutually fascinated, the possibility of heaven on earth did not seem at all like a vision or a fantasy or an ideal or a dream. It *was* a reality.

We hovered there, like secret lovers in the blue-green ray-beamed void, all equally aware that we were strangers greeting each other across enormous gulfs of experience, but our *classes* of consciousness were not really so very far apart. It even seemed for a few seconds that the adult dolphins and I were sharing our appreciation for that milk-white youngster, so similar to a harp-seal pup, yet so much larger. It seemed to me at that moment that the entire world was a temple and everything I met was a manifestation of God. I was not the only crew member from the *James Bay* who emerged, dripping and shivering after thirty or forty minutes in the water with the Risso'ses, who felt that way. There was exuberance on board, after that particular experience, and high energy— but there was also a quiet. It was not enough to babble about being in the water with those creatures: each and every one of us had to also spend a little bit of time being silent. Finding a private corner and thinking. Opening ourselves up to full recollection of the event. Letting its possible meanings come sifting through our minds. Anyway we looked at it, we had met the dolphin-people on their own turf, and they were *people*. Strange people. Different language, different world, different bodies, different minds. But *people*, damn it! PEOPLE! Possibly there was no way we could have come back from that experience without being affected, however greatly, however lightly.

We emerged from the Mendocino Ridge dolphin experience with a clear sense of purpose. We were united in the way that religions and revolutions can sometimes be united, at least until they get into a position of power. It was clean. It was something that few other people had gone through. For a day there in the North Pacific in July 1976, there was a boatload of us who understood what joy could be expected to come from interspecies contact—the *real feeling* of it. So I suppose we were pioneers, overcoming our fears and plunging ourselves completely into a whole new environment, not only a physical environment, but a psychological, perceptual, emotional, and experiential environment as well.

The rendezvous with the *Phyllis Cormack* and Byrd Baker and the Mendocino Whale War Committee a few hours later was definitely anticlimactic, although it was great theater. Camera crews from both vessels were in the water with Zodiacs to record Byrd Baker and me exchanging hugs as our own two Zodiacs met, with the minesweeper and the halibut seiner in the background, and the available light was good enough to make a cameraman give praise to the Great Lighting Technician in the Sky. There was a good feeling all around simply because the Russians had not so far entered this area of ocean—for the first time in decades—and we all felt we could take credit for the fact that any whales in the vicinity were free to travel without fear of being harpooned. The only reason we could see for the Russians breaking their pattern of whale hunting off the North American Continental Shelf was the presence of two antiwhaling ships, and the potential they represented of drawing unwelcome public attention to the hunt itself.

Moore radioed a statement back to shore, saying that "as a result of international pressure we helped to generate, half the North Pacific Ocean is now free of whaling ships for the first time in modern history." It seemed clear that our next step was to go deep sea, whether we wanted to or not. If we could make it to Hawaii and refuel, we could strike right at the heart of the Russian fleets. But first we'd have to go into San Francisco, top up the tanks, and add more ballast.

By this time a number of us, including the captain, were in a cold sweat at the thought of trying to travel that far in our wobbly aluminum-and-wood steed. One good storm out there, and the ship would probably fall over on its side.

We had been out of Vancouver for eighteen days—and only four of those actually at sea—yet the captain seemed to have aged several years. His temper was foul. His patience was thin. His normal European charm was down to just about zero. Also, he was sensitive to the fact that many in the crew doubted his ability. There were not just lives, including his own, at stake—there was his pride itself. Neither Korotva's pride nor anyone's confidence in him was helped much by the landing at San Francisco. We had to put into a rotting old dock beside a warehouse below Fort Mason. There were a dozen reporters and cameramen as well as a couple of dozen supporters, all cheering and waving. It was a twelve-scream landing, in which the captain took several runs at the dock, none of which worked, having to repeatedly back off and come around again. In the end, the *James Bay*'s sleek bow sank itself slowly, elegantly, and with a terrible tearing sound, into the dock, sending a horrified crew of cameramen who had been standing in the way scurrying in every direction. Fortunately, the dock was moldy enough that there was no damage to the cedar hull. The only real damage was to the self esteem of a panting, red-faced Captain Cruel, whom no one dared go near for at least an hour after the ropes were fastened. There were no jokes. Everyone steered clear.

San Francisco was hot—and crazy. It was the peak of the American Bicentennial celebrations. The ship quickly became an ant heap. Hundreds of people flowed through each day on guided tours. Volunteers from the Greenpeace office fanned out along with our own crew members into the streets, hawking merchandise, urging people to come down and see the whale-saving boat. Jerry Garcia and The Grateful Dead did a benefit for us at which the Hell Angels appeared. They dug the idea of roaring around on Zodiacs. Work parties were organized to go out in trucks to a nearby quarry and shovel some thirty tons of sand into burlap bags. A human chain sweated by the hour, passing the bags up from the dock, and down three stories to the bottom of the hold. It was back-breaking, arm-snapping work, but it was the only solution Korotva could see to the problem of the vessel's apparent instability. It would have been much easier to lower bits of old railway track, but because the ship was made of aluminum, contact with any other kinds of metal would cause electrolysis: the superstructure would eventually rot. Sand it had to be. And later, we all knew, we'd be glad to have that extra weight at the bottom. Very glad.

Not everything went smoothly. Alan Wade dropped his watch from the deck of the ship. It landed on a cross-beam down among the pilings. Before anyone could stop him, he had slipped over the side. Trying to wiggle his way around a piling to get to the watch, he got caught just as a slight swell came in and the *James Bay* lurched against the dock, pinning Wade between the hull and a wooden piling. All he suffered was a couple of broken ribs, but had the ship moved even two inches more, his entire rib cage would have been popped like fresh fruit. It was time for some crew-changes. Taeko Miwa and Bree Drummond both had to be asked to leave too, because of their pregnancies. Don Webb's holiday time was over.

In the middle of a benefit in Union Square, Melville Gregory got carried away and stripped off all his clothes. The police hauled him off to jail. We decided to leave him behind bars overnight rather than post bail, which we could ill afford.

For some reason, Korotva and I had fired Patrick Moore as first mate and replaced him with Paul Watson. But then I had turned around and screamed at Watson because he refused to agree with a decision on another matter. It was clear to everyone that El Presidento had not completely recovered from his bow-case state in Sidney, and was behaving like a power-crazed madman.

The problem came down to money. We had arrived in San Francisco in exactly the condition we had left Vancouver: broke. In order to refuel, we had to have money. It was strictly a one-way ticket. If we could raise the cash to get to Hawaii, then we could try again to raise the cash to get us out of there, and if we could get out and confront the Russians, then we'd worry about getting back home. Getting home was a long way down the road. Right now, the entire struggle was to maintain our forward thrust, hold our shaky crew together, and generally beg money from wherever we could get it. This also involved approaching entertainers, politicians, well-heeled conservationists, and environmental- or animal-welfare groups. It involved renegotiating our movie contract from the year before, which was still a sore point with veteran crew members and new recruits alike, many of whom did not want to sell our "rights." Furious board meetings ensued in the wardroom of the *James Bay*. One school held that we needed every penny we could get, therefore the contract was worth renewing since it would give us another desperately needed couple of thousand dollars. The other school called the deal a sellout and still refused to have anything to do with it. To make matters worse, the personality of the lady agent from New York drove almost everybody right up the wall, especially the women on board.

It took twelve days in all before we were able to scrape together the necessary money to allow us to pull away from the dock. By that time, everybody was bleary-eyed and depressed, we were at each other's throats, and Captain Korotva's Honeymoon Cruise was beginning to look like nothing much more than a long dragged-out marathon group therapy session. Maximum disagreement on every point. The leadership— especially myself—was discredited in the eyes of half the crew. Everyone was nervous about the deep-sea run. I was contemplating suicide. And probably quite a few other people were contemplating murder. My bride was threatening to leave me, and all the other remaining couples were experiencing personal-relationship problems.

John Cormack and the Mendocino Whale Warriors had returned from their patrol, even more disunited than Korotva and his Rainbow Warriors. The charismatic Byrd Baker was last seen slumped on the edge of the dock, staring blankly into space.

By the time we did finally pass under the Golden Gate Bridge en route to Honolulu, our voices were raw from hustling T-shirts on the street, yelling at one another, making speeches, and talking all night, night after night, with potential donors. There were several people on board who were not on speaking terms.

327

Ironically, at that point our image was better than it had ever been. There seemed to be the equivalent of a natural law at work: the better things looked in the external world, the worse they felt internally. The only common denominator it was possible to find on board was a mutual desire to save whales. Otherwise, it seemed we should all have been on separate boats.

We had acquired a couple of new shipmates: a frail-looking white-haired physicist named Dr. Norman Seaton, who had been included in exchange for a donation from Cleveland Amory's Fund for Animals. We had also taken on Barry Lavender, a Vancouver musician and actor whose energy was ceaseless, who never argued, worked like a horse, and gave off such positive feelings that people had real difficulty being nasty-minded in his presence. He was also an artist of considerable talent whose cartoons began to adorn the walls, making us laugh even when we didn't feel like it. He wore his dark hair in a long pigtail and frequently sported a black cowboy hat.

We passed under the Golden Gate just before dusk, July 12, divided, exhausted, quarreling as usual, but at least in motion in the direction of the Russian whaling fleets, and it did remain that however imperfect we were, individually and collectively, ours was the only vessel on earth heading out to save whales, not kill them. Perfection could wait for later. The whales could not.

The Golden Gate was barely behind us when the winds began to pick up. Within hours, the *James Bay* was lurching and shuddering. By midnight the wind was up to forty-five knots. Several times during the night, despite the thirty tons of sandbags in the hold, the old minesweeper leaned over so far that it seemed we had, indeed, collapsed on our side and were only floating for a few seconds before gurgling down into the deep. In the bunk rooms, a mood of fatalism set in. In the wheelhouse, Captain Korotva became so alarmed he altered course, putting the *James Bay*'s bow directly into the wind, which was coming from the northwest. This meant that instead of heading toward the Hawaiian Islands we were now heading for the Aleutians. By morning, the waves were running at fourteen to sixteen feet. The sea and sky were gray. There were few signs of life on board. Our Japanese interpreter, Kazumi Tanaka, could be seen from time to time staggering down the hallway to the deck, letting fly with a thin stream of vomit, then lurching back to his bunk. Norman Seaton was the second-sickest person on board, followed by Michael Manolson and Mary Lee from Montreal. For Rod Marining, this was nothing new: he had been in worse weather in a minesweeper in the Gulf of Alaska and he still retained his sense of invulnerability. The camera crew, housed in the starboard bunk room on the main deck, which we had dubbed the "media room," were soon singing a song called the "Media Room River Blues" because there was so much water pouring through ventilators and leaks in the roof. Fido the iguana became so upset she slithered out of her nest and disappeared somewhere into the depths of the ship. By lunchtime, only ten people out of thirty-one were willing or able to get out of their bunks.

Role playing now became very important. Korotva, for instance, became every inch the wartime skipper. Under pressure, he was magnificent; his beard was trimmed, his blond hair was clipped in a short, almost military style. He prowled the wheelhouse, scanning the gauges, and went on inspection tours of the engine room, aware that everyone's eyes were on him, probing for any sign of uncertainty. Until then, the crew had not been particularly respectful. But now that we really were at sea, and in a storm, the realization come that it was all up to Korotva's judgment and capability. A subtle wave of relief passed through the ship that first gray day at sea because it was so obvious that the captain had risen to the challenge, that he was steady, confident, stern, calm, alert—everything you wanted from the man who held your life in his hands. There had been fear that he would get into the sauce, and we'd drown. Instead, it was his finest hour. We admired and loved him enormously.

With Korotva setting such a good example, the rest of us felt called upon to rise to the challenge too. That is, those of us who could function at all. The chief engineer, Haggarty, went about his business with the conscious superiority of an old navy veteran toward amateurs, only grudgingly acknowledging that his engine-room team was doing a flawless job, even though half of them—Eileen, Marilyn, Bobbi, and Susi—had to tie their hair to avoid having it caught in the gears as they serviced the V-12 engines. Despite their overalls, they were like sisters in a religious order tending to a shrine. Overlooking the engines was a soundproof booth where you could take off the earphones that otherwise had to be worn. We referred to it as "Mission Control." And since most of the crew had never even *seen* twelve-hundred-horsepower engines at work before, let alone oiled them, it was an experience tinged with wonder just to go down there. Inside the booth, Gary Zimmerman sat at a control panel, wearing a yellow T-shirt with a huge green ecology symbol on the chest, which seemed delightfully out of place among those roaring gas-gobbling machines. He moved about with the nonchalance of Technological Man at home with his power. In this way, he was much like Al Hewitt, whose Draconian features managed a smile each time one of the younger seasickness cases stumbled by. Hewitt, too, had the aura of authority, of being completely at home and at peace among the machines, even though the wooden starship in the galley was hitting Warp Factor 8.

Our two Oglala Sioux warrior-brothers were, of course, calm and cheerful. Watson moved about serenely, making the occasional John Wayne joke, taking his watch in the wheelhouse with the oh-hum air of the seasoned seaman. Walrus Oakenbough shuffled food supplies from the storeroom to the huge walk-in refrigerator, pausing only to make entries in his logbook. He was cheerful, meticulous, ready to help anyone at any time. Around him buzzed David Wise, who had now been nicknamed "Fonzie" after the famous television character whom he vaguely resembled. Barry Lavender did his Mr. Clean routine, ceaselessly scrubbing, polishing, washing, mopping, sweeping, moving like a nimble elf.

In the role of helmsman, Lance Cowan rose to the occasion every bit as well as his skipper. He sat behind the automatic pilot, saying very little, his earrings jiggling as the ship shifted and banged about, with a serious slightly tired look, but no sign of fear. The only giveaway of his inner tension was his habit of chewing on sunflower seeds. He had chewed so many that he had worn a groove on his left front tooth. Working with him, in the role of navigator, Matt Herron poured over charts and studied the loran, quick to volunteer information on anything from the weather report to the sea conditions, and the only sign of nervousness on his part was the fact that he spoke out of one side of his mouth, which left the other side of his face disconcertingly immobile.

The other veterans of previous expeditions who were on board—Rex Weyler, Precious, Easton, Gregory, Moore—all seemed to me to shine with a kind of nobility. I felt proud just being among them. Weyler retained the air of spirituality he had picked up in India, which took the form of an almost beatific smile. Precious sported a white Panama hat and a flamboyant purple tie-dye shirt. Half balding, with a kind of reddish flapjack cowboy beard, he looked much older than he was, like a gentle uncle. He seemed the most polite man on earth. Easton took everything

330

noncommittally, tending to his cameras with the same absorption that Gregory brought to bear on his guitar. Both Easton and Gregory wore headbands, giving them a native Indian look that contrasted strikingly with the generators and diesels around them. Gregory took on as his role the task of staying up all night in the wheelhouse with the skipper, singing songs and clowning, driving out "uptight vibes" by calmly sipping on rum-laced coffee and letting his own totally confident sense of our good karma seep out into other people's heads. Gregory's strength was that he had no fear at all. The feeling was: *So long as we have Saint Melville with us, how can anything go wrong?* Moore took a daylight watch and helped out with the navigation. A writer had described his hair as looking like a "Byzantine nimbus," and as he strode about wearing a Levi's jacket with a Tibetan symbol embroidered on the back, he seemed from behind to look like a visitor from another planet. Even in the worst of the weather, he ate enough for three people.

New recruits—Mike Bailey, Bob Thomas, and Chris Aikenhead—went about their tasks industriously with no complaints. Bailey hung around mostly with the chief engineer and his uncommunicative hired assistant, Bruce Kerr, while Aikenhead fussed with his sound equipment in the sea-drenched media room, his eyes twinkling and a grin fixed almost permanently on his face. Thomas radiated boyish innocence. The storm lasted two days.

We emerged from it united again, purged. The aftertaste of the battles in San Francisco, when we had been at each other's throats, was just about completely washed away. We were back to being a family. The ladies were all beautiful; everybody was beautiful. Throughout the long awful nights of swooping and pitching and seeming hour after hour to be on the very edge of tumbling sideways into the sea, no one freaked out, so we were all able to emerge on the deck on the third day, looking each other in the eye, proud of our captain, proud of our vessel, proud of ourselves for having made it through. Now the pavement of clouds above broke apart, the sea gave a few last heaves at the hull, then backed off like a pack of wolves who have tested a buck and found him too strong to bring down. We were able to get back on course. And now the *James Bay* moved swiftly and neatly, her knife-edged bow making a giant zipping sound as she sliced through the water. Dr. Seaton rigged two three-foot-wide speakers on either side of the wheelhouse, and the ship trembled from the volume of the music. If the tape recorder wasn't on, there was almost always the sound of Gregory or Weyler playing their guitars and any number of people playing flutes or beating on tom-toms or singing or dancing on the deck. The sun brought us outside like a magnet. After just two days of being trapped inside—with no portholes, only the aluminum walls—we were all suffering from one degree or another of claustrophobia. Now we could relax in the rubber comfort of the Zodiacs mounted on top of one another on the back deck. To ride on a bed made up of three wobbly, jellylike rubber inflatables, like air mattresses, while the ship went swinging vigorously but comfortably through the waves was one of those experiences that makes the twentieth century seem worthwhile. There was also a V-hulled double-engined Avon rubber speedboat with a windshield, a steering wheel, and foam seats, exactly like a little seagoing racing car. It was mounted on the starboard stern. Sitting in it, riding on the back of the tremendous plunging projectile we called the *James Bay* was more fun than most kids could dream of having.

If both stacks of Zodiacs were full of people, sitting around having intense, excited conversations, and the Avon was being used, you could always climb up to the flying bridge and view the world around as though from the top of a wide castle tower, a castle that was weaving and dancing amid waves. Or climb back down and pull yourself against a blast of wind out onto the forward deck. Usually you had to wait your turn to hang on to the front light pole, at the very point of the bow, and become a human figurehead for a while, as though flying effortlessly through the sea. "Bow tripping" was by far the most exquisite of all the experiences available. And everyone was very fair about it. Nobody hogged the bow for too long. It was so good it had to be shared.

I think during that period, July 15 to 18, we experienced a collective high that was as good as any a group of human beings has ever shared.

It reached a kind of peak on the evening of the seventeenth. A large object was spotted ahead, just as the sun was setting. A quick slight course change, a great deal of commotion between the wheel-house and the engine room, and the *James Bay* was suddenly slowing down, zeroing in toward some kind of extraordinarily large creature: a whale?

"No," came the cry from the flying bridge, where the binoculars were being trained, "it's a giant squid!"

The veterans of the previous summer's voyage all exchanged shocked looks, remembering the giant-squid press release I'd cooked up to get us some desperately needed coverage. Had the press release been prophetic?

Apparently. Within minutes, Watson and Zimmerman were in their wet suits, streaking toward the thing. Everybody was on deck. The minesweeper's engines idled peacefully. No one seemed particularly concerned that two maniacs were racing directly into the embrace of a giant squid somewhere out there in the middle of the ocean. It was an unconscious collective decision to investigate. It crossed my mind then that what this odd group of people did share was a mutual nearly complete fearlessness in the face of what most people thought of as Monsters of the Deep. It was either fantastic courage or a wholly new type of consciousness that simply did not include fear instinct.

The giant proved to be in the last throes of dying. A few of its tentacles still writhed in the water. Nearly a third of its harpoon-shaped head was missing—bitten away by something that could only have been a sperm whale. Its saucer-like eyes still contained a glint of awareness, even as Zimmerman and Watson began to pull it up into the Zodiac, its limbs *slishing* around them, the sucker pads grasping feebly.

"It's only a small one," Zimmerman yelled. It was duly hauled aboard—a nineteen-foot mass of translucent tubes of jelly, the most alien creature any of us had ever laid eyes on. It was eventually carved up and packed into plastic pails and placed in the freezer with the idea of selling it as a delicacy in Honolulu or maybe even trading it for other food. Watson cooked some of it up, Japanese style, and ate it, but nobody else could force it down—and Watson didn't go back for seconds. We never did sell it, and eventually, when the freezer stopped working, the stuff gave off a stink that made even strong stomachs convulse.

I radioed back a press release to the Vancouver office about the discovery of the giant squid, claiming that it was evidence of the depletion of the whales in this area of the Pacific. A giant squid on the surface, even a youngster, was a rare thing and probably an indication of an unusual abundance of the creatures. With fewer whales around to eat them, the squids would naturally multiply. A vision of the future came to mind—where giant squid grown larger than ever for lack of a natural predator would have moved up to inhabit the surface waters.

The press release never got past the tape recorder in the Vancouver office. We had a new press officer back at home base, a bright young university graduate named John Frizell, who had enough of a general scientific background to figure that my reasoning was pretty shaky. He decided to check with at least one veteran from a previous Greenpeace trip: Carlie Trueman. The summer before Carlie had objected to my original squid story more than anyone else. Now, she listened to the tape of what I had just reported, and said: "Forget it. It's just Hunter up to his old tricks. Don't believe a word of it."

There was a kind of justice to that.

But it also revealed a subtle change that had come over our operation, a sign that we were growing dangerously fast. For the first time, there was a stranger—rather than an old friend and ally like Rod Marining—at the other end of the radiophone. And there were other people holding down the fort—not the same small tribal unit that had worked together before. The fact that several of our women were with us made a huge difference in itself. On the first whale run they had been at home, hustling the money to keep us going, pushing, butting, working fifteen and twenty hours a day, week after week. Their vital energy had been removed from the home fortress, leaving their spaces to be filled by others who were not so intimately bound to us. In terms of full-time participants, our small band had tripled if not quadrupled in just the last year. There was already a first and second generation of Greenpeace whale crusader. Most of the first generation were out on the boat, leaving the second generation in command at home. In fact, there were really three generations involved. For there had been a point in time—as far back as 1973—when there had only been two of us in the great whale conspiracy: Paul Spong and myself. We were technically, therefore, the gurus of the movement, or at least the patriarchs. It made us feel like old men, and in a sense we were treated like old men. Then came Rod Marining, who was the next one to have been let in on the plot, and then the others, each of whom was consciously or unconsciously perfectly aware of their seniority and the ill-defined but nevertheless real authority it gave them. My wife, Bobbi, was, for instance, the matriarch among the females, even though she was younger than Eileen, simply because she had been sweating away licking stamps and making Xerox copies of pamphlets long before any of the others. At the very bottom of this pecking order was Bruce Kerr, who had had to be hired because he wouldn't volunteer and who had not helped out in any way before.

As the old men of the expedition, the architects of this bizarre notion of going to sea to save whales, Spong and I shared a sense of responsibility not quite felt by any of the others. If anything were to happen to any of them, it would be our ultimate blame. If we'd just kept our mouths shut, no one would be out here in the first place. And at a final level, because the idea of the trip itself had been mine, it was me who agonized the most. God help me, I felt paternal. I tried to laugh at myself for taking things too seriously. But the truth was that the strain of bearing—even if only in my own head—the responsibility for the fate of all these lives was just about more than I was psychologically and emotionally capable of handling. By this time, I had a feeling of love toward everyone, a real sense of brother and sisterhood. At least I felt that whether they recognized it or not, I was their brother.

Brotherhood I could handle, but being an Elder Brother was where the weight got my head down.

It got Spong too. And Korotva, whose responsibility was not so long-term but far more immediate. Each of us was a high-energy free spirit, full of ourselves and very much bound up in the wonders of our own lives, none of us particularly enjoying the rude discovery that he who rules must serve by worrying and fretting and planning and scheming and thinking harder than anyone else, getting ulcers and ruining your stomach in the process. It was no coincidence that it was the three of us who smoked and drank the most and who were probably the least pure souls on board.

Our sanity itself was constantly in question. There were people on board who thought of Korotva as a "crazy Czech," meaning a *really* crazy Czech. Spong himself was known to go into black rages where he said absolutely nothing, yet gave off such furious vibrations that people backed away from him as though he were an electric eel. The moods were hard to predict, but sometimes when he paced the deck, it was easy to think of him as a kind of Captain Ahab in reverse, so intense was his desire to reach the side of the whales. As for myself, it was conceded by everyone, including my wife, that I was in a trance half the time, either meditating on my knees as I cleaned and scrubbed the latrines and showers, or wandering about the boat, staring into space, perceiving everything with the open-eyed astonishment of a child. Perhaps fifty percent of the time I was lucid in the ordinary sense and up to playing the role of old man or eco-commander. The rest of the time I was far more interested in cosmology—at least until word came from our intelligence sources that the Soviet whaling fleet had left the vicinity of Kauai Island and was now moving purposefully toward the California coast.

The Russians' timing was rather exquisite. Obviously, they were monitoring our movements as closely as we were monitoring theirs. It could not just be an accident that they had waited until we were committed to a deep-sea run before launching themselves eastward. Clearly if we engaged them close to any shore we would have a greater capability in terms of fuel to stay with them and harass them. But if we were to meet halfway between Hawaii and North America it would be at the point where we had the least possible flexibility. This, too, was a legitimate tactic of naval skirmishing. By July 17, we were eleven hundred miles out from San Francisco on a southwest course. That night, we got a radio message from Vancouver that gave the Russian fleet's position in a code we had worked out in advance. Unfortunately, there was a "clerical error," which left us staring in shock at the charts. If this information were correct—which it wasn't—it meant that the *Vostok* and her pack of killer boats were only fifty miles directly ahead of the *James Bay*. Yet try as he might, Al Hewitt could not pick up a trace of Russian voices on the airwaves. Of course they might be maintaining radio silence, hoping to slip past us on their way to the North American coast. Just in case, we called a full crew alert. Zodiacs were tested and retested, along with cameras and outboard engines and tape recorders. Nobody could get any sleep that night. We tripled the watch, with the result that there were at least two or three people prowling the flying bridge with binoculars scanning the darkness for a glimmer of lights that would tell us the fleet was approaching.

Toward dawn, the calculated interception point came and went. Still no Russians.

Tempers flared. Morale collapsed. Bitter looks were exchanged. It was amazing how swiftly our sense of unity vanished.

By noon, however, we were able to confirm the error through Vancouver. The Russians had not slipped past in the night. In fact, they were still some five hundred miles ahead of us. We were on course for an interception within twenty-four hours.

The jockeying for positions in the Zodiacs now began in earnest; practically everybody wanted to be in one. Some—like young Mike Bailey—were more eager than others, but eagerness was not going to be the criterion.

There was an interesting moral dilemma here: two years ago, when I'd begun to assemble a crew for the first expedition, the problem had been to find someone willing to get in a Zodiac with me in front of a harpoon. This year, I could easily step back and let any number of them rush forward. But then that would put me at the level of a general who allows troops to go forward as cannon fodder. "We'll do it the way they did in the old days," I insisted. "The leader goes first."

So I would be in the kamikaze Zodiac again. But who else? Paul Spong was adamant that he be the driver, and because of his seniority, no one could argue. Everybody agreed that there should also definitely be a woman aboard, and on the grounds of her seniority among the female crew, my wife, Bobbi, laid claim to the spot. "If Bob gets killed," she argued, "then I get killed with him, and neither one of us is left behind to miss the other."

Early the following morning we picked up Russian voices on the radio.

"Yeh, dat's de *Vostok* all right," said the captain.

An RDF fix followed almost immediately. Bodies erupted from the radio room.

"We got 'em," said Weyler breathlessly.

There was a slight discrepancy between the position we had calculated and the direction in which the RDF needle was pointing, which indicated the Russians had changed course farther to the northeast. As far as Matt Herron could figure it, they had altered course within hours of our own most recent course change. They were dodging northward and speeding up. It was interesting to speculate how much they knew about our fuel range. Certainly they knew where we were—their satellites would have told them that. But there was more to it, because if their intention was really to try to avoid us, they were making all the perfect moves: delaying before moving into the gulf between the Hawaiian Islands and the continent until we had reached a point of no return, altering their speed, and now angling northward, as though they had calculated our fuel capability down to the last gallon and intended to remain just a couple of teasing miles out of our range.

"That's gonna be awfully close," muttered Moore, shaking his head.

Dawn came gray and cold but by 7:00 A.M. the shutter of clouds had opened ahead on the horizon, letting long shafts of golden light spill toward us across the waves. It was against this background of gold and blue that the tiny black dots finally appeared.

The minesweeper seemed to be racing now like a train, streaking across the remaining miles of water. By the time we reached the outer perimeter of the area in which the Russians were operating, the sky had cleared almost completely, and the scene that came into focus around us was like a replay of the previous year. The available light was precisely the same. The water was exactly as choppy. And again the dark bulk of the factory ship lay silhouetted on the horizon while the killer boats worked zigzag patterns in the water all around.

Then just about everyone's hair stood on end.

The nearest killer boat, NK-2007, switched course and began plunging toward us. Directly ahead of the ship were the faint vapor puffs that betrayed the presence of fleeing whales. Once again, the whales were running directly to our side, bringing the whalers along in their wake. Many on board had heard us describe this scene from the summer of 1975, but most had dismissed it as exaggeration. Now there were no doubters. The whales *were* running to us for help.

It had been Paul Spong who, years before, had insisted "you can count on the whales to help." He had said it at a time when even his best friends thought he was slightly mad for attributing that much intelligence to the creatures or expecting any such science fiction-like encounter with an alien mind. So, for a moment, the diminutive scientist from New Zealand stood by himself on deck, his face twisting with emotion. There, only a few miles away, were the cousins of the whale he had come to know so well back in the Vancouver aquarium and whose impact on him had changed his whole life. And there, beyond, was the enemy he had sworn to destroy. Now everything he believed and felt would be put to the test.

Within minutes, the *James Bay* was slicing to a halt and the Zodiacs were being lowered swiftly over the side. Bobbi was first down into the Zodiac, followed by myself. Spong had disappeared. Everybody was yelling for him. The metal hook attached to the ropes that had suspended the Zodiac became snagged and I had to wrestle with it furiously, panting and cursing, to get us free. But no sooner had the hook been pulled loose than I realized we were left sitting motionless in the water while the *James Bay* drifted inexorably away from us and I couldn't remember how to start a damned outboard engine. I started to scream in rage, but Bobbi said: "Just sit down." Deftly, she started the engine and steered us back to the minesweeper's side. Spong came bounding over the side with his camera. We were ready to go.

But the engine on the camera crew's Zodiac was stalling. And the rope was snagged on the other. The whole operation was sloppy, and would have been comical—all the shouting and recrimination and Korotva bellowing over it all—except for the fact that less than a mile away, whales were being hunted while the whale savers stumbled and fumbled about. Any second now that harpoon might go off.

"That's it," snarled Spong through clenched teeth. "Can't wait. We're going." He opened the engine full throttle and we leapt away from the *James Bay* rapidly leaving our back-up crews behind.

During the long moments while the rainbow-hulled minesweeper dwindled away and the rusty killer boat grew larger and larger, I felt an amazing sense of detachment, or at least an almost complete lack of feelings. At one point, I wanted to laugh. Bobbi and Spong's expressions were so grim. They were evidently feeling the same things I'd felt the first time—but now I was just curious. Bobbi had wrapped a white scarf around her head to keep her long hair out of her eyes and hung on against the fierce pounding of the Zodiac against the waves as we finally came up parallel to the racing whaler. It was evident right away that the men on the deck and on the bridge could tell that there was a woman in the rubber craft that was now maneuvering into position in front of their bow, directly in the line of the harpoon's fire. I felt my first emotion—a surge of pride in her that made me want to cry.

"Well, we're here," yelled Spong over the buzz-saw noise of the outboard.

I was sitting at the bow, looking back at the ship as it rose and fell above Spong's head. Bobbi was in the middle, looking ahead over my shoulder to watch for whales. Spong's head swiveled back and forth as he throttled down, letting us fall into a position scarcely ten feet ahead of the whaler's bow, the gunner perched so high above us that I had to crane my neck to look up at him. He did not look back, merely spat out of the side of his mouth.

It was Bobbi who noticed that the ship's number—2007—was the same as the address of our old office back on Fourth Avenue. Was there some hidden meaning in that? It seemed to me that there had to be, but this wasn't the time or place for a lengthy discussion about it.

Then the gunner stepped back from the harpoon.

"He's moving away from the gun!" I yelled, daring for the first lime to hope that they might not shoot after all.

And a moment later, the NK-2007's engines ceased to roar and the ship abruptly slowed down, her bow wave pushing out ahead of her and picking us up. Dumbfounded, Spong throttled down. The next thing we knew, the whaling ship was dead in the water and we were sitting there, bobbing up and down, the only sound being the spluttering of our engine as it idled.

Minutes later, three other Zodiacs arrived and pulled in on either side of us, forming a floating picket line in front of the stalled Russian ship. It had worked! We had blockaded them! The decision not to fire a harpoon had probably been taken months in advance, perhaps at the level of the

Kremlin. It had to be a political decision. Yet at that moment, it seemed to Spong and me that the power of sisterhood had played a crucial role. It was one thing for a Russian skipper to play macho games with other men, but he could not have had any stomach for running down or blasting apart a woman. Two of the crewmen waved at Bobbi. One whistled.

While our camera crew recorded it, we moved our Zodiac up to the very edge of the rusting steel bow, I got up on my feet and put my hand out to press it against the metal. At that moment, the NK-2007's engines went into reverse, and as I pushed against the hull the boat backed slightly away. An illusion, of course! But it did seem then that my hand had become an instrument for some force greater than myself.

We were too low in the water to be able to see what had become of the whales the ship had been pursuing, but back on the *James Bay* there were wild cheers when NK-2007 came to a halt and the pod—maybe five, maybe seven—moved off rapidly toward the horizon.

Our hearts were of course pounding, and it was still difficult to believe that this was all really happening. It was the feeling that Pygmies must have had when they finally brought a mastodon to bay. NK-2007 sat in the middle of the ocean, surrounded by tiny pellet-shaped rubber boats, completely pinned down. The silence was stunning. The only activity on the ship was the gunner wrapping a canvas cover over the harpoon, as though to hide it from our cameras. Otherwise, the Russians stood about in small groups, smoking and ignoring us. After about twenty minutes, the engines came to life. Immediately, our small flotilla of Zodiacs darted back in front of the bow. But NK-2007 was merely heading back to the *Vostok*. After several minutes of pacing her, we broke off and swung our own noses back toward the *James Bay*.

Once the equipment was all back on board, Korotva spun the wheel and took us rapidly across the water to within five hundred feet of the starboard side of the *Vostok*. As before, two killer boats were flanking the factory ship's stern and dead whales were being winched up the vessel's grim gray rectum. We put one Zodiac and our speedy V-hull Avon into the water. Watson took the Zodiac, with cameramen Kazumi Tanaka and Matt Herron, and maneuvered in between the killer boats, right up to the *Vostok*'s stern. With characteristic impulsiveness the Oglala Sioux warrior-brother fired his engine, driving the craft right up on the back of a whale that was still half submerged in the water, with only its tail already being pulled up the metal chute toward the flensing deck above. He said later: "I thought that mighta stopped them." But it didn't. Whether it was because the Russians were willing to dump the Zodiac regardless of the risk to its occupants—none of whom were women—or because the winch operator was simply unaware of what was happening, the great chain continued to drag the whale up out of the sea, and as the slippery bulk of the creature rose, the Zodiac that had been aground on its back slipped sideways some four feet back into the churning wake behind the massive propellers. The Zodiac came within inches of going right over on its side. Kazumi Tanaka only managed to hang on by one hand with a split second to spare before he would have gone under, probably to be chewed to mincemeat by the props. Caught off guard, Herron crashed to his knees as wash surged over the rubber sides, soaking—and ruining—some two thousand dollars worth of camera equipment.

While this was happening, Moore had taken the Avon over to the side of one of the killer boats. He had Weyler and Eileen Chivers with him. As they pulled alongside, a long-haired blond young seaman leaned over the side and yelled, in perfect English:

"Got any acid?"

It was not exactly what they'd expected to hear from a Russian whaler.

Weyler threw up his arms in a gesture of helplessness. "No man, sorry."

Instead, Eileen handed up buttons, T-shirts, and pamphlets. The crewmen scrambled to scoop the stuff up, then scattered the moment the captain roared down at them from the bridge.

When the rubber boats were both back on board the *James Bay*, the skipper moved us to a position parallel to the factory ship, and we turned on our three-foot-wide speakers to deliver a lecture. First I addressed the *Vostok*'s crew—most of them lined up almost obediently on the deck—pleading with them to stop killing the whales, then turned the microphone over to Korotva who had at them briefly in Russian, then at greater length in Czechoslovakian. Next came Kazumi Tanaka, talking mainly to the two Japanese observers we could see among the Russians. After Tanaka had said his piece, Mary Lee spoke in French, and finally Walrus Oakenbough practiced his surprisingly fluent Spanish on them.

The captain of the *Vostok* spurred his massive vessel on from the four knots he'd been doing when we arrived up to eighteen knots. He also changed course some eighty degrees so that instead of heading any longer toward California, he was heading now toward Alaska.

The killer boats moved away at an even faster rate, fanning out ahead over the horizon. Within a couple of hours, there was only the *Vostok* and the *James Bay* racing side-by-side across the ocean. On our decks, the crew exulted. We had stopped a killer boat in its tracks; the entire fleet had splintered apart and taken, it seemed, to their heels. Now even the monstrous factory ship was following. It was as though a Texas Ranger had arrived, sending a gang of surprised cattle rustlers galloping for the hills. We allowed ourselves to believe for a while that we were the planetary eco-police and that we were escorting the criminal Russian whale killers out of the whaling grounds.

But up in the bridge, Korotva, Haggarty, and Hewitt were shaking their heads.

"Dat captain," rumbled Korotva, jabbing a big finger in the direction of the *Vostok*, "he's no dummy. He's taking us in exact da goddamn opposite direction from Honolulu. He knows dat's where we gotta go for fuel. He don't give a shit. He's got tankers come out and feed him. Pretty soon we gotta turn back. Simple like dat. He knows it."

Up until the last day, we had been cruising at a steady thirteen knots. But now we were doing eighteen. We could go faster—but for each knot faster, the cost in spent fuel was enormous. Nervously, the old chief engineer eyed the gauges. Al Hewitt was making his own private calculations. The two men disagreed. Hewitt said that we had a larger margin than the chief was saying, but of course the chief had been trained in a navy tradition that demanded that a wide safety margin be maintained at all times. An angry debate followed, during which the fuel supply was checked and rechecked and checked again. We finally agreed upon a point of no return where we would simply have to let the *Vostok* go and turn toward the fuel pumps of Honolulu. The chase lasted forty hours. We never caught another glimpse of the killer boats. Just the *Vostok* lumbering along like a bear beside us.

Finally, at 11:00 P.M., July 20, Korotva gave the order to come about 180 degrees. The lights of the *Vostok* were suddenly on our stern as we headed away, gentling the V-12 diesels back down to thirteen knots, and then the lights were gone entirely. Shortly afterward, the *Vostok* swung back onto the course she'd been following before we arrived: heading once more toward the coast of California.

We'd cut the fuel safety margin to a sliver. Any bad weather at all would mean that we wouldn't make it to land.

Yet as costly as the race had been to us in terms of fuel, it had been far costlier for the Russians. We estimated that the fleet had been forced

342

to burn up some fifty thousand gallons of diesel fuel. And if the pod of whales we'd saved had indeed numbered five—some thought seven, but we tried to be conservative—then the loss, calculated at twenty thousand dollars per whale, was one hundred thousand dollars.

The next day we got our reward.

Melville Gregory had just turned on the speakers and begun to sing a song when the cry came from the bow: "A whale! A sperm!" While people scrambled out of their bunks and up onto the front deck, Gregory continued to sing—the sound booming out over the sea.

Another whale. Another. Two more—and one of them was a baby. Five whales! We were at that moment passing back no more than fifty miles from the spot where we had saved the pod of sperms only two days before. It was perhaps unlikely, given the vast distances involved, that this might be the same pod. But on the other hand, we had not seen any other whales at all during the entire voyage from San Francisco.

"It's the same guys!" shouted Spong, literally jumping up and down with excitement. Whether it was or not, we'll never know, but the whales did behave oddly for creatures who are hunted routinely by ships roughly the size of the *James Bay*. Instead of fleeing, they closed in around us, breaching and blowing, coming as close as two hundred feet, maintaining their pace so that they cruised along beside us for close to an hour, and it was impossible to avoid the impression that their movements were definitely synchronized with the music that Saint Melville was sending across the waves. Finally, the whales pulled ahead of us, then, as if on a signal, they all halted, allowing us to glide right through their midst, while several enormous heads lifted out of the water to stare at us as we passed—silent now, everyone tingling, all of us choked with the sense that we were taking part in an event whose meaning was probably beyond our understanding. It was as though the whales were saluting us. The whales *were* saluting us. It was a conscious and deliberate action on their part.

Then, one by one, they sounded. And disappeared. When they surfaced again, it was far away and out of the line of our direction. Less than a minute after they vanished, a sudden minor squall hit us, pelting rain down on us as though we were in a needle shower, and, of course, a rainbow appeared. No one went inside or below to escape the rain. We were far too ecstatic to be bothered. We were also gripped by a vast sense of tragedy, for we could understand better than ever what John C. Lilly had meant when he wrote that it is the very size of the sperm-whale brain that "leads toward non survival of their species." Although hundreds of thousands of whales were killed during the 1800s, there were only six cases recorded where the sperm whale lost its temper and destroyed the ship that was attacking. Dr. Lilly had argued that the larger the brain, the better able it is to control its lower-brain reflexes, especially those that precipitate rage and trigger the attack mechanism. Sperm whales—with brains six times as large as our own—were undoubtedly better at overriding their primitive urge to fight or slaughter than human beings, yet for the very reason that they had larger brains, they were helpless against a smaller-brained creature that had as yet scarcely begun to *dream* of controlling its urge to kill, to exploit, and to dominate. The mood that prevailed afterward on the *James Bay* was a mixture of joyousness and sadness deeper than anything we had yet felt, for it seemed so clear to us that in the midst of wars and the innumerable calamities human beings bring down on themselves, the real horror and tragedy of the century was not to be found in man's inhumanity to man, regardless of the scale, but in humanity's blind destruction of beings so far ahead of us that we might only hope—someday—to reach the state of peacefulness they had long since achieved.

We did not realize it then, but in terms of immediate feedback relating to an action, Greenpeace had just had its peak experience. It had been good on the ice off Labrador to halt a sealing ship for even a short period of time, and it had been better, more satisfying, to send a whole whaling fleet running toward the horizon in the North Pacific—and to be able to see with your eyes that it was running. But nothing until then—and so far as I am concerned, *since* then—could compare to that hour afterward with the great sperms leaping and blowing around us. It had been a source of pleasure and pride to have been involved in exercises that succeeded in bringing nuclear-weapons tests, both below and above the surface of the earth, to an end, but you never got to *see* the bombs stopping: they were simply no more. There was no direct feedback. But at that moment, the whales seem to have spoken eloquently, to have confirmed our newly awakened belief systems, to have made everything that Paul Spong had said was true *be* true.

In centuries to come, I was absolutely certain, people would look back on these records of ours—the films and tapes and notebooks, the charts and readings—and say: "Look, here, this is where the turning point occurred. It was not just that humans went out to save whales, but the whales came back to acknowledge the effort. It was the real beginning of brotherhood between living creatures."

*

The only docking facilities in Honolulu were in the midst of concrete warehouses where freighters and tankers normally load and unload, next to the military base at Pearl Harbor. Jet fighters rose screaming into the sky almost directly overhead. Ships so vast they dwarfed even the mighty *Vostok* were constantly in motion around us. Every couple of days we'd have to shuffle the puny little *James Bay* to another dock to make way for an incoming supergigantic freight carrier. There was not a palm tree in sight—only echoing concrete caverns, empty warehouses and depots, barricades, walls, and swinging wire-mesh gates. Police cars prowled, making scarcely a whisper of noise, in and out of the block-long warehouses and along the wharves. It was a tense, paranoiac atmosphere, a constant battleground between smugglers and cops. Everywhere there were tough-looking men in Hawaiian shirts and sunglasses, and it was impossible to tell if they were plainclothesmen or Mafia. Or, for that matter, CIA. Everywhere one went, one had the spooky feeling of being watched from the shadows. We felt like mice lost in an abandoned furniture store, afraid any moment that a giant might come along and step on us. Our vessel, which had seemed too huge at sea, now looked like nothing more than a toy compared to the Brobdingnagian brutes in whose shadows we were moored.

"Well this sure isn't St. Anthony," remarked Eileen Chivers.

Had it really only been four months since we were zipping around over the ice off the coast of Labrador?

If our stay-over in San Francisco had been hard on the nerves, it was an idyll compared to what happened now. Eighteen days were to pass before we were finally back out to sea, heading once more after Russian whalers. With only a couple of hundred dollars left over from T-shirt sales, we were completely dependent on the good will of the islanders—or for that matter on anybody who was willing to help. Help came: within hours of our arrival, the ship was swamped with visitors all volunteering to supply us with food, set up television and radio and newspaper appeals, and head out into the streets to hustle merchandise. Cash donations began to pour in. Carloads of bananas and papayas arrived. The harbor master waived moorage fees. The mayor's office sent us his greetings. Whaling was apparently highly unpopular in Hawaii, and anybody who was opposing it was automatically a hero. Still, as enthusiastic as the response was, it did not come close to solving our main problem:

345

the need to completely refuel the *James Bay*. It was our U.S. liaison guy, Peter Fruchtman, who came through again by putting me in touch over the phone with a colorful maverick airline-company owner, Ed Daly, whose main claim to fame was his spectacular last-minute flight out of Saigon with a planeload of Vietnamese orphans, just before the American withdrawal. There were grumblings from people like Paul Watson, who had supported North Vietnam in the war, that we shouldn't have anything to do with Daly. Basically, Watson was told to stuff his politics: we needed the fuel and there were no strings attached. So, shortly after our arrival, a World Airways car pulled up beside the ship, and Daly's Hawaiian agent came aboard to arrange for the fuel to be delivered, free of charge.

With food and fuel assured, life should have been simple. Instead, it got more complicated by the day. We were just about at the point where we figured on being able to get back to sea when the Honolulu newspapers ran a report from the U.S. Coast Guard stating that a Japanese whaling fleet had been spotted near French Frigate Shoals in the leeward Hawaiian Islands. An exquisite tactical dilemma now faced us. We knew that the *Vostok* fleet was operating some eight hundred miles west of California and could be expected to swing back toward us before too long, leaving us in a good position to hit them a second time. The Japanese fleet, led by the factory ship *Kyukuyo Maru #3*, had taken off within hours of being spotted by the Coast Guard, and could be anywhere by now. We had to choose between heading west after a Japanese fleet that could easily elude us or moving eastward after Russians whom we knew we could find. Still, everybody agreed, it would be much better if we could hit a Japanese fleet at least once, thereby driving the whaling issue into the mass media in Japan.

Most of the crew was impatient and wanted to plunge after the Japanese immediately. They were all psyched for action and could not endure the tension of waiting. Besides, the *James Bay* had been turned into an oven by the fierce steady sunlight. The decks were so hot no one could walk barefoot on them. The ship had been designed for cold northern ventilation. Between the heat and the restlessness, the frustration of not knowing which way to go, and a certain degree of simply being tired of each other's company, the crew quickly broke down into a half dozen dissenting factions. The Greenpeace leadership— Korotva, Spong, Moore, and myself—did not want to commit ourselves to an expensive fuel-consuming wild goose chase after the Japanese unless we had at least a chance of finding them. From the information Spong

346

had acquired in Norway the year before, we knew the rough pattern of Japanese movement, but it was spread out over so large an area that it might take us months to find them. There was one more factor that worked against a successful chase to the west. While the Russians used relatively old-fashioned radio equipment that allowed us to track them with our RDF, the Japanese were known to use sophisticated telex-like facilities which left nothing for us to zero in on. The message would be totally incomprehensible, even to a Japanese listener. We would not be able to distinguish between whaling fleets or any other kind of seagoing operation.

A compromise was finally reached: we would move the *James Bay* westward to the island of Kauai, putting us closer to the last known Japanese position, but remain in port without consuming fuel while several Greenpeace spies fanned out to carry on clandestine reconnaissance. Taeko Miwa—who had flown in from San Francisco—set out with Kazumi Tanaka to visit every Japanese ship in port, trying to pry loose information about the whaling fleet's radio frequencies. If we could at least be sure of the frequencies, it would give us something to go on, even if not very much. Fred Easton and Rex Weyler stripped off all their GREENPEACE buttons, necklaces, and T-shirts and boldly walked into the offices of the U.S. Air Force, posing as free-lance cameramen who wanted to take advantage of a free weekly press flight out to Midway Island. Several islanders were approached to take us up in their private planes to scout around the outer edges of the islands. Peter Fruchtman managed to talk some navy pilots into letting him stow away on a high-altitude surveillance flight heading southward. Within less than a week, Greenpeace members had succeeded in flying over some 170,000 square miles of ocean, peeking over pilots' shoulders to get any glimpse of Japanese whalers below. For the record, it was easily the largest undercover ecology aerial action ever undertaken. By the time Weyler and Easton got back from Midway, it was clear that the Japanese had moved completely out of the area. A report carried in the Honolulu newspapers indicated that the U.S. ambassador had suggested to the Japanese ambassador that with antiwhaling feeling running so high, it would be considered unfortunate if there was to be a confrontation between whalers and antiwhalers in nearby waters. Not only the whaling fleet had been asked to withdraw, but all Japanese fishing boats as well.

The waters around the Hawaiian Islands were suddenly safe for whales.

During the time that the reconnaissance flights were taking place, the *James Bay* had remained tied up at a dock in the community of Nawiliwili. It was a bad time for everybody. A heat wave—phenomenal even by Hawaiian standards—had settled like a great weight over the islands, leaving everyone breathless and exhausted. A "security watch" had been established, so someone would always be guarding the gangplank. Mostly, in Nawiliwili, the main task of the security watch was to keep the cockroaches from coming aboard. Most people simply crushed them when they came near, leaving hundreds of squished cockroach bodies on the baking pavement of the wharf. A handful of mystics refused to kill them and instead tried to shoo them away. Herding a cockroach away across a dock is slow, laborious work. The damned things are quite independent-minded. Saint Melville got to the point where he would simply eyeball the cockroach and send telepathic messages telling it to "get out of here." It did appear to be true that cockroaches did not try to come on board when he was on watch. But the presence of the cockroaches set the mood for our collective state of mind. It was as though we were all suffering a real-life case of the DTs.

We moved to Lahaina, and there bickering broke out worse than ever. Korotva punched Walrus Oakenbough, sending him flying several feet through the air. Watson left, taking Marilyn Kaga with him. Paul Spong ended up in the blackest depression yet. The chief engineer had become unbearably crotchety. Bobbi and I wound up under a huge banyan tree, shrieking at each other. There were plots afoot to murder Peter Fruchtman and drop his body overboard when we got back to sea. Kazumi Tanaka raged with frustration because we weren't getting to the Japanese fleet. On and on and on. What had happened to the spirit with which we'd left Jericho, newly blessed by Cree medicine man Fred Mosquito? Where had all our karma gone? Why were we getting into these knots?

Rod Marining and Mary Lee both decided they had to get back home for various reasons. What with the other losses we had sustained, we had to pick up some new crew members. We got three: a long-haired bearded radical named Ross Thornwood; a short-cropped UPI reporter, Hal Ward, who had had some navy training; and a young lady from Maui named Juliette who took up with one of the males on board, and who was expert at giving massages.

Only one "magical" coincidence occurred while we were at Lahaina. It was the discovery that we were not the only boatload of Greenpeacers anchored just outside the harbor. Only a few hundred yards away from where the *James Bay* had come to rest was a thirty-six-foot sloop, *Wings II*, with two doctors from previous Greenpeace missions on board: Lyle Thurston and Myron MacDonald, my wife's former husband. They had just sailed in from Vancouver. It was the stuff that makes reality so much more interesting than novels. Along with them was MacDonald's new lady, Leslie, and a lawyer friend of ours, Davie Gibbons, who had helped us out several times in the past. So there was a fine reunion, and before we left Lahaina, we rechristened their sloop *Greenpeace VIII*, gave them a Greenpeace flag, and told them to keep an eye out for whalers.

It was not intended as a joke, nor did it work out that way. Although almost a full year was to pass before *Greenpeace VIII* got a chance to go into action, the opportunity did finally come. They were in Suva, Fiji, at the time, when they looked up to see a vessel they recognized from our photographs as being a Russian killer boat. Running up their Greenpeace flag, they had charged across the harbor at it. By that time, the Russians were so sensitive to any kind of public exposure about their presence anywhere that the ship had gunned its engines and fled from Suva Harbour, where, presumably, it had been in the process of picking up supplies. The little sloop had bobbed triumphantly in the Russian ship's wake. By that time—1977—Greenpeace had launched an "eco-navy" that was tackling two Russian fleets in the North Pacific and going after Australian whalers in the southern ocean as well. For yet another Greenpeace vessel to show up in the waters around Fiji had no doubt left the Russians shaking their heads.

At last we found ourselves moving steadily northward toward the very heart of the North Pacific Ocean, roughly as far away from any point of land as you could possibly get while still remaining on Planet Earth. Reports indicated a steady northwestward movement by *Vostok* and company. From the time we left Lahaina, if we maintained our northward course, we could expect to connect in six days, thirteen hundred miles north of Honolulu. We were going to be very far off the regular shipping lanes, in a part of the Pacific where whalers had not been known to go so far. Spong emerged from his hermit shelter behind the smokestack long enough to observe that "this means they're going into the last whale refuge in the whole bloody ocean." With a wild, furious look, he turned and marched back out through the hatchway, leaving us with the feeling that it was somehow *our* fault.

It was only the gray murmuring of the sea herself, the sensuous stroking of the hull, that finally restored us. Gray upon gray upon gray, laced with sizzling toothpaste lines, the ship making a motion that gentled you out of almost any mood, gentled you to sleep, gentled you awake ... what a lover the ocean could be! Like a hostile beast being soothed, our collective negative mood slowly moved down from a twitching snarl to that blank opiate-like space of noticing everything but not being inclined to anything except what you absolutely must do. That way, we did not interfere much with each other. Life could become a tolerable routine, surely one of the unheralded states closest to heaven itself. Or so—after the hassles of being in port—it seemed then. A relief.

Rex Weyler decided to start an evening entertainment program called "The Trapper Dan Hour," during which he would act out the role of a Texas DJ "comin' at ya from the North Pacific," while sitting in the radio room, using its facilities to broadcast to the rest of the ship. He was soon joined by Gregory, Barry Lavender, Chris Aikenhead, and anyone else who could even begin to play an instrument or pretend to act. Skits, songs, tapes, and "interviews" flowed outward, keeping the rest of the crew in hysterics. Kazumi Tanaka became radio interviewer "John Rennon," who wandered about the ship using his dry Japanese wit to elicit the most absurd remarks from people, just so the regular "Trapper Dan" audience would have something to laugh at that night when the show came on. Let's laugh: that seemed to be the general desire. And the shows were so funny we came close to laughing ourselves sick.

To buttress this, there was now a daily gossip column being published and pinned up on the bulletin board. I started writing a newsletter called *The Daley Bulletin*, which dutifully reported our position, current plans, inside information, and all of that. I also became a saloon-type gossip column, capitalizing the names of the people mentioned. Frequently, *The Daley Bulletin* would do reviews of "The Trapper Dan Show." But mostly it would fire snide barbs at everyone in sight. When the *Bulletin* got posted, a happy crowd of crew members would gather around to laugh convulsively at the reports of what had happened to other people, only to pause, frowning and suddenly embarrassed, at an item concerning *them*. It was a classical journalistic situation. Becoming the shipboard columnist had one particularly attractive feature: until then, I couldn't move ten feet without someone coming along and demanding to know what was the latest news. I could never get a minute alone. But now that they knew anything they said to me might wind up in tomorrow's *Daley Bulletin,* they tended to shy away. I was finally allowed to be alone almost as much as I wanted.

The collective headspace was definitely improving. On August 12, it took another one of those instant nose dives.

We were entering a fogbank late in the day when Hal Ward took a look at the radar only to spot a blip on the screen, a rather large blip. It indicated some kind of craft four miles off the port bow. A craft at least three hundred feet long, maybe four hundred. What was odd was that we were already well off the normal shipping lanes: what would a freighter or tanker be doing out here? He called the captain.

Korotva paced nervously around the wheelhouse, muttering to himself. A whaler? No, it was the wrong size. Who the hell else would be out there? Moreover, the blip seemed to be pacing us rather than following its own course. It was not long before the fog lifted and there, in the binoculars, was a dark rectangular shape. Ward stared, then passed the binoculars over to Korotva, who saw the vessel's running lights come on, and then watched, still not willing to accept what he had seen, as the submarine began to move swiftly, effortlessly away. At that point, we were roughly one thousand miles north of Hawaii. An odd place to run across a submarine. And any submarine operating that far out in the ocean could reasonably be assumed to be nuclear powered.

A hasty meeting was called. "It was definitely a submarine," Korotva said.

The question was: Whose?

Well, it could only be American or Russian.

For the first time, the phrase *Monsters of the Deep* had meaning. No one had been afraid of dolphins, whales, or even giant squid. But now there was a slightly electric buzz to the edge of everyone's conversation. Out there, beyond our hull, was a monster capable of sinking us faster than Moby Dick might have taken out a canoe. There was no natural creature left in the ocean whom we really had to fear, but this unnatural creature now presumably circling around out there somewhere nearby gave us the same feeling that ancient mariners must have had toward the whales and great white sharks themselves. If it was a Russian craft, then for all we knew the torpedo that would sink us instantly might already be on its way. But who knew for sure that the Americans might not decide to sink us while we were closing in on a Russian fleet so they could blame it on the Kremlin? The general feeling was that we were somewhat out of our

351

depth, yet—again—there was nothing to do but keep going.

In all, the submarine was spotted seven times in the next ten days, never coming close enough for us to make out any identifying marks or get any decent pictures. At most, the object was only a black rectangle scarcely visible to the naked eye, one whose running lights could occasionally be seen. The only time it appeared was when we were in the vicinity of the whaling fleet. Once, our radios, RDF, and loran began acting strangely and a prickly feeling came over us. "We're being scanned by some pretty heavy-duty equipment," commented Al Hewitt. From that point on, we posted someone in the radio room at all times, ready to put out an SOS, although it seemed somewhat futile in the face of a steel shark that made every other creature that had ever inhabited the ocean look like a rag doll.

By August 13—the day after spotting the submarine—we picked up Russian voices on the radio, still distant, but the first clear sign that we were once again on track. Now began a nerve-racking game of blind man's bluff.

The voices seemed to be coming from all around us, some faint, some harsh and booming. A faint fog hovered over the sea. Between the fog and the echoing, crackling voices, we had the feeling of having entered a ghostly realm—a feeling that was not diminished by the fact that since leaving Maui we had not seen a single porpoise, dolphin, whale, or even a bird. There was only the gray of the sea and the water and the throb of our engines, as though we had entered a huge echo chamber in a world stripped of life except for ourselves and the mysterious black science-fiction beast that was shadowing us. Ghostly alien voices called from other dimensions.

Captain Korotva was finally satisfied that we had a fix on the Vostok herself. He altered course to bearing 240 degrees. It was the first of dozens of course changes as we frantically strove to engage a fleet that seemed determined to avoid us. The Russian effort to stay away seemed so intense that there were moments when it was possible to imagine that they were afraid of something—but what did they have to fear? Our cameras? Or was there much more going on here than met the eye? Was it impossible that the nearby submarine was, in fact, American?

If it was, it might mean that some sort of superpower chess game was being played out here in the lonely fogs of the mid-North Pacific, with only ourselves as witnesses, or rather as pawns caught in the gravitational field between two enormous entities. There could be no doubt any longer that the *Greenpeace VII* had become involved in the Cold War. It was fascinating to speculate on the possibilities. But in the meantime, regardless of whatever games were being played by the Pentagon and the Kremlin, we had one task to perform, and that was to get back between the harpoons and the whales. And it was to this task that everybody bent with surprising energy, trying not to think too much about the titanic twin military-industrial complexes whose presences we could sense, like icebergs, just beyond the horizon. In our helplessness, we felt very much as the whales must feel, aware that there is absolutely no way they can resist such forces, nor escape.

Whether the Russian whalers were trying for some reason to avoid an American sub that was pacing us, or whether they were trying to avoid further exposure in the media and it was their own sub that was prowling along beside us just to make sure we didn't try anything funny, was finally a moot point. It was clear that whatever the reason, *Vostok* and her pack of killer boats were looping about, zigzagging, changing speed, giving every sign of taking evasive action.

"God damn," said Korotva, "dey working around us!" Late on the afternoon of August 15, Matt Herron emerged on the bridge with the worst news yet: the old minesweeper had been using more fuel than anyone had expected. Since Maui, fuel consumption had increased by thirty-one percent, indicating that something might be seriously wrong with the big diesels. Unless the situation could be corrected, it meant it would already be impossible for us to make it to North America, and if the problem got much worse we might not make it back to Honolulu, either.

"We've gone from using twenty-nine gallons an hour to fifty-one," said Herron, looking shocked and weary.

Korotva stared blankly into the fog, shaking his head, knowing that *Vostok* was somewhere ahead, within less than two hundred miles. But now we were only two direct steps away from disaster. If we were to be left lifeless at sea and a storm were to occur ... well, we were roughly twelve hundred miles from shore. In theory we could always send a distress call to the Russians, but maybe we'd be better advised to simply go ahead and drown. "Cut da engines," Korotva ordered.

Ironically, we were now at a position where Japanese whalers had been operating on the same date—August 15—two years ago. Already, the Japanese had retreated westward, and the Russians themselves had fallen back to the previous Japanese position. It was impossible not to perceive in their movements two great fleets slowly being forced back across the Pacific, away from the rapidly expanding North American centers of whale awareness.

Toward the end of the day, a short-term solution to the fuel crisis was devised. Haggarty and Hewitt worked out a system of temporary compensation for the loss by operating the engines in tandem. Korotva could get back to work trying to reach *Vostok*, but he had to do it more slowly.

We were close enough to the fleet to know that they were working in a wide range directly ahead and on either side. We knew from Korotva's eavesdropping on their radio calls that they had killed twenty whales that day. The *Vostok* had ordered them to deliver the bodies immediately, a sign that the factory ship may have already picked us up on its radar, putting her less than fifty miles ahead.

Agonizingly, all we could do was crawl across the ocean toward her, not daring to go any faster than nine knots. Our previous fuel couldn't be wasted on one wild shot in the dark. It was better—although far more painful—to move in slowly, monitoring, waiting for the moment when the zigzag movement of the fleet might bring *Vostok* back toward us.

August 16 was the most unbearable day yet. While the engine-room crew measured and calculated and thought furiously of possible solutions, the *James Bay* came again to a dead stop, engines shut down. We decided to break off communications with the outside world, something we knew was going to worry the folks back home, whose voices we could hear calling over the radio, pleading for us to answer. But it was the only way we could avoid being pinpointed by a radio fix ourselves. Even without satellites, the Russians could easily track us now. They had ten ships—ten positions—from which to triangulate our own position. We shut down all our outside lights, allowing ourselves to drift silently, invisibly, in the heavy fog that had rolled in. We listened and listened over the radio, recording every minute change in their movements, but not one came within the twenty-five-mile range of our radar. All that day and through the night the waves lapped against our strangely lifeless ship. Visibility dropped to zero. You could not see one end of the minesweeper from the other. Once, the fog drew back some three hundred feet to reveal an oil slick, perhaps fifty feet across.

"That means there's a ship near us," said the old chief engineer. We could also glimpse bits of garbage floating in the water, egg cartons, a few bottles. The phantom fleet was scattered on all points of the compass around us. Late that night, all ten vessels swung westward, leaving us drifting helplessly behind.

By morning, hanging on to the thinnest thread of fuel, we started the engines and moved slowly, gingerly westward after them, not wanting to waste a single drop on anything less than the most direct course. In the navigation room, Patrick Moore and Matt Herron peered blearily at the lines they had drawn on the chart, knowing that if they had made any errors in calculation, our last chance for a confrontation would be blown.

It was not until shortly after midnight, on August 20, that we picked up a blip on the screen, eight miles dead ahead. It was coming toward us. Then it turned and began racing away. After half an hour of chasing, Korotva cursed and throttled the engines down again. An open high-speed pursuit at this stage would be a disaster. Again, we drifted. Again the navigation room and the radio room became the centers of activity. Again we were left with nothing more to do than pace the decks, straining our eyes to penetrate the fog. Despite our best intentions, a feeling of despair was beginning to set in again.

In the end, it was a combination of forces that launched our final stab toward the fleet. While Herron and Moore had been busily working out their own calculations, Paul Spong had been riveted to the RDF unit for days, meticulously noting the direction from which every radio impulse was coming. Finally, on the morning of August 20, a squad of mystics was perched up on the flying bridge, composed of Melville Gregory, Rex Weyler, and Bobbi Hunter. For hours, Gregory had been saying: *"That's the way, go that way!"* He was basing his instructions on the fact that he had been watching petrels all through the previous day as they winged their way out of nowhere, hovered over the *James Bay*, and then headed back in the direction from which they'd come. Gregory could remember John Cormack describing the petrels as "gooney birds," and so the only way he could express his notion was to say: *"That's the way the gooney birds are going!"* It was not until roughly ten o'clock in the morning that Weyler and Bobbi finally realized that Gregory was pointing his finger in the direction of the most obvious phenomenon in sight: the petrels were looking for food. Having found little or none around our minesweeper, they were heading back to the one place this far out to sea where they could be sure of a feast: the factory ship.

Down in the radio room, Spong had just come to a conclusion about where the *Vostok* had to lie, based on the RDF readings. Excitedly, he dashed into the wheelhouse, on his way to rouse the captain. En route, he collided with Weyler, who was on his way down from the flying bridge to tell the captain, too. To their astonishment, they were both pointing in the same direction.

Once awakened, Korotva took only a couple of minutes to agree with the conclusions. Also, he knew that we only had five hours of full running ahead of us, and if we didn't try it now, later would be too late. Besides, on top of the ravings of the mystics about gooney birds and Spong's sense of the RDF results, there was Korotva's own intuition. He had been prowling the wheelhouse, the flying bridge, the decks, restless, moody, glowering, his head moving this way and that, sensing. Finally, at 10:40 A.M., he made up his mind and shouted to helmsman Lance Cowan: "Damn it, *Vostok*'s dead ahead! I *know* it! Maybe just over the horizon!" The *James Bay* surged up to fourteen knots.

In the water, the amount of garbage increased. Albatrosses appeared. The Russian voices stabbed through the loudspeakers, coming from every direction.

At noon, the fog began to lift, revealing nothing except a sea burnished by shafts of gold light.

"Damn it! Open her up!" Korotva snarled. "They're just ahead!"

No one knew quite what wavelength the captain had tapped. All we knew was that he was gambling our entire remaining fuel reserves on this wild charge forward, and if he was wrong, we would never get another chance.

"Hey George, how do you know?" Moore demanded.

"I sniff dem, brudder."

At 1:30 P.M., the first blip appeared on the screen.

"We've got them!" screamed Lance Cowan. Yet it was not until late afternoon that Bob Thomas of Ottawa finally saw the ships themselves through his binoculars. By then, everyone was in a frenzy to get into the water.

For the first time, I would not be in the kamikaze Zodiac: there had been a mutiny on the way up from Maui. Almost the entire crew had assembled to announce that they each wanted a turn in front of the harpoon and that I couldn't keep "hogging" the front lines. For the sake of appearances, I argued that I had a "moral responsibility" not to ask anyone to do anything that I wasn't willing to do myself. But the truth was I did not argue with much passion. I was secretly relieved beyond measure that I wasn't going to have to get back out in front of that harpoon. Others were willing—*eager*—to do it. I found myself looking at them all as though they were crazy. To *want* to do a thing like that was insanity. And then I realized that this was indeed how I had been viewed by everyone else for quite some time now. Crazy.

I retreated to the latrines to wash the toilet bowls and scrub the floor, alternating between quietly laughing and even more quietly crying—with relief, with amusement at the absurdity of human existence, and with a surprisingly deep sense of loss, because I knew my period of office as Greenpeace eco-commander was over. I was being kicked upstairs.

So it was from the wheelhouse—shoulder to shoulder with Korotva, who as captain could not leave the ship—that I watched the Zodiacs fling themselves out across the water, moving like tracer bullets toward the rusting hulks ahead.

Moore and Spong had taken over, Moore in the racy little Avon with the deep V hull and Spong in the slightly slower moving Zodiac. Spong had taken Rex Weyler and Susi Leger. Moore had taken Peter Fruchtman and Fred Easton. The *James Bay* was sitting still, two miles from the nearest whaler. The light was almost perfect, and once the sound of the outboard engines had faded away, it was quiet and almost peaceful on the ship. There was an unusual feeling of confidence that could be traced to the fact that we had installed a radio on Spong's rubber boat and so could stay in verbal contact. Voices as clear as taxi drivers' came through.

"*James Bay* to Zodiac, do you copy? Over."

"Roger, Roger. Read you loud and clear."

The two craft were reduced to the size of mosquitoes, halfway to the horizon, when Korotva suddenly boomed:

"Dat ship! Where'd it go?"

A moment ago there had been nine killer boats and the distant brute shape of the *Vostok* spread out in a wide arc ahead. But now there were only eight killer boats and something was happening to the *Vostok*. She was vanishing.

The fog that had lifted earlier in the afternoon was drifting rapidly back toward us.

The captain wasted no time seizing the microphone.

"Dis is a command! Return to *James Bay!* Return *to James Bay!*"

But all that came back for an answer was static and the sound of something heavy bumping.

"God damn," said Korotva.

"Ah, this isn't gonna be real, is it?" asked Lance Cowan, digging into his bag of sunflower seeds.

"I repeat, Zodiacs back to *James Bay!* Dat's an order!"

It might, indeed, have been an order, but we were neither an army nor a navy. Instead of responding to the order, Susi Leger's reply was to the effect: "Don't worry. We can still see the *James Bay*. We're all right." Before anyone could think of a way to respond to the independent-minded Newfoundland lady, it suddenly dawned on us all that the Avon, driven by Moore, did not have a radio, and since it was in the lead, if Spong's Zodiac stopped, Moore and his crew would be left to keep on flying blindly into the fog.

At the controls of the Zodiac's outboard, Spong could see the problem all right. There *was* a fogbank advancing ahead. Now that he had been alerted, he could detect it. But Moore, Fruchtman, and Easton were ahead, and they weren't looking anywhere except at the ship upon which they were targeted. Spong knew the danger of continuing on into a fogbank, but he also immediately knew that he could not turn around, saving himself and his crew, only to leave Moore and his companions to get lost. Desperately, the New Zealander cranked up his outboard. Normally, he might have been able to overtake the Avon, since Moore had not yet opened up to full throttle. But the Zodiac began to buckle badly. It had not been properly inflated. The floorboards popped and the rubber walls bent like soggy sausages. He couldn't catch up.

Before either Spong, Weyler, or Susi Leger could get the attention of the three men on the leading rubber speedboat, both craft found themselves suddenly enveloped by grayness, as though it had sprung down out of the sky on top of them. Only the dark shapes of a few killer boats could be made out ahead. The *James Bay* had disappeared completely.

The two rubber craft pulled in together beside a boat whose crew was in the process of lashing a dead whale—probably thirty-five to forty feet in length—to the side. In the eerie silence, cameraman Easton and photographer Weyler focused on the whalers, and the whalers in turn brought out their own cameras, clicking back. One Russian even had what looked like a Super-8. Hurriedly, Moore and Spong consulted about how to get back to the *James Bay*. The full dimensions of the disaster had not yet struck them. They were nervous but could still imagine that they knew—roughly—where the minesweeper was.

Unfortunately, back on the *James Bay* we were having other problems. From the instant the rubber boats vanished, the Russian fleet had begun to move erratically. It had been functioning in a coordinated pattern at the time of our arrival, but since the launching of the rubber boats, the pack had broken up, as though scattering. Suddenly, there were 150-foot steel harpoon boats in motion all around us in a fog that quickly rolled over our own decks. Our film team was in the wheelhouse trying to make their movie, Ron Precious with his Bolex and Chris Aikenhead with his shotgun mike, but the bright lights they were using partially blinded the captain, who was meanwhile becoming extremely upset about the ghostly movement of ships in the blankness around him. Uttering a rare Czechoslovakian oath, he ordered them off the bridge. For a second they tried to appeal to me to get the captain to let them capture some priceless real-life dramatic footage, but at that point I knew Korotva would kill them if they didn't get out of the way. It might make a great movie, but right now there were six people's lives on the line and a possible deep-sea collision imminent. The cameras had to bow out.

Worse, Korotva had to start dodging and ducking the ships around him. That meant moving. And it meant that within no time at all, we had traveled quite some distance from the position in which we'd been the last time the crews on the rubber boats had seen us. One split-second decision was having to be made after another, each one putting us all a little further down the road toward a serious problem.

From the wheelhouse, we told the rubber boat crews to raise their radar reflectors. But it turned out Mike Bailey had failed to put the radar reflectors aboard. Someone else had forgotten to stow the emergency survival kits. And, of course, someone had failed to make sure the Zodiac had been properly inflated. This was adding up to just too many mistakes.

From the wheelhouse, I radioed: "Now listen, we can't see you, and you can't see us, so we'd better all sit still and figure this one out."

But now even the killer boat beside which the Zodiac and Avon were poised was in motion.

Back in the wheelhouse, we could at least see the phosphorescent outlines of the fleet on the radar screen. Without radar reflectors, the rubber boats were impossible to detect. We *thought* we knew which blip represented the ship beside which our people were parked. So Korotva yelled over the radio:

"*Stick* with the killer ship!"

But the Russians weren't making it easy. They were up to full throttle within minutes. The sea had joined the conspiracy, and with it, the wind. The waves were coming up with that appallingly sudden change of mood so common to the ocean. The temperature was dropping rapidly. Bounding along in the wake of the mist-streaked killer boat, surrounded by nothing else but fog, both Moore and Spong and their crews were beginning to find themselves dwelling rather more than they wanted on the thought of just how far it was to the nearest land. At that stage, some fifteen hundred miles. While, normally, the rubber boats could easily have kept pace with the killer ship, they were handicapped now not only by rising seas but by the tendency of Spong's Zodiac to buckle and bend. Finally, it couldn't maintain the pace and began to fall back. The killer boat vanished. And for a bad few moments, the two rubber boats were lost from each other. A plaintive Susi Leger, her voice shivering, called over the radio to the *James Bay:* "We've lost them. Can't see them at all. We cut the engines. We've stopped, trying to figure out what to do."

In the wheelhouse there was dead silence. Somewhere out there in the void around us were two boatloads of our crew, one with a radio, one without, the two of them separated, with the Russian fleet speeding away, leaving us all behind, blind and lost. And the wind coming up.

It began to rain. The darkness coagulated.

On the rubber boats, five minutes passed before Moore realized he'd lost visual contact with the crippled Zodiac, and managed to do an immediate 180-degree turn, taking him back almost directly into the nose of the other craft.

Susi's voice crackled across the radio, a whisper: "We've found them. We're together." The problem now was: Together where?

On the *James Bay*, we sounded the foghorn repeatedly, only to hear Susi's weary voice calling out over the speaker: "Would you *please* sound the horn? We can't hear you."

In the blankness of the wet, cold, darkening North Pacific universe, Paul Spong was desperately twisting and turning a forty-dollar FM radio he had picked up at the last minute in Hawaii, trying to pick up timed-impulse signals being sent from the minesweeper. In a nighttime test only a few days before, Spong had found that if you held the FM at right angles to the direction from which the electronic signal was being beamed, the sound would come through more loudly than at other angles. The radio could therefore be used as a primitive radio-direction-finding device. The trouble now was that he couldn't tell the difference in volume, and so still didn't know which direction to face.

As he described it later, "We were all starting to get fairly agitated. Fruchtman was extremely agitated, Moore was agitated, Weyler was agitated, Susi was agitated, I was agitated, even Fred Easton was agitated."

It was Weyler, finally, who came up with the observation that so far, they'd been speeding along with the swells. If they were to turn around and travel against the swells, that should take them roughly back in the direction of the *James Bay*. It was a long shot, depending on how far the minesweeper had moved and whether or not their recollection about traveling with the swells was accurate.

A tense two hours followed before, in response to our by-now despairing and monotonous blasting of the foghorn, Susi's voice suddenly called through the static: "We hear you! We hear you!"

Half an hour later, just as the fog had pulled back to some five hundred to six hundred feet and a faint silver-red glitter of light cut through the gray banks to signal the last flare of the day, the rubber boats appeared in the distance, and a tremendous cheer went up from the *James Bay*.

There were no recriminations about the goof-ups and breakdowns in communication that had led to the near-disaster. It was enough that everyone was back safely, and that, after all, the usual guardian angels had been at work. The main thing was that no one had panicked, either on the minesweeper or on the Zodiac and Avon. And however much we all might come to be tangled in political conflicts, emotional knots, or even sexual binds later on, for one evening—after the *James Bay* had retrieved her lost children—we were brothers and sisters again, everybody in love with everybody else. Korotva put us back on course to chase the *Vostok*, still visible on radar.

By morning, the fog had melted away and the *Vostok* could be seen lumbering along at four knots, heading back toward North America. The killer boats had sped away during the night, and only the rare trace of smoke on the horizon gave any indication that there were any other vessels within hundreds of miles except for this multicolored minesweeper and its enormous prey. For a while there was the good feeling that we were herding the fleet before us again, that the camera was, indeed, mightier than the harpoon—but toward noon, two killer boats appeared several miles off our starboard beam.

"Hurry!" Bob Thomas screamed from the flying bridge. "They're on whales!"

This was to be our last attempt at interception for the year. Within moments we had three craft in the water, piloted by Spong, Moore, and young Mike Bailey. Bailey's Zodiac swept into the lead, carrying Matt

Herron and myself, streaking for the nearest killer boat. But we were only halfway across the intervening space when the harpoon made its giant firecracker sound and a puff of white smoke told us we were too late. By the time we got there, the whale was a floating gray-blue island in a red slick, its tail thumping heavily but spastically against the steel hull of the killer boat while the crewmen leaned over the side, slipping a loop over the tail. Then the ship was in motion, dragging the whale away toward the factory ship even though the creature was still stiffening and going slack, convulsing, obviously not yet dead. The ship was soon up to lull speed and another harpoon had been fixed in place. Bailey cranked up the outboard and sent the Zodiac whirling across the waves, staying directly in front of the harpoon. After fifteen minutes of bone-bruising pounding, we saw three men surround the harpoon and cover it over with canvas. The ship was heading to the *Vostok*, having abandoned the pursuit of the pod it was chasing when we arrived.

Several miles away, Spong's Zodiac, with Michael Manolson and Kazumi Tanaka aboard, had managed to put itself in position before the harpooner could fire. The moment they swung in front of the bow, the Russian captain altered course, likewise abandoning the pursuit of any whales and plowing away to the south. We seemed to have saved one whale family and all but one of another. The one dead whale was quickly winched aboard the *Vostok*, which picked up speed and started barreling to the northwest, exactly the same tactic used to shake us off before. We gave chase for two hours at sixteen knots and then, when we reached our final point of no return, we shut down the engines and sat there in the late afternoon sun, watching the outline of the factory ship's stern diminish, grow faint, and finally become nothing more than a puff of smoke on the horizon. Several of us lounged around on the flying bridge, feeling good that we had routed the bastards.

We had been at sea nonstop for twelve days. We only had two days worth of food left and we were at least six days from shore. Yet no one objected to the idea of staying put for the time being, and going back to radio silence, in the hope that the Russians might think we had headed back to Hawaii. If they were to think that, they might swing back into this area, and we'd have another chance to sting them.

At 1:00 A.M., August 22, we got our last glimpse of the Monster of the Deep. It appeared as a cigar-shaped blip on the radar screen, eleven miles away, seventy degrees south of our bearing on the port side. When it vanished beneath the sea again, we were left still not knowing whether it had been sent from Russia or America, whether its purpose was to guard the whaling fleet or to track it—but now that the fleet was moving away, the submarine lost interest in us. We never caught a glimpse of it again.

By now we were $140,000 in debt, and, of course, the real problem lay in the fact that most of our movers and shakers were on the boat instead of on the shore raising money. Reality—in the form of the lack of hard, cold cash—reached down and clamped its vice-like grip around our mystical little necks. The vision might be as clear as ever, but the path was blocked by an avalanche of debt. My own will caved in at that point, I know.

Matt Herron emerged from the navigation room with the news that there were only four possible places we could reach, and any one of them would be out of reach if we ran into any bad weather at all—so the sooner we started the better. Using a great-circle route, we could either continue on to Midway, go back to Honolulu, head for Dutch Harbor in the Aleutian Islands, or aim for the one place on the entire West Coast of North America that lay within our reach: Patrick Moore's home town of Winter Harbour on Vancouver Island.

From almost the moment we did a final 180-degree turn and started toward the coast, the weather changed. By the dawn of the first morning of our northeastward haul, we had picked up a following sea with ten- to fourteen-foot waves sloshing sometimes above the stern, sometimes below. It was not particularly serious, even though the wooden starship in the galley was hitting Warp Factor 5 at times. But had we gone on toward Midway, we would have found ourselves running directly into the storm. It was not surprising that now we took turns standing alone at the stern, staring back into the thousand jaws of a sea that was chasing us, always on the verge of making the leap that would seize us and drag us down. The wind had turned cold. Sleet danced lightly on decks that were wet gray mirrors in which you could see a brooding silent version of yourself.

A few days before we reached Winter Harbour, Rod Marining, by then back in Vancouver, read us a news report over the radio quoting the Russian ambassador to Canada as saying that the Soviet Union would quit whaling within two years. Some of us refused to believe it. But most of the crew celebrated wildly, leaping down ladders, running about the deck, breaking out last hidden stashes of booze, laughing with a frenzy that was vigorous but forced—as though their subconscious was still completely aware that the struggle would be longer, harder, and more draining than anyone dared to imagine. Not so easily would we be let off the hook, said one's subconscious. The following day, when the inevitable denial from Moscow came through Associated Press, it was the crew members who had celebrated the day before who suffered most.

By the time we reached Winter Harbour, nobody knew what to believe about anything. We suffered the common fate of people stepping off a boat upon which they have been isolated from the rest of the human community for long periods of time: we had trouble communicating with the land dwellers. We had been so involved in our group trip that most of us quite forgot that other people had been living their own lives, only marginally aware—even if at all—of the adventure we had been experiencing. It might have seemed like the most important journey in human history to us, but to the rest of the world it was just another event.

From Winter Harbour, we moved to Alert Bay, partially to pay our respects to the Kwakiutl again, partially because Spong's wife, Linda, had arranged a benefit concert for us, and not only did we need money to pay the bills, but we needed a good party to shake loose all the muscular knots that had developed during the long, uneasy crawl toward shore. The entire crew turned out at the dance hall to join with some three hundred other people, flinging themselves madly about the floor, hollering and leaping into the air, not quite believing the huge sign that hung over the West Coast Indian-village stage: ALOHA! The dancing seemed to take our internal psychic bruises away—for a while at least.

On September 6—a Sunday—we were roaring ashore toward Jericho, with the Indian symbols on the walls of the hangars towering over the heads of the assembled crowd—and there was *still* magic in the air! For all the psychological pits we'd been through, for all the arguments and conflicts and contractual hassles and politics and bad vibes and interpersonal conundrums, ideological differences, competition, ego trips, sexism, betrayals, counter betrayals, and otherwise natural organic bumpings-together, there was *still* the fundamental sense that what we were doing was more than anything any of us had individually done before. By Paul Spong's calculations, our presence had cost the Russian whalers the loss of as many as one hundred whales in terms of direct intervention in their operations: time lost moving quickly instead of hunting, actual pods not hunted because of the presence of cameras and protestors, reduction in efficiency because of the unwillingness of whaling skippers to leave dead whale bodies floating in the water with electronic beacons so long as we were around—plus the knowledge that for the first time in decades, no whaling fleet had ventured closer than seven hundred miles to the West Coast of North America. Up until the previous year, the Russians had taken thirteen hundred sperm whales annually within that zone. What was it now that kept them away? If it was political pressure, and we had helped to generate it, then we could count another thirteen hundred whales saved, for a total of fourteen hundred. We were not coming home empty-handed. Only a year before, I'd been able to say that we had saved eight whales—now I could say that we had saved at least one hundred, if not fourteen hundred.

"Maybe next year we can save them all!" I yelled.

Three thousand people roared their approval. The hangars echoed.

On the surface, it was a triumphant return. We were the heroes returned from battle, having conquered more territory than they had the year before. There was nothing to be ashamed of, even if we had not saved all the whales in the world at one stroke. I even remember admonishing some people who were expressing frustration at not having been more effective, saying: "Look, you guys, everybody wants the revolution to happen by the weekend so you can go back to goofing off on Monday. The Industrial State can't be overthrown in a day." But the fact was that I was as frustrated as anybody. Back at Alert Bay, I had taken a rainy afternoon off to go out to a secluded beach and furiously smash rocks against the ground, howling obscenities, panting, trying to work out the complex of knots that had developed through my brain and body. Maybe passivism was just a crock of shit. Maybe the only thing to do was to buy some explosives and sink the bastards. This business of "bearing witness" must have an end point in it somewhere. I had borne witness to the point of nausea and soul-sickness, yet nothing in the world had really changed. Gandhi's approach may have been appropriate for India, but hadn't India gone on to detonate her own nuclear bomb, and didn't that make a mockery of Satyagraha, the philosophy that does not contain such a word as enemy? A hell-pit of doubt and confusion fell upon us all. The limitations we'd run into—fuel limitations, money limitations, and worse, limitations of the will—had seemed implacable, like a Great Wall blocking our path from horizon to horizon. This business of stopping a whale ship or seal ship for a couple of minutes here, a couple of minutes there, was so far from the solution we wanted that it seemed, in that incredibly bummed-out period after the return of the *James Bay*, that everything we'd done had been nothing much more than a diversion.

Maybe it was time to take up guns. Maybe I was just overwrought.

Bobbi could see that I was falling apart, so when an urgent request came from David McTaggart for me to go over to Britain to sit down with him for a couple of months and ghostwrite the story of his adventures at Mururoa and the subsequent court cases in France, she insisted I go. It would give me a chance to "dry out" after spending most of the summer on Valium. Going to Britain would also be a chance to get away from the seemingly interminable pressures of holding the burgeoning offices together despite their centrifugal tendency to fly off in every direction on their own. As for the financial and political situation at home: "Forget it for a while," McTaggart advised. "If it can't hold together without you, it isn't worth preserving. It'll fall apart anyway, sooner or later."

Paul Spong agreed to take on the job of acting president, a task that left him facing a mountain of bills and internal dissension on every side.

The two months I spent in North Wales with McTaggart, writing his book, was a relative paradise. We stayed in a little village called Corwen, in a rented cottage overlooking a meadow where sheep and cows were feeding and a river where local fishermen caught trout. While the relief of not having to deal with administrative problems was tremendous for me, it left a situation back home where rebel elements and friends alike had a chance to organize themselves and get ready for a coup d'etat upon my return.

Just forcing myself to go back to Canada—to face it—was one of the hardest things I'd ever done.

The Year of the Dragon ended with a fire that took the life of Fido, our iguana, and killed Melville Gregory's two children, Nathaniel and David. After the *James Bay* voyage, Saint Melville and his wife, Sybil, had moved up the coast and were living in an A-frame near the port of Powell River. Sybil had gone into town to do the laundry and Gregory had gone across the street to visit some friends. It was December 27. Some twenty minutes to half an hour later, the musician stepped back outside only to see his A-frame ablaze. The fire had started under a floor heater and smoldered undetected for several hours before suddenly turning the house into a torch. Gregory tried to throw himself into the fire but was held back. When the RCMP showed up, he tried to grab one of their guns to shoot himself. Nathaniel had been four, David three. Somewhere in a corner of the flaming house, the little dragon who had been the Greenpeace

366

mascot also died.

Within a matter of days, the boat on which Bobbi and I had been living sank. Someone had retied the lines improperly at low tide with the result that as the water rose, the boat was unable to struggle free and was dragged down into the murky depths of the Fraser River. We had been living on the boat since before the first antiwhaling voyage. It was our only home.

Chief engineer Haggarty decided to sue us for wages he felt we owed him and which we didn't think we did. The upshot was that a sheriff went down to the *James Bay*, moored only a few miles upstream from where my own boat had just sunk, and impounded the vessel on the spot, with the announcement that if the chief engineer's claim was not satisfied, the old minesweeper could be auctioned off.

John Cormack learned that his beloved old halibut seiner was going to have to be sold out from under him because he had insisted on going off whale saving the previous summer. The Mendocino Whale War Committee had failed to pay him the charter fee they'd promised, so the veteran skipper was left without enough funds to maintain the mortgage payments. Cormack was shattered.

Paul Watson had begun to lobby vigorously with the various branch offices to get another seal campaign going—one that would leave him the undisputed leader. Walrus Oakenbough and his Japanese lady, Taeko Miwa, had taken advantage of a donation of money from a Florida-based conservation group to go to Japan to organize a "grass-roots movement" against whaling, but had come into conflict with Paul Spong over Greenpeace's position concerning calls by other groups for a boycott of Japanese goods. They ran out of money in Tokyo and had to survive in a one-room shack, along with Kazumi Tanaka, and were enraged when the Vancouver office refused to send them any more money. George Korotva decided to head over to Honolulu to connect with the people we'd met there in order to set a new antiwhaling mission in motion, one that would be bigger, better, more effective than the "media-oriented" type of operation we'd engineered the two previous years. By the end of the Year of the Dragon, Greenpeace had effectively splintered into more groups than most of us could name. The American groups were fiercely squabbling over jurisdiction and control of funds. Most of the Canadian groups outside of Vancouver were starting to wither from lack of support or direction.

Some strange law of simultaneous increase and decrease seemed to be at work. Membership had shot up to eighty thousand, our credibility in the media was generally good, yet, internally, we were more disunited than ever. And it was not just that we had the usual differences of opinion: we had always been a fractious organization, but now there was an unexpected element of bitterness and personal resentment. In the drive to launch the *James Bay*, a lot of egos had been bruised, a lot of people had been overpowered by the leadership of the time. There did not seem to be much goodwill left. Was this the inevitable price of any kind of power? Did it have to be? Somehow, in the rapid and casual creation of dozens of different boards of directors, we had created a situation that more closely resembled a loose coalition of city-states—each with its own princes and guilds and loyalties—than the integrated ecological strike force of our fantasies. Much of the blame for the chaos, of course, was directed at me, and not without justification. My style of leadership might have been appropriate in the context of a small band of borderline outlaws, but it was completely unsuited to a miniature multinational corporation struggling to bring itself into existence. The idea of simply saying okay to anybody and everybody who wanted to form a Greenpeace group did give us a long list of addresses and names of contacts scattered at great distances—but in the absence of any recognizable structure, there was no check on individual ambitions. Virtually anybody could set themselves up as a Greenpeace office, taking more or less full credit for all the achievements to date, and appoint himself or herself to a position, using no formula more elaborate than the one we had used ourselves in Vancouver: simply, you get a bunch of your friends together in a room and proclaim yourselves. By just such a method, with only two other people present, Gary Zimmerman had elected himself to the awesome position of president of the Greenpeace Foundation of America, Inc. Similarly, presidents, vice-presidents, and directors were coming into existence like moths throwing themselves into the perimeter of bright light. And with the creation of positions of authority, based on bylaws and incorporation papers, bounded by federal, provincial, and state regulations, we had drifted far, far away from the original Fellowship of the Piston Rings. Bureaucracy had arrived, full-blown and grinning, mocking our every attempt to resist it. And with the arrival of not just one, but more than a dozen bureaucracies, the ability to make decisions became a liability rather than an asset. It now required a particularly crippling process known as "getting a majority vote" at the board level. "Unilateral actions" were out—passé. It did not seem to matter that everything that had transpired so far had been built on unilateral actions. With the passing of the Year of the Dragon, for Greenpeace at any rate, such freedom was a thing of the past.

By the time we came up on our annual general meeting in January 1977, politics—not ecology—had come to dominate our minds. In view of how badly our luck seemed to be running against us, no one dared to think anymore in terms of miracles. Instead, an atmosphere of hard-nosed politicking had taken its place. Watson had maintained his role as "rebel leader" and had pulled together a coalition of people who were disgruntled with the way things had gone before. Their first objective was to depose the current president; yet maybe I was addicted: I decided to fight to stay on. Surprisingly, it was my old comrade, Patrick Moore, who opted to run against me, backed by Paul Watson. It seemed inconceivable, but there it was. Something inside me went sour and never did recover. It had been a long time since Rod Marining and I had deposed Ben Metcalfe, and now that it was my turn to have to put my back to the wall and fend off those who were closing in on me, there was very little joy to be gained from the effort. I won the election by one vote—and was left with a board that was openly divided between financial and legal people whom I had brought in with the promise of a directorship and a "youth element" who had taken advantage of the confusion left in the wake of the *James Bay*'s long and expensive journey to mobilize support for themselves.

Behind and beneath all this surface turmoil there lay the outlines of a possible universal paradox: no matter how disgusted we might each feel about ourselves and the others around us, no matter how furious we might be with all the "extraneous bullshit," the worse it got the more deeply we found ourselves being involved, as though we intuitively understood that if there was any meaningful test in all this, it did not come in the finest hours, it came in the darkest. Exactly at the point where we most detested our brothers and sisters and they detested us—*there* we were most deeply involved with each other. It was not just love that bound us together, it was loathing too. One could think: *Am I actually putting up with his jerk? My god, I really* must *love the planet!* So the deeper into the quagmire of struggling with each other for control of the whole turf, the richer the possibilities became for discovering one's own private sense of commitment to the real cause.

Although the process was brutal, it did leave us with one distinct advantage: the only people who stayed were those whose commitment was fierce. We had addressed ourselves to some of the most desperate issues on earth, and so it was not surprising that we felt desperate ourselves. We were desperate people in a desperate time. We did not simply jostle with each other: we tended to claw and bite.

Jet Johnson summed it up best when he said: "It's a dirty business, but someone's got to do it." That became our motto. There we were, after trying so hard, so innocently, so idealistically to "defend the environment"—faced with so many bills it took a full-time accountant just to keep track of what we owed, even the best friends among us pitted against each other, struggles going on between the various branch offices that rivaled the intrigues of Machiavelli's Italy, our karma gone rancid, the essence so badly lost in the particular—most of our energy seemed to go into fighting one another rather than the common enemy. The more confused it got, the harder everyone tried to make it work, thereby confusing it even more. The more fragmented it got, the harder everyone tried to pull it together, tearing it further apart with each decisive yank. Whatever "it" was, this Greenpeace thing, it was obviously precious to us, it obviously represented some means to a common end, or we would not have clung to it so ferociously.

It might be that we entered 1977 with an organizational situation that could only be described as a nightmare, a financial situation that was ludicrous—with Vancouver carrying the debt and all the other offices exploiting the name that had been built up at such cost—but none of these factors had the effect of slowing anything down at all. Rather, the agitation and friction seemed to act upon us as though we were molecules in a pan of water being brought to the boiling point. Activity was up, not down. For every tender psyche that couldn't stand the heat, two or three more had appeared to take its place. Even though it seemed, at home, that we were pinned down and frustrated at every turn by debts and dissension, none of the other offices felt that way. Instead, they gave every sign of possessing the kind of energy we had enjoyed ourselves two years before when we'd opened our first office on Fourth Avenue. Moreover, because the head office looked helpless in the face of its debts, they knew that further campaigns would have to be financed by themselves.

By February, the broad outline of where the money was coming from could be seen in the list of crew members who had been picked by Watson and his Seal Steering Committee to head back to the Labrador Ice Front. There were people from San Francisco, Seattle, Portland, Detroit, Montreal, Thunder Bay, Winnipeg, Kitchener, Oslo, Bergen, London, and only a minority from Vancouver, these being Watson, "Generalissimo"

370

Mike Bailey, newcomer John Frizell, and Walrus Oakenbough, freshly returned from his unhappy stay in Japan.

The first act of the new Seal Steering Committee was to make sure that I was excluded from the crew. They were determined to run a "pure" campaign, which meant that there would be no further talk of negotiations or compromises with the Newfoundlanders. The fragile alliance we'd initiated during the first antisealing campaign with the fisherman's union and the outport landsmen, in league against the big commercial hunt, was thrown out the window. Greenpeace's official policy was now absolutely rigid: no seals were to be killed by anybody, not even by Eskimo or Indians. Watson had whipped up most of his support in the new offices around the argument that I had "sold out" to the sealers the previous year, and was therefore basically a traitor. And while the antisealing expedition was coming together in North America around the desire to go further, to be more radical, to take a hard line, the same process was well under way in Hawaii, where George Korotva had linked up with Ross Thornwood, a disgruntled radical who had shipped out of Honolulu on the *James Bay* the summer before. They too were going to be tougher, more aggressive, more unrelenting.

It was March 8 when the antisealing crew climbed aboard the train and headed east, a handsome collection of young eco-hawks that included three women from Norway, Vibeke Arviddsson, Elizabeth Rasmussen, and Kristin Aarflot. From London, England, came Alan Thornton—who had moved there from Vancouver—and a lady with the delightful seal-saving name of Susi Newborn. From the U.S. had come veterans Jet Johnson and Gary Zimmerman, Margaret Tilbury of Portland, Ingrid Lustig of Seattle, veterinarian Dr. Bruce Bunting from Michigan, and cameraman Patrick Ranahan. The rest were Canadians: Corre Stiller of Winnipeg, David Drainville of Thunder Bay, George Potter of Kitchener, Andrew Pines of Montreal. The remaining crew members consisted of the film crew, a lawyer, and a number of advance scouts from Montreal who had gone ahead with Vancouver organizers to set things up in advance. On balance, it was the youngest crew yet.

At the last minute, I had decided to pass my pouch containing the piece of cloth from Tibet to Watson who, as expedition leader, spokesperson, and squad leader, was going to need all the help he could get. As before, wherever the train stopped, press conferences were held. Watson set the tone from the beginning. It was a hard line: "This year we are determined to physically stop the hunt. Some of us may not be coming back." He had left himself almost no room for maneuvering.

Yet while the stamp of Watson's personality—a simple, straightforward bulldozer drive from Point A to Point B—was deeply imprinted on the style of the campaign, he lacked the absolute control he wanted. In the creation of his own "party," he had been forced to seek out allies. And now they demanded their share of the decision making power. Worse, from his point of view, he had not been able to totally deny the authority of the Old Guard back in Vancouver, so he was saddled with the presence of Dr. Patrick Moore, who as vice-president technically outranked him, and who was there to keep an eye on him. Politics would not just go away.

Among the last to join the main party was a red-headed lawyer from Vancouver named Peter Ballem. Ballem was an ex-football player who, at age twenty-nine, was already a partner in the city's biggest law firm—and a close friend of Marvin Storrow. Ballem had a broken nose and a disarming smile. He was present strictly as a legal advisor. All he knew about Greenpeace was what he had read in the papers or seen on television.

At first, everything was more or less as Ballem had expected. Chaotic. High energy. A pregame atmosphere. There was a staggering amount of equipment scattered over the reception area at Quebec airport, and he wondered whether it would all be able to fit in the plane or not. From Quebec City, the crew flew on a regular Quebecair run to Sept Iles, on the edge of the frozen St. Lawrence. Watson's action-oriented crew didn't want to waste time dickering with the Newfoundlanders, so they had opted to travel to the French-Canadian village of Blanc Sablon, on the Labrador border, within helicopter range of Belle Isle. This time they would set up a major base on the desolate ice-cased little island and leap out from there to the Front ice. There would be more tents, more people—a bigger operation all around.

So big, in fact, that, upon arriving in Sept Iles, it was to discover that the equipment couldn't be loaded on the two helicopters and fixed-wing Otter, which were all the aircraft that had been chartered. A DC-3 had to be rented. While the main party lifted off in the DC-3 and the Otter, Watson and his closest advisors made the four-hundred-mile flight to Blanc Sablon by chopper. The lead pilot this year was our old friend, Bernd Firnung, the NATO-trained Red Baron who had flown during the days at St. Anthony. By 11:30 A.M., March 13, everyone was off the ground and heading eastward toward the Gulf of St. Lawrence. The DC-3 and the Otter got through without trouble, but the two helicopters were forced down in a sudden blizzard some thirty miles short of their goal. Firnung's craft swept perilously close to an outcropping of rock. After about half an hour on the ground, they were able to get airborne again and make a dash into Blanc Sablon.

The village was far smaller than St. Anthony, and the predominant language was French. A Montreal Greenpeacer, Laurent Trudel—who had sailed on Jacques Longini's *Vega* during the 1975 whale voyage—had gone into the village ahead of everybody to check out the vibes. Stopping into a motel bar and speaking perfect Québécois, he was punched in the eye shortly after it became known he was with the seal savers. Another Montrealer, Michael Manolson—a *James Bay* veteran—had followed Trudel into town to help out with the advance work but almost immediately came down with pneumonia. A third veteran Greenpeace seafarer was also waiting for the main party: Bob Cummings, who had sailed on the *Phyllis Cormack* on her aborted run to Amchitka. He and Patrick Moore were, in fact, the two "old sea dogs" of the expedition. But whereas Cummings had originally served as underground-newspaper writer during the 1971 voyage, he was now serving as media coordinator, a role that had originally fallen to Metcalfe and later, to me. Cummings was the seal committee's choice as my successor.

Certainly he would have his hands full with this one. Not only had the small Greenpeace army descended on Blanc Sablon, but a horde of mainly European journalists was already beginning to arrive, demanding to be taken to the ice. Media? Soon there would be media coming out of everyone's ears.

The next day was spent ferrying fuel drums by Otter out to Belle Isle. The only place for the craft to land was on a frozen lake in the middle of the island. It was such a hazardous exercise that the pilot—someone we'd never hired before—came back to Blanc Sablon that night, got loaded on cognac, and quit. Another pilot had to be sent in by the company to replace him.

Some forty-five cameramen and journalists from Europe had arrived, one of them so ill prepared for the blizzard conditions that he came dressed in a pair of blue jeans with his mother's full-length fur coat. Most of them had been promised accommodation and free rides out to the ice front by a Swiss conservationist named Franz Weber, who had vowed to stage a "mass happening on the ice." Weber had made headlines around the world in the months prior to the hunt by offering first to buy off the sealers with money raised from the sale of furry toy harp-seal pups, and then, when that was refused, by promising to build a multimillion-dollar synthetic-fur factory in Newfoundland. That offer, too, had been rejected, since it was made only on the condition that Newfoundland give up the hunt. Weber's final move had been to promise to bring six hundred journalists from around the world to witness the slaughter. Of these, only forty-five made it to Blanc Sablon, which was just as well, since the village's two hotels were already booked solid.

Dropping out of the sky from places like Germany, France, England, Denmark, Norway, Switzerland, and Italy, the journalists had come expecting to find their rooms waiting for them, along with regular daily flights out to the ice. Instead, they could count themselves lucky if they were able to share a single hotel room with as few as four other equally frustrated, angry journalists, most of whom were ready to lynch Weber by the time he arrived. The fleet of helicopters he had promised had dwindled to just three machines, all of which arrived late. Across the Gulf of St. Lawrence, back in St. Anthony, Brian Davies was landing in the midst of a howling mob of Newfoundlanders held in check by some forty RCMP officers.

The stage was cluttered this year. Between Weber, Brian Davies, plus Watson and Greenpeace, there were a record number of protestors, each of whose story was being scrambled together by a thoroughly confused media corps. Davies was being mixed up with Greenpeace, Watson with the IFAW, Weber with them both, and both with him. To compound the confusion, there were only so many helicopters available for charter on the East Coast at that time. Each protest group wound up constantly accusing the others of stealing machines and pilots.

In the twelve-room Alexander Dumas Hotel at Blanc Sablon, there was a twenty-four-hour lineup for the two public phones, with reporters spilling over into the kitchen to use the family phone. Sleep was impossible because each newsman had a different deadline, and because they were mixed up in the rooms together, someone always had to have the light on, and the sound of typewriters went on through the night, even in the bar, accompanied by blaring jukebox music. One German film crew became so desperate for film that they borrowed a stuffed toy seal and hired a local man to pose on the ice, as though getting ready to strike it. They borrowed the exact angle from a year-old Greenpeace photograph. The bar was open from before breakfast until 3:00 A.M., with at least three tables filled with local youths, gaping in amazement at the scene. In a village of eight hundred people, an event on this scale was a once-in-a-lifetime experience.

And it had only begun.

By the fourteenth of March, a day before the hunt itself was scheduled to begin, twenty-one Greenpeace crew members were airlifted out to Belle Isle, where they proceeded to set up camp. On the first run out to Belle Isle to set up the base, the advance crew unpacked the boxes and bags of equipment only to find that the tents were missing. By then, the helicopters and Otter had gone. But for the fact that a French biologist had left a tent on the island after a trip out to the ice to examine the way in which the seals were bashed to death, the advance crew would undoubtedly have frozen to death overnight. As it was, six of them had to crowd into a tent intended for three at the most.

With the arrival of the rest of the forward camp crew, there was a sizable encampment, although it was not clear exactly what everyone was doing there, since it was unlikely they'd all get a chance to travel to the ice. Still, it looked good. It was too bad that hardly anyone in either the United States or Canada got a chance to see it. While Blanc Sablon was packed with European media, the North American press contingent was back in St. Anthony, covering Brian Davies and several counterdemonstrations being staged by such people as Roy Pilgrim, the man who'd led the roadblock against us the previous year. Pilgrim had abandoned his old organization, the Concerned Citizens Committee of St. Anthony Against Greenpeace and had started up another one called the Society for the Retention of the Sealing Industry. He and about 250 other wily Newfies were getting prime-time coverage across North America, while the Greenpeace forward camp at Belle Isle remained almost completely out of the picture. As long as the mainland newsmen had stories to cover in St. Anthony, without having to move more than a few hundred yards from the beer parlors, it made no sense at all to them to spend outrageous amounts of money flying northward to film a crowd of Greenpeacers sitting around in their tents. In terms of how the story got out to the mass viewership, it could not have worked out much worse. The coverage about Greenpeace, originating from Blanc Sablon, was going straight to Europe, little of it feeding back to North America, where most of the Greenpeace offices and members were. The very people who had put up the money to finance the expedition were the ones who were seeing lots of film footage of protesting Newfoundlanders swarming around Brian Davies in St. Anthony, but almost nothing of the brave little band of shivering eco-hawks camped just thirty miles north, so remote from the cameras and mikes that they might as well have been on the moon.

While the Belle Isle crew was snuggling into their sleeping bags the last night before the seal hunt began, a meeting was being staged in the school auditorium in St. Anthony, with some three hundred people attending and the eyes of North America's mass-communication system looking over their shoulders. They were addressed in turn by Newfoundland Fisheries Minister Walter Carter, the grandson of a sealer, who described the seal-hunt protesters as "a bunch of almost crackpots," and by Reverend Stewart Payne, an Anglican, who announced that all the churches in the province—Roman Catholic, United, Presbyterian, Pentecostal—and the Salvation Army were united in favor of the hunt. Standing in the sidelines, watching, was a low-profile group of Greenpeacers: Peter Ballem, Patrick Moore, Michael Chechik, Fred Easton, and a new photographer, Errol Baykal. Remarked Moore, after Reverend Payne had spoken: "To listen to him you'd think Christ was a sealer."

Then it was the morning of March 15. With the first light across the jumbled landscape of ice, the first hakapiks crashed down on the heads of the seal pups, and the blood began to flow.

Scarcely minutes later, the Bell Jet Rangers were lifting from Belle Isle and whirling northward across the jigsaw-puzzle ice pans toward the Front, and the mood in the base camp as the machines soared skyward was electric. A shiver that had nothing to do with the cold passed through everyone. The battle was on. Even those who were indignant or angry because they had not been included in the first confrontation squad were moved to the edge of tears.

In the helicopters there was an almost blissful, fatalistic sense of the moment. Below, the ice was loose and soft, and even as the Front slid into view over the horizon there was not much change. Loose and soft it remained. There were the dark specks against the early light, the ships, and there were the long claw marks of blood. For Michael Chechik, Jet Johnson, and Watson, it was a familiar sight. For Fred Easton, Alan Thornton, Peter Ballem, Mike Bailey, and *Time* photographer Arthur Grace, it was new. There was the initial shock once the sheer physical scale of the slaughter taking place below registered—then there was another, worse shock. The ice was so broken that there were gaps of two and three feet between the pans, many of which were no bigger than the floor space of an average house. The swells of the Atlantic pulsated underneath, heaving the pans about like chips, a dizzying, ceaseless swooping motion. The choppers were unable to find a pan large enough for the two of them to land side-by-side and so had to settle gingerly down on pieces of ice separated from each other by open water.

The *Time* photographer refused to even get out of the helicopter. Said he later: "The entire ocean was heaving when we landed on an ice floe. The ships kept disappearing up and down below the horizon. It was like a roller coaster." It was across this shattered, writhing mass of ice that Squad Number One would have to clamber if it was to reach the ships. No one knows what was going on in Expedition Leader Watson's mind at the time, except that he must surely have known that he was facing an almost-impossible situation, far worse than anything we'd gone through before. A single slip, and he and his crew would be thrashing in the water, trapped between uncountable tons of grinding ice. What was required at that stage was a kind of ultimate broken-field run, throwing any thought of getting back to the helicopters out of one's mind and just concentrating on going forward. It was as Gray Wolf Clear Water that Watson started his run once again, making no attempt to shepherd or mobilize his troops. He was seized by telescopic vision, his senses fixed on the distant ships, just as high-wire artists fix their own senses on the platform ahead, never the street below.

The swell was running at roughly twelve feet, which meant that there were moments out there in the middle, with the helicopters a mile behind and the ships still at least another mile ahead, when the pan over which Watson was loping was down in a trough and neither choppers nor ships could be seen—just the sky and rustling canyons of ice all around. The other crewmen were scattered, just as lonely, each on their own pan. Jet Johnson had twisted his ankle the day before and now found that no matter how badly he wanted to make it, his leg would not respond, so he had to fall back to the helicopter. Generalissimo Bailey found himself faced by too much water to cross, decided that the scene was crazy, and turned back. Of the others, only Easton, Chechik, Alan Thornton, and Peter Ballem were able to follow Watson, and only Ballem, the former football player, was able to catch up with him before he reached the ships. Ballem had the presence of mind to drag along a small rubber inflatable, which had been brought along in case of an emergency, but which Watson had decided to ignore.

It was Ballem and Watson, almost neck-and-neck, who came charging—

it must have seemed out of nowhere—onto the scene of the carnage, panting and sweating inside their gear. Watson hardly paused. Ahead of him, a sealer had just finished bashing one seal on the skull, leaving the animal twitching, not yet dead, while he went over to another pup, killing it, and was bending down to skin it, leaving his club on the ice beside him. Watson scrambled across the floe, scooped up the club, and heaved it into the water. Snarling, but making no effort to attack Watson, the sealer retrieved his club from the water and moved—determined to carry on business as usual—back to the first seal, finishing it off and reaching down to flop it on its back and begin the skinning process. While he was doing that, Watson picked up the remaining pelt and threw it, too, into the water.

"I don't think you should do that, Paul," Ballem warned. "It could be construed as theft under the Criminal Code." The lawyer found himself thinking of how absurd it was, in this situation to be worrying about *animi ferae*.

Ballem tried to ask the sealer if he was Norwegian or a Newfoundlander. The man refused to answer. Watson pointed at him, and in a sneering tone of voice, said: "Well, can't you tell he's a Newfie; only a Newfie would wear a sticker on his sunglasses." The man still had the price tag on his lens. He flushed at the insult but turned away.

The ship they were coming up on was the *Martin Karlsen*. Her decks were lined with sealers whose faces were smeared with blood, many of them no older than Watson, and most looking as though they would like nothing more than to get down on the ice and start throwing punches. They jeered loudly as the small party advanced. A large stack of pelts was piled up in a bale beside the ship, with a cable attached, ready for them to be hoisted aboard. Without warning, Watson produced a pair of handcuffs, marched up to the cable, and hooked his left wrist to the wire. Belatedly, he yelled: "Where's the camera?" But he had not given Fred Easton enough lead time, and the frustrated cameraman was down on his knees on the ice, frantically reloading the magazine. Easton had been the one who, against all odds, had captured the harpoon in motion over our heads at Mendocino Ridge, but now there was no such luck. By the time he raised the viewfinder, the action was all over.

Watson's account:

> The sealers stare at me speechlessly. The sealers on the deck of the ship begin to chant, "Haul ta bye in" and "Give ta bye a right cold swim." I feel elated

that the ship is unable to proceed. Surely they won't attempt to haul in the line and risk killing me. The sealer beside me mutters, "Ye are right daft bye, we's will kill ya fo sure."

My heart almost stops when I feel the first tug. The wire line is taut. The two dozen shipboard sealers are cheering and urging the winch man on. I am being pulled off my feet and across the ice. My parka and pants have been ripped on the sharp ice. Suddenly, beneath me there is no longer solid ice. I am being hauled through a thick sludge of slush and water. I feel the line pull upward, and my body leaves the water and slams against the side of the steel hull. Ten feet above the water I'm hoisted. The line stops, slackens, and I fall back to the water, jerked to a stop while waist deep. Immediately the line tightens again, and once more I am hoisted upward. Once more the line slackens and I fall, this time plunging into the water up to my neck. Again I am hoisted, again I'm dropped. I'm a mouse on the end of a string being treated as a plaything. On the fourth pull, my belt breaks beneath the strain, and I find myself falling five feet into and under the frigid waters of the North Atlantic. The shock immobilizes me. I can't move my legs, my arms are numbing, my chest feels as if it's on fire.

Into his tape recorder Ballem said later:

I hadn't given much thought to the danger to Watson in this maneuver, and he certainly hadn't consulted me before doing it. My impression was that he would swing over the water and then be hauled up the side of the ship. I was very surprised when the line came to a stop with him in the water. I remember him being in the water and me yelling to the ship's crew, "For Christ's sake, haul him up the rest of the way." I took some pictures of him in the water, but I didn't really think a rescue attempt was going to be necessary, because I felt certain that the crew would haul him into the ship. When they didn't get him out of the water and aboard the ship quickly, I was yelling at Al Thornton to get the raft, because Al was closer to it, throw it in the water, and get over to Watson. I suppose the reason I didn't do that myself in the first place was that Thornton was closer to the raft, and I wasn't particularly anxious to go in there. The next thing, Thornton did get the raft off the ice and into the water and was skimming toward Watson, but he wasn't hauling him in or getting to him, so I jumped into the raft. In doing that I got half soaked from my waist down, but I was able to reach Paul and haul him half aboard. Once we had him hauled onto the raft, I was able to reach with the paddle to Chechik, who grabbed it and pulled the raft in.

It was only when they had the dripping, shivering protester safely laid out on the ice, with Ballem's coat under him, that Fred Easton was finally able to get his camera working. There is some very interesting footage of Watson lying there, shouting at the sealers, and the sealers shouting back, but it was not quite the stuff of which dramatic newsreels are made. Watson looked as helpless himself as a seal at that point.

It was now that the "objective" lawyer found himself taking charge of the operation.

A great deal of yelling back and forth occurred between the protesters on the ice and the sealers on the ship. When Ballem identified himself as a "lawyer from Vancouver," one sealer shot back: "That's too bad for you." They seemed content to leave Watson there in his soaking survival gear in below zero weather. It took Ballem thirty minutes to convince a fisheries officer to convince the captain to let Watson be taken aboard before he died of exposure. A stretcher was thrown down, but no sooner was it hauled aboard than the sealers deliberately dragged Watson through a pile of bloodstained pelts. Three of them tried to prevent Ballem from climbing on board after his client, pounding their hakapiks against the iron railing and swearing. But the burly ex-footballer was in no mood to be turned back. He shouldered them aside, strode across the deck, seized the stretcher from the men who were gleefully dragging Watson through the blood, and snapped: "All right, you've had your fun. This fellow could be hurt. Smarten up."

He ushered Watson down to a cabin near the stern, got his clothes off, got him into bed, and checked him over. Aside from some tenderness in his back, there was no damage. But it seemed best to Ballem to keep him on board overnight, rather than risk further exposure now. Through negotiations with the fisheries officers, he arranged for the faraway Greenpeace choppers to be allowed to come in close to pick up Easton, Chechik, and Thornton, while the lawyer stayed overnight on the ship with Watson, guarding him.

By the time the remaining members of Squad Number One got back to the base camp at Belle Isle, the light was waning. Alan Thornton put a hurried and garbled radio message through to Blanc Sablon, saying that Watson had suffered back injuries and broken his arm. The story went out immediately—especially to Europe. By the time it appeared in one particular German newspaper, it had gotten so distorted that Watson was actually reported dead. By the time Ballem's radiophone call from the ship on the real situation got through to Moore, it was too late to head off the reports that were being carried on the wire service. The result was that when Watson showed up in Blanc Sablon the next day, without his arm in a sling, our credibility in the media took a swift nose dive. It did little good that media coordinator Cummings assured Watson that he was "the news of the hour." In fact, his moment had come and gone. In the credibility lag that developed, the assault on the ice by Squad Number Two got lost completely.

While the expedition leader was being taken to hospital in Blanc Sablon to be given a checkup, Gary Zimmerman and Walrus Oakenbough led another six people from Belle Isle back out to the ice. The conditions had not improved. It took the weary band close to an hour to pick their way across the heaving pans, jumping from floe to floe. But this time, they stayed close together and tried to act as a team. They placed themselves between seals and hunters, turning back more than a dozen attempts to kill the pups. Laurent Trudel and Zimmerman marched into position in front of the sealing ship *Theron* in an attempt to block it—but made the mistake of facing the captain. Instead of bringing his ship to a halt, he pushed right through, smashing the ice apart, forcing both Zimmerman and Trudel to jump for their lives at the last minute. Determined, Trudel dogtrotted along beside the ship until it halted, then marched back in front of the bow. This time, the skipper decided against another run. Instead, he put his ship into reverse and backed away to go look for another patch of seals where there were no protesters in the way. A thin wind was whipping the ice as Squad Number Two stood there, watching the *Theron* retreat, leaving some one hundred pelts behind, unclaimed. In practical terms, the assault by Squad Number Two was the most successful to date: not only had they saved individual seal pups, but they had forced a ship to back out of an area where pups were still numerous and still alive. When they finally clambered out of the helicopters into the embrace of their comrades back at the base camp, they were ecstatic with the success of the operation.

The media at that stage could not have cared less.

For the reporters at Blanc Sablon had what they really wanted: a phenomenon.

French actress Brigitte Bardot had arrived to protest the hunt.

In fact, not only was the bravery of Squad Number Two totally swamped in the wave that preceded the actress, but so was the return of Watson from the battlefield. A press conference had been scheduled at the hotel, but it was a bit of a flop. Lights and a backdrop had been set up, and Peter Ballem remembered thinking to himself: "Gee, they're going to be terribly efficient in their press conference with us." But there were only a few perfunctory questions.

The preparations were all in readiness for the arrival of the star of such films as *And God Created Woman*.

Brigitte had not been having an easy time of it herself. She had flown from Paris in a Corvette chartered by Franz Weber but ran into the same problem getting helicopters as the Swiss conservationist himself. Stopping in St. Anthony to hold a press conference, she found that the reporters were uniformly hostile. They nudged each other in the ribs every time she tried to speak and laughed in her face. Blanc Sablon proved to be no easier. In her journal she wrote:

> At 5 o'clock I try to pull myself together for a press conference, but my eyes have circles under them and my clothes are crumpled from two days in my bag. The hotel where the press conference is being held looks like a trapping lodge—the place where I am going to spend one of the most painful hours of my life.
>
> Several hunters have slipped in among the journalists.

Franz [Weber] tries to speak softly into the microphone, to explain why we are there, but no one listens. At this moment, a strength rises up in me and I get up. I look them all in the face and order them to shut up. There are some rumblings. I am still looking them straight in the eye when finally a silence falls over the room. I approach the subject directly. I am not here for fun and I am speaking for the entire world. I will say for the umpteenth time what I have to say. I am strong willed and relentless. Insidious questions are hurled at me. They ask me if I will attend the massacre as if they were asking about a premiere at The Lido. I am sick and tired of all this. Since they don't understand anything, I use the strongest means. "In Europe," I say, "you are called Canadian Assassins." The word is out. After a frosty silence, the room again becomes agitated. I am at the edge of my nerves. I look out into a sea of black camera holes pointed at me.

During this terrible tumult, a voice comes out from the crowd, "Miss Brigitte, would you like to show the journalists a baby seal who had been freshly killed this afternoon?"

This is too much. There is the little body of a still-warm baby seal in a plastic bag. I feel like vomiting as tears come to my eyes. I get up, thank the audience, and make for Franz Weber's room, my eyes blurred by my tears. I will cry a long time there, desperately. ...

The wind is howling outside—the house is cut off from the rest of the world. It is impossible to try and get a helicopter on the ice fields. Life has come to a standstill.

Two men arrive on foot, covered with snow. They are members of the Greenpeace Foundation, and have learned that we are having trouble. They are putting two helicopters at our disposal. ...

The two men were Patrick Moore and Peter Ballem, both of whom had come to the conclusion that there was no way of competing with Brigitte for the attention of the press. The actress's love for animals was well established. She was a known opponent of the fur industry. But more to the point, she was a living legend. She had not been outside of France for several years. There could be no doubt that her movements were monitored like those of a queen. Just in making the move to the East Coast of Canada—albeit surrounded by an entourage of her own cameramen and her lover, Mirko Brozek, who seldom moved more than a few yards away from her—she had generated worldwide coverage of the hunt, more than had yet been generated by the combined efforts of Brian Davies, Franz Weber, and Greenpeace. If the objective was to focus attention in the mass media on sealing, there could scarcely be a better way—at the level at any rate of sheer volume of coverage—than to bring the world's most famous "sex kitten" into action on the side of the milky-furred pups. Whatever clout or fame Greenpeace might have earned in the previous years, it was a drop in the bucket compared with the publicity tidal wave that was automatically displaced by her mere presence. Until her arrival, the seal-hunt "story" was all blood and death, but now it was blood and death and *sex*. No more potent combination could be put together.

Of course the decision to exploit Brigitte's presence by putting the two helicopters at her disposal was not reached without there being a bloodbath of another order. On Belle Isle, there was outrage—especially from the young women involved. The Greenpeacers were all very much caught up in their own drama, suffering already from a degree of tent fever, and quite thoroughly isolated from the rest of the world. Media coordinator Cummings argued that we would "look bad by associating Greenpeace with Bardot." Watson and the rest of the steering committee did not much feel like being upstaged by the forty-four-year-old actress, but the fact was that they had already been swept inexorably into a back corner of the stage anyway.

The fight that broke out behind the scenes between Moore and the others came close to blows. In Jet Johnson's view, Moore "succeeded in making a fool of himself and he alienated even the gentlest of our crew members." But in the end he managed to overrule the steering committee, and Brigitte was given her chance to fly to Belle Isle.

To Bernd Firnung fell the honor of piloting the lady's helicopter. Brigitte's entry in her journal for Friday, March 18, went like this:

The alarm goes off at 5:50 A.M. It is cloudy, and we get dressed for the ice fields where the temperature is dipping to minus 20 degrees Centigrade. I am a bit worried—it's a difficult expedition and I am not used to this difficult existence. But my love for the baby seals gives me strength. At 6 o'clock there are two helicopters outside the house with two members from Greenpeace.

As we approach the Greenpeace camp on Belle Isle, the weather becomes more and more cloudy and the snow starts to fall. There is absolutely no visibility. Just above Belle Isle, the wind whistles and the snow swirls. Our pilot can't see anything anymore. I am afraid. It would be so silly to die like this. Our helicopter is jostled. The pilot decides to land without knowing where he is. Here we are lost in the middle of a deserted ice field without a radio in the middle of nowhere—no gas, no other life. The Greenpeace camp is around this area—somewhere, but the island is relatively big and the camp relatively small. We are thrown about in the storm, the wind so strong that we have to shout to be heard. The pilot sends out an S.O.S. message, hoping that someone will pick up the call and guide us to the camp, but the wind is blowing in the wrong direction and nothing can be heard anyway. I am scared and numb with fear. But then again—I think of the baby seals and the confidence that so many people have put in me to fight for their lives. I think of my life at home—my animals—my friends—my home—all which is so comfortable. I am not used to this life, but I must think of the lives of the seals.

One of the two helicopters leaves without any passengers. The pilot has decided to be responsible himself in the risk for reconnaissance. Five minutes later there is a replacement. I am the only one to stay in the helicopter, the others take turns warming up. A miracle: the second helicopter comes back, flying low. The camp is only one kilometer away and we leave, flying so low that you would think we were in a car. Suddenly, I see about a dozen tents and a flag through the thick fog!

The Greenpeace people welcome us with open arms. They are all between the ages of 20 and 35 years old with 4 women among them. They are all nationalities and working as volunteers. They are like apostles.

They show me their kitchen—a Bunsen burner and an ice wall protecting it from the snow. There is a casserole covered in snow which is cooking—for soup. They offer me a cup of hot chocolate. Never before has hot chocolate tasted so good to me.

I am cold. My feet are freezing and I go over to one of the tents. Even on my hands and knees, my head touches. The wind whistles around, and there is no heat source. They sleep four to a tent in sleeping bags. At night the temperatures dips to -40 degrees.

They lend me special shoes with enormous socks. They make me drink rum from a bottle—their way to get warm. The tent breaks, the wind is always strong, but they tell me that today it is warm in comparison to other days. I admire them. What courage and devotion ...

Time passes by but nothing changes while we are waiting. If we have to spend the night here, I ask myself with terror if I will have enough courage to test myself again. Laurent, one of their members, plays the flute. He plays well. He tells me about the whales who love the sound of the flute when their group goes to protect them. We will fight together.

The wind has died down a bit outside. The pilots decide that we can afford the risk of going back to Blanc Sablon. It would be unthinkable to go out on the ice field. We leave the encampment in an atmosphere of warmth. They are fabulous people.

Vivent Greenpeace!

By the weekend—March 20—it was clear that the camp would have to be evacuated. The forecast was for winds up to eighty miles an hour, enough to sweep the camp, occupants and all, over the glassy edges of the cliffs, just as the entire camp had been destroyed the year before. The evacuation turned into yet another near-disaster itself, when one of the helicopters was forced down by a whiteout some five miles from Blanc Sablon. Local snowmobile owners mobilized a rescue party to pull them out of the deep drifts in which they were trapped, freezing. To the very end, the expedition had played on the edge of tragedy. It was not just Brigitte Bardot who had moments of terror. All the others felt it too. The helicopters seemed as fragile at times as eggshells.

The second Greenpeace campaign to save the seals was over.

Out on the ice fields, Norwegian ships had taken 24,043 of their quota of 35,000 harp seals and were moving eastward in preparation for the opening of the season on hooded seals. Canadians had killed another 15,688. They were not quite done either. But now they would be free of even the slightest degree of harassment. Brian Davies had been arrested and charged with violating the Seal Protection Act by landing too close to a seal and flying under two thousand feet. Brigitte had gone back to France. Franz Weber had retreated, licking his wounds, to Switzerland, pursued right to the last minutes by furious European journalists who never did get to the ice. And Greenpeace's helicopter contract had run out. In the American and Canadian offices, there were demands for Media Coordinator Cummings's head because the coverage at home had been so poor: the Newfoundlanders had come out of it looking good.

Elsewhere, in Hawaii, a cauldron of energy was brewing around plans for the "biggest ever" antiwhaling voyage. It did not seem to occur to anybody that we had just been through our "biggest ever" antisealing voyage, and the results were not exactly up to expectations. Did we really need another "big" campaign? Could we survive it?

Watson's seal campaign was supposed to have been a test of the new network, to see if the offices really could cooperate on a simultaneous effort. It was supposed to have been the trial run of the "super group" concept: a multitude of Greenpeace groups striking as one force. All things considered, it had worked out not too badly. If the postmortem criticism was harsh, it was also very positive in at least one way: everybody pledged that they just wanted to make sure these mistakes wouldn't happen again. The drive was still toward carrying on, getting better and bigger and stronger, acquiring more and more brute political force.

For the original Kitsilano tribe, it meant an entirely new life, a horizon that had expanded from the mere abstract thought of a global culture to the actual experience of moving about in one, living in it daily, communicating almost at whim with people tens of thousands of miles away. A whole new tribe—and yet, the *same* tribe. There seemed to be certain characteristics by then that were common to young new Greenpeacers I kept meeting in places like Honolulu, London, Montreal, even Los Angeles. Although it was a wild, wild blend of old folks and young folks, tough, practical technicians and visionary animal freaks, dropouts, conservationists, action junkies, traditional Quaker types, and a few flinty-eyed businessmen, it was predominantly what I might get away with calling the Post-LSD Generation. That didn't mean it was made up of acid freaks—it meant simply that it roughly reflected the attitude that had finally begun to emerge out of the apparent wreckage of the psychedelic revolution, which had, in its own way, affected society at large. What its potential was—God only knew. But it was an electrifying kind of consciousness to keep running into in cities that were sometimes halfway around the world from each other. It was proof to me—there it was in front of my eyes—that a genuine planetary consciousness was coming into existence, and these issues Greenpeace was tackling were the flash points of its awakening.

There could be no doubt that there was a special thrill to be had for a Vancouver kid to walk down Whitehall and find the small GREENPEACE, LTD., sign on a narrow door right in the center of London. Of course there were five floors to climb: no elevator. The office was tiny, with a dirty gas stove in the kitchen and a jar of honey beside the coffeepot. But there were close to a dozen intense young men and women, pounding typewriters, keeping accounts, clipping newspapers, drawing lines on navigation charts on the wall, heatedly discussing the "correct policy" to be adopted toward North Sea oil, nuclear reactors, the dumping of chemicals in the Mediterranean. ...

Or to stand on the cracked garbage-strewn pavement outside the San Francisco office on Second Street, looking up at the whale banners and Kwakiutl symbols hanging in the windows on the second floor, where maybe two dozen busy souls were doing inventories on Greenpeace T-shirts and whale bracelets, and polishing Greenpeace position statements on everything from boycotts to dolphins to the two-hundred-mile limit. It was particularly enjoyable—and privately amusing—to pick up a glossy press package, complete with photographs and even a biography of Gary Zimmerman, that included statements on both "The Greenpeace Philosophy" and "The Greenpeace Ethic." The philosophy was:

> Just as Copernicus demonstrated that the earth is not the center of the universe, so ecology teaches that man is not the center of life on this planet.
>
> Through the study of ecology, man has embarked on a quest toward understanding the great systems of order underlying the complex flow of life on our planet. Ecology has taught us that the whole earth is part of our "body," and that we must learn to respect it as we would respect ourselves. As we feel for ourselves, we must feel for all forms of life: the whales, the seals, the forests, and the seas. The tremendous beauty of ecological thought is that it provides a pathway to an understanding and appreciation of life itself. That understanding is vital to the continued existence of life itself. That understanding is vital to the continued existence of our environment and ourselves.

The ethic was:

> The Greenpeace Ethic is one of personal responsibility and nonviolent confrontation.
>
> According to the ethic, a person who witnesses an injustice becomes responsible for it. He must then decide whether to act against the injustice or to let it continue. The choice is a matter of personal conscience.

Personally, I found the ethic to be a bit corny, but the philosophy was nicely phrased. ...

Or to take an elevator in Montreal and ride some twenty-four stories up through the innards of a soaring silver tower, padding along with a bunch of obvious hippie types through faintly hissing glass doors, over plush blue carpets, into a small back office looking out across the city to the faint thread of the St. Lawrence, while everybody giggled nervously about how weird it was to be operating out of a space that had been donated by a big corporation executive. Yet there was the typewriter, also donated, an immense electrical job, even a donated secretary to go with it, and there was the Greenpeace newspaper the group had just published—in French! And next door was the telex, which had also been donated. ...

Or to go down into the depths of the most seedy and worn-down-looking section of Toronto, surrounded by narrow greasy restaurants in brownstone buildings, with the CN tower a faint scintillating needle shape rising beyond rows of tenements, watching while another dozen young men and women noisily debated the best way to go about infiltrating a nuclear power plant to demonstrate the vulnerability of such installations to sabotage. Above their heads, tacked and taped to cracked yellowish plaster walls, were the usual blown-up photographs of Greenpeace Zodiacs closing in on the *Vostok*, of Jet Johnson kneeling over a harp seal with a hunter looming behind him, and there, on a warped wooden table, was the usual display of Kwakiutl T-shirts and buttons. A cold wind blew through a hole in the floor, rustling the pamphlets and newsletters. ...

The same pamphlets and newsletters that would be fluttering under the old-fashioned electrical fans in the office at 404 Piikoi Street in the middle of a sagging, semi-abandoned shopping complex in downtown Honolulu. Despite the fans, the heat was staggering for a Northerner. Hastily hammered-together plywood walls separated the cubicles where some two dozen deeply tanned young people were jabbering over the phones, typing, counting merchandise, tabulating bills, chewing on pencils, swirling together in little groups for one meeting after another, somehow managing to ignore the allure of the surf making its muffled booming sounds on the sands just a two-minute walk away. Entering the office, a brilliant burst of color hit the eye from all the aloha shirts, as though so many butterflies were hunting together. ...

These scenes were repeating themselves in places like Portland, Victoria, Eugene, Detroit, Paris, Tokyo, Wellington: the same posters, although perhaps in different languages, the same stacks of publicity photographs, the same symbol on the letterhead—a symbol that had been designed by Barry Lavender back in 1975, depicting the globe inside an ecology symbol—and, as often as not, the same blown-up black-and-white photo of an orca named Charlie Chin leaping out of the water just at sunset. Wherever that particular photo appeared, you knew that Paul Spong had been there at some point in the previous few years, as he had wound his way around the world, turning people on to the whales. Even in the midst of exponential growth, there were a few cables that linked the organization, however tenuously, to its own recent past. The Charlie Chin posters were one. The Kwakiutl connection was another. The Greenpeace symbol. Pictures of the *Phyllis Cormack*. Striking oval-shaped whale buttons. Ornate olive-green bumper stickers. And pictures or postcards or paintings of rainbows. The rainbow was still our basic unifying theme. It was exciting, amazing, electric. ...

But it was too much for me, man.

On April 20, I submitted my resignation as president and chairman of the board.

After years of feeling so passionately about the Greenpeace fantasy, I suddenly felt nothing but apathy. The ongoing energy of those around me seemed repugnant, embarrassing, silly. It was as though a gyroscope in my brain had shifted, directing the focus of my consciousness to another window. Looking out through it at my old friends and comrades, I found that I did not really have anything to say to them anymore. Out of all the experiences in crowded boats, crowded offices, crowded cars and apartments and homes, all the thousands of intense conversations and soul-baring night-long rap sessions, the shared moments on the high wire, the brotherhood and sisterhood had come a surprising feeling of loneliness that was in most ways deeper than it had ever been. Overload—it was just plain overload, I guess.

There was no doubt who should succeed me: it was obviously Patrick Moore's turn. He had been my own ecological guru for so long, it seemed inevitable to me that sooner or later he would run the organization anyway. Whatever the outcome of his regime, it *had to be*. Everybody else seemed to have a special focus, whether whales or seals or bombs or nuclear reactors. He was the lone inter-disciplinarian. Whether he would be able to lead people or not would depend on himself and the flow of events hurtling around him. He was eager to try it. I felt oddly guilty passing it on to him, knowing that it would be harder than he thought, and that the path would not likely change from a toehold on the edge of a precipice into a four-lane highway; it was doomed, I was sure, to remain nothing much more than a toehold. There was something in the nature of the exercise that suggested it wouldn't change. Moore's role would be to see if theories of ecology could mesh with the realities of human dynamics. He seemed to think they could. That was excellent, because I was beginning to have my doubts, and no organization needs a doubter at its head.

Moore had not been president more than a month, however, before we came up against a political crisis that also seemed to involve leftover karma. The karma, in this case, was Paul Watson's.

No one doubted his courage for a moment. He was a great warrior-brother. Yet in terms of the Greenpeace gestalt, he seemed possessed by *too* powerful a drive, too unrelenting a desire to push himself front and center, shouldering everyone else aside.

His "crime" at one level was straightforward. By tossing the sealer's club and the pelts in the water, he had, indeed, violated the Criminal Code, and within less than a month, the San Francisco office was finding that its application for status as a charitable organization—an application whose success might have opened the door onto hundreds of thousands of desperately needed dollars—was being blocked by the Internal Revenue Service on the grounds that "illegal actions" had been taken by Watson during the seal campaign. The action had been unilateral. Even though there had been offices across North America depending on him, as our most visible spokesman of that moment, to stick within the framework of nonviolence and non destruction of property, he had blithely ignored the larger obligations to the survival of the organization as a whole. He had behaved like a young bull in the china shop of our rapidly expanding network of delicate relationships with governments.

Internally, he had committed a worse crime. He had consistently gone around to the other offices, acting out the role of mutineer. Everywhere he went, he created divisiveness. The experience out on the ice front had not changed him. He came back from the trip and promptly flew to Hawaii to get involved as the "anti-Vancouver" rebel leader, leaving his former allies, Frizell and Bailey, behind to pay for the bills that had been run up, far in excess of the seal-campaign budget. Basically, he had declared open war. It left us, back in Vancouver, with no choice but to remove him—by an unanimous vote—from the board of directors. No one felt good about it. We all felt we'd gotten trapped in a web no one wanted to see develop, yet now that it had, there was nothing to do but bring down the ax, even if it meant bringing it down on the neck of our brother.

We had evolved into a political party whose scope was both planetary and as immediate as one's own backyard, for it was one of the first real glimmerings of an *ecological* party—almost a contradiction in terms.

Yet ecological thought demands that you think simultaneously about the entire sphere of life in the world and the goings-on of the garbage man in your own backyard. It requires a positive kind of doublethink. You have to think big, very, very big, and small, very, very small, at the same time. What is happening is that the macro-picture has to be dealt with as directly as what is happening in the microcosm within. Trying to think that way makes your head sore, even though you know it is necessary. If this is the kind of attitude that ecology is forcing upon us, it has some amazing parallels with Zen. For most people in Vancouver, the hustle and bustle of the preparations for the "biggest-ever" antiwhaling campaign down in Hawaii was distant, removed. What we needed was an event to pull us back together in our new configuration with Patrick Moore as president and myself retired to a position as just one of the directors.

It was almost a classic example of the wish fulfilling itself instantly. It was also a chance for us to pause in the midst of our global worrying and get back for a moment to our roots. "Our" roots meaning the roots on the West Coast, dating back to our first encounter with the Kwakiutl on the way to Amchitka, long before the exponential growth curve had hit us or the super group had begun to take shape, back when we were fresh. ...

We'd known since the late 1960s, of course, that sooner or later the supertankers would be nudging their massive blunt noses down along our coastline from Alaska, looking for a place to discharge the stuff they'd sucked up out of the bowels of the tundra. One of the places that was being considered as the major West Coast oil terminal was Kitimat, at the head of Douglas Channel, in the very heart of the British Columbia shoreline. It was adjacent to ancient Kwakiutl territory, in the land of the Tsimshian Indians.

The annual North Central Municipal Convention was being held in early May on board the M.V. *Princess Patricia*, a three-hundred-foot pleasure cruiser that regularly plied the Inside Passage to Alaska.

Normally such a voyage would have attracted hardly anyone's attention, except the one-hundred-odd mayors, aldermen, and their wives from the municipalities, villages, and towns involved, who were being sent on the cruise by the taxpayers to exchange notes, have a few drinks, and enjoy the view. But this year, the event was being extended two days at a cost of twenty-five thousand dollars to allow the delegates to the convention to have a "firsthand" look at Douglas Channel, the route being proposed by an oil consortium called Kitimat Pipeline Company, Ltd., for future supertanker traffic. The expenses were being covered by the oil consortium, and the politicians, after having viewed the route for free—and having a scenic cruise out of it in the bargain—would all go back home thinking how nice it would be if the supertankers were to be allowed to use Kitimat as a pipeline terminal. On board the vessel during the extended cruise would be the president of the consortium and a half dozen aides, on hand to answer any questions. It was a buy-off if there ever was one. For the Vancouver Greenpeace group, it was as though a red flag had been waved in front of our eyes. Debt or no debt, internal squabbling aside, we had a boat ready to go within two weeks, with the purpose in mind of going up to Douglas Channel to blockade the *Princess Patricia* when she tried to get through to Kitimat with the politicians and oil men on board. The vessel was the sixty-seven-foot M.V. *Meander*, renamed *Greenpeace IX*, skippered by a young seaman whose face looked familiar. ... We had connected somewhere in the past. His name was Dennis Feroce. And where we had connected had been a fine spring afternoon in 1975 at Jericho when he had appeared out of nowhere with his landing craft to whisk our crew from the beach out to the waiting *Phyllis Cormack*, saving us the embarrassment of having to stand around helplessly in front of some twenty-five thousand people, unable to get to our boat. So far as Feroce was concerned, as long as enough T-shirts and buttons could be sold en route to Kitimat by stopping in at little ports to cover expenses, the *Meander* was at our disposal. And a beautifully appointed craft she was, intended for luxury cruises, complete with stained-glass windows, stereo, a liquor cabinet, and a lounge on the stern deck where you could sit underneath a canvas canopy. By the end of the first week in May, the *Meander* was heading north from Vancouver with a crew that included veterans Rod Marining, Melville Gregory, and Walrus Oakenbough. It also included Bill Gannon and Linda Spong, both of whom had been working behind the scenes for a long time but had not yet taken part in a voyage. The boat was loaded not only with T-shirts but speakers and amplifiers and wah-wah pedals, electric guitars, and Linda's violin. At each port, the *Meander*'s crew would sing songs, show films, and generally beg for their supper.

The day before the actual confrontation, Moore and I flew to Prince Rupert and hopped on a United Church missionary boat, the ninety-foot *Thomas Crosby V*, which was on its way down to Douglas Channel to join the blockade. Also en route were the top executives of the United Fishermen Food and Allied Workers Union, traveling in their own high-speed motor vessel. The rest of the flotilla of protest boats would be made up of fishing boats manned by Indians from the village of Hartley Bay, on the edge of the channel. For these people, the stakes were the highest. As one of them put it when we finally arrived there, "Just the wake from one of those supertankers will ruin our life." At Hartley Bay—a "dry" community, meaning no booze allowed—the shore was still so free of pollution that the people could live mainly off of clams, oysters, abalone, and seaweed picked up on the tidal flats in their front yards. Perhaps a dozen other West Coast conservation groups had also sent representatives to join the blockade.

Long before the *Princess Patricia* emerged from the fog at the mouth of the channel, the protest fleet knew she was coming. Fishermen had been reporting her movements for days, relaying the information from boat to boat along the coast. Each house in Hartley Bay was equipped with a CB radio. Reports on the *Patricia*'s latest position had been coming in on an almost hourly basis. The high speed UFFAWU boat had pushed ahead into the fog to do the last-minute scouting.

Hail was clattering on the decks, and visibility was down to three miles when the *Meander* and a Dunkirk-like fleet of little boats slipped away from the dock at Hartley Bay to ambush the big cruiser. There were whoops and hollers. Huge helium-filled balloons were wobbling above several of the boats. Engines were revving up as vessels ranging from the *Thomas Crosby V* to herring skiffs and seiners chugged out into the channel. The Warriors of the Rainbow felt very good about this one. We were surrounded by our brothers, the Tsimshian, by fishermen and missionaries and raincoast ecology freaks, and even though most of the high treed slopes of the mountains around the channel were lost in a fog cloud, we could sense them all around us—in some mysterious mountain way *watching* this little rerun of the old Wild West days, except that this time the Indians carried no weapon more fearsome than a Super-8 camera, and our own weapons were musical instruments and loudspeakers. It was also a magnificent relief to be back in action, able for a few hours to forget the horrendous politics of international environmentalism. There were some twenty to twenty-five boats out in the water now, strung out across the channel, with messages flying back and forth on the CB radios.

The tension along the blockade line was something like a football team bracing itself for the onslaught. The United Church vessel had taken on the task of radioing the *Patricia*, asking the cruise ship to stop so that the village chief and a delegation from Hartley Bay could go aboard to deliver a statement opposing the construction of an oil terminal at Kitimat. It did not seem too much to ask since the whole purpose of the *Patricia*'s cruise up Douglas Channel was to show the conditions a supertanker would have to face. But the cruise ship's skipper refused to acknowledge any of the appeals from the *Thomas Crosby V*. When she finally heaved into sight, a great wash of white water was boiling around her bow. She was coming at full speed—which surprised just about everybody.

The consensus that morning had been that the *Patricia* would stop, rather than risk a collision with the little boats strung out in front of her. There was a definite pause apparent all along the line of protest boats as the realization sank in that the six-thousand-ton cruise ship was not only not stopping, she had turned the corner into the channel and was picking up speed. She was already up to about fifteen knots. It was Dennis Feroce who broke the paralysis by barking into the radio: "Everybody in front of her bow!"

The protest boats became unfrozen. They began to converge, slowly at first, but picking up speed, until many of them were leaping clear out of the water, moving like chips drawn to a magnet. Most of the boats until then had been concentrated on the southern side of the channel just in case the *Patricia* tried to do an end run into another channel beyond. But instead she was slicing directly along the northern side where the blockade line was thin. It was apparent by her burst of speed and the smoke belching from her stacks that the *Patricia*'s skipper had calculated he could outrun most of the fishing boats by striking briskly along the north side of the channel.

His calculation was almost right. Except by the time he hit the line, at least a dozen of the smaller, faster skiffs and outboards had caught up and were swarming around him. The larger protest vessels had pulled abeam. As the pack closed in, several skiffs could be seen leaping directly in front of the *Patricia*'s high-heaving bow. The *Meander* had fallen back into a position directly in front of the cruise ship's path. Captain Feroce had cut his engines, radioing to the *Patricia* that we were "dead in the water," meaning that in navigational terms the onus was on the larger vessel to avoid a collision.

Directly under the *Patricia*'s bow, actually surfing on the bow wave, was a Greenpeace Zodiac with Rod Marining and Melville Gregory aboard, looking almost straight up at the passengers crowding around the rails.

They had set out from the *Meander* twenty minutes before to push ahead into the fog. There had been some debate about letting Saint Melville take the controls, because of what had happened to his kids back around Christmastime. In the aftermath of the fire, he'd been openly suicidal for a while and no one could blame him. But then he seemed to come out of it. He'd lost weight and looked as though half of his insides had been torn out, but within a month of the deaths he had been able to pick up his guitar and sing again, even though there was a ragged edge of pain to his voice. Yet this was only five months after the tragedy, and there was some concern that he might be looking for a way to get himself killed. We loved him. We didn't want him to do that. So we were hesitant about letting him take the Zodiac out in front of the *Patricia*. What had offset the fear was the fact that Rod Marining was willing to go with him, and while Gregory was a father who had recently lost his kids, Marining was now a father himself for the first time. He and his lady, Bree, who'd discovered she was pregnant back on the *James Bay*, had had a baby girl. So whatever lingering suicidal impulses the one man might have, we figured they were more than balanced by Marining's own desire to keep on living. Gregory would never take someone down with him. Still, there was a bad moment on the *Meander* when we spotted the Zodiac in the distance, directly under the towering bow.

"Christ," said Moore, "Gregory's cutting it awfully close. I hope he knows what he's doing."

Then the scene became too confused, with so many other little boats zigzagging across the cruise ship's bow, for us to make out what was happening to the Zodiac. It was getting just a trifle too confused for the cruise-ship skipper's taste, too. Begrudgingly, he slowed down to roughly six knots—but kept pushing determinedly through. At the last minute, Dennis Feroce had to gun the *Meander*'s engine and throw her into reverse, swiveling on her side, to avoid being run down. On came our loudspeakers mounted on the decks, subjecting the passengers to a blast against the pipeline terminal, while TV cameras on the big ship rolled, newsmen's microphones recorded every word, and a Canadian Broadcasting Corporation helicopter swooped down out of the sky to catch an overview of the action. Afterward, one of the TV-network cameramen joked: "We shot enough film to remake *Gone with the Wind*." From a media standpoint, it could scarcely have been more perfect. The oil consortium had paid for everything—only to have their stage taken over completely by a ragtag bunch of Indians and eco-freaks, whose blockade would be the big item on the news for the next couple of days. The finest irony was that it had been the municipal politicians and the oil men who had arranged to have a small army of press people on board, all of whom were now leaning over the railings, frantically poking their cameras in every direction and recording every bit of our antipipeline propaganda. In fact, by the time the *Patricia* was finally surging free of the encircling protest boats, we were starting to feel rather smug on the *Meander*, knowing we had just pulled off another— rather classical—media coup.

We didn't know then that Gregory and Marining had been run down.

Either their Zodiac hadn't been properly inflated or it was just one of those accidents that can happen anytime. At the last minute, just as the *Patricia* was breaking free, the floorboards of the rubber craft had cavitated—buckled—and even though he desperately tried to recover his momentum, there was nothing Gregory could do to stop the craft from momentarily floundering. The cruise ship's bow caught the rubberized side, the Zodiac bent in two, twisted, and vanished within seconds beneath six thousand tons of swiftly moving ship. Out of all the cameras present, only one—the Super-8 held by Chief Clifton of Hartley Bay—actually caught the scene, although the footage was too blurred for either Marining or Gregory's expressions to be made out. They could be seen being tossed out of the Zodiac and disappearing like stones into the water. Gregory was not even wearing a wet suit. Marining, who was, said afterward:

"It was all white under there, and I could feel myself being sucked right under. The first thing you think about is the props, and the instinct is to start kicking like mad to get out of the way. Once I came up and the hull was going past so close I could have touched it. I was looking straight up. Then I got sucked down again. I have never held my breath so long in my entire life."

When he did finally come up again, the *Patricia* had passed, and he found himself in her turbulent hissing wake. The first thing he saw was the Zodiac, its engine somehow still running, roaring around—riderless— in a circle. Then Gregory surfaced slowly, not moving, like a bubble breaking the surface. Young Oliver Clifton, son of the chief, had seen the accident, wheeled his skiff around, and now swiftly moved in to haul the limp Gregory aboard, then pick up Marining.

On the *Meander* the first we knew that anything had gone wrong was when another skiff roared up to us and someone yelled: "Your boys got run over!" The smug smiles were whipped from our faces and Linda Spong cried: "Oh no, not Rod, not Mel!" Bleak, empty moments followed, with no one speaking. Then Clifton's skiff appeared with our two soaked, shivering kamikazes huddled together on the floor. Gregory threw up for a long time, getting all the salt water he'd swallowed out of his system. Marining had a bruised hand. That was it. The floorboards of the Zodiac were smashed. It had been as close a brush with death as any of us had yet experienced.

Later, at Kitimat, the *Princess Patricia*'s passengers were subjected to angry picket lines and demonstrators who called them fools and battered the windows of their tour bus with placards. Altogether, the oil consortium-sponsored tour up Douglas Channel had not been a happy affair. From a public-relations point of view, it was a disaster. The top oil-consortium executive found himself being badgered by reporters, and the municipal politicians were not amused at having been dragged into the limelight. The captain of the vessel had not scored any points either for his reckless charge through a cluster of small boats whose owners basically wanted nothing more than a chance for their voices to be heard. The oil consortium did not win many friends in the neighborhood that day. Where they had hoped to cruise with dignity and cool self-assurance, they had been forced to bolt and run.

There was to be a long series of hearings into the pipeline-terminal proposal for Kitimat. West Coast politicians, both provincial and federal, whose jobs were on the line, were to become involved. In fact, a year was to pass before the final decision came down: that there would be no terminal at Kitimat. The supertankers would never enter Douglas Channel. Gregory and Marining had not taken their long dip in vain.

Our feeling was that Greenpeace had in part paid its debt to the Indians of the Pacific Northwest who could now go on fishing without the danger of a massive oil spill in Inside waters. There was also a sense of satisfaction that stemmed from knowing that the very waters through which the tankers would have traveled were those where our old friends, the orcas, lived in their greatest numbers. The relatives of Skana would not now end their days covered with oil.

There was one more local issue that seized our attention before we could get down to concentrating on the summer's biggest-ever campaign against whaling, and that was triggered by a creature almost as tiny and ugly as the whale was large and beautiful. It was a creepy little pest called the spruce budworm, an infestation of which was reported to be destroying trees in the Fraser Valley, a couple of hundred miles from Vancouver. The provincial government had announced that it intended to spray a fifty-thousand-square-mile area with an insecticide that would incidentally have poisoned almost every bird or small animal in the area as well. Our new president had studied forest biology in the process of getting his degree and knew that the budworms actually function in a seven-year cycle. This particular infestation was in its last year and would begin to decline—so there was no need to dump tons of poison all over the landscape, except, of course, that it was good business for the chemical companies, whose lobbyists hung around the provincial capital in droves. It was Moore who decided to launch the "Great Budworm Crusade." It came off magnificently.

The scheme he worked out was to send a convoy of volunteers into the valley where the spraying was to be done and set up camp under the very trees that were to be sprayed. That meant that the helicopters that were to do the spraying would have to dump their loads of poison on people as well as budworms, trees, birds, and squirrels. The volunteers would split up into a dozen different camps so that the chopper pilot would have no way of knowing exactly where they were. The volunteers would be equipped with helium balloons to be floated at random above the trees, high enough to be in the way of the choppers. That in itself would give the pilots reason to pause. It was an idea borrowed from the use of dirigibles above England during the war.

The British Columbia government responded by sending two dozen Pinkerton police into the area to remove the protesters and to block the roads to prevent any more from getting in. Moore's plan for getting around this was to ask a local Indian council for permission to use reservation land as a base from which to launch the campaign. The Indians agreed, leaving the government in the position, short of invading Indian land, of not being able to do anything to head the protesters off. It turned out that one of the Pinkertons was a card-carrying Greenpeace member, who phoned in daily reports to our office, telling us the exact movements and plans of the police.

Twelve hours before the spraying was to begin, the premier of the province called a special cabinet meeting and ordered a complete halt to the spraying program and the withdrawal of all Pinkertons. By our calculation, Patrick Moore had saved fifty million spruce budworms. "Now," he laughed, "when someone tries to tell us we're only interested in saving cute animals like seals, we can say: Have you ever seen a spruce budworm? Yecch!"

Think big, very big. Think small, very small. We were getting the hang of it.

While all the other offices were occupied with the seal campaign and the Vancouver office was busy dealing with Kitimat and the budworms, it was the Honolulu office where the main thrust of energy was developing, being driven along at a relentless pace by the Czech and the Kiwi, Korotva and Spong, with perhaps three dozen hard-core Hawaiian whale freaks backing them up. Ostensibly, the president of Greenpeace Foundation, Hawaii, was Ross Thornwood, a broodingly handsome dark-haired radical who had been so unhappy because we had not done "more" to harass the whalers from the *James Bay* the summer before. As a veteran of a previous voyage, he had lots of stature within the new group. But the real boss was Korotva, whose role as Greenpeace captain put him right up there with the heaviest of the old-time heavies. When Spong arrived to help out, he came with the title—ordained in Vancouver—of whale-campaign coordinator. The local folk had lots of titles too, but in practice it was Korotva and Spong who made most of the decisions. Ross Thornwood's sense of organization left something to be desired: what he really wanted to do was to climb a flagpole and refuse to come down until both Japan and Russia stopped whaling. This was presumably his idea of a radical action. As a Hawaiian, however, Thornwood was the logical local spokesman: he was photogenic, he spoke with a deep voice, and most of the workers in the office loved him. They had an affectionate feeling about him precisely because he was so obviously a dreamer.

There could be no doubt that the Hawaiian operation possessed the kind of collective overdrive energy that had not been seen in the Vancouver office since the time of the UN conference and the launching of the *James Bay*. With his monomania to see a ship—any ship—get out after the whalers, Paul Spong mostly reveled in it and thought nothing of working sixteen and twenty hours a day, day after day, week after week.

Korotva, too, was functioning on daily rushes of adrenaline. He had to. He'd taken on the central load—namely to find a ship, get it ready, and take it out to sea. In the pursuit of the ship, he sky hopped from Japan to Los Angeles and back again, making deals with every kind of hustler and broker he could find to track down a ship that would go faster and farther than the *James Bay*. In the end, though, he'd had to settle for a 176-foot tin hulk called the *Island Transport*. She was narrow-beamed, rode low in the water, and had an ornate little wheelhouse that was somehow reminiscent of a prop in an early Flash Gordon movie. She was dirty, caked with rust, trailing a green mantle of seaweed, with a hull so thin and dented that it was possible a good kick would stove her in. A broken-down hydraulic "cherry picker" reared up amidships, like a bent Mecanno set. The decks were a sand pile of rust turned to powder, and the space below decks was an empty echo chamber where the only sound was the pinging of drops of condensation dripping from the hull, and the smell was of musk and bilge. "With a little work" the engines would probably run. The *Island Transport* had been built twenty five years before, in time to go out and sink two Japanese submarines before the end of the war, and as a subchaser she had been built to go fast. She was supposed to be able to do twenty-six knots. And she had enough space so that extra fuel tanks could be built in, giving her an incredible theoretical range—providing the engines didn't burn more fuel than expected.

The other great feature of the vessel was that she was the only one available in the Hawaiian Islands that the Honolulu office could afford. She cost seventy thousand dollars, with twenty thousand dollars worth of spare parts thrown in. The spare parts were important, because nobody made spare parts for ex-navy subchasers of World War II vintage anymore. On top of the initial payment, another twenty thousand dollars would be needed to get her going, although at the time of the purchase no one quite realized that.

With the purchase of the *Island Transport, Greenpeace X*, the big drive was for money. In this area, the Honolulu Greenpeacers proved themselves to be geniuses. They staged walkathons, sponsored baseball games, put out public-service announcements, talked entertainer Elton John into doing a benefit in Lahaina, ran ads, got a famous local artist, Maui Loa, to do a special painting depicting the whale, ran off tens of thousands of T-shirts using the design, hustled Greenpeace belts, necklaces, badges, posters, and booklets to the tourists crowding around Waikiki Beach, enrolled all the local politicians, and generally made

Greenpeace's presence in Hawaii so high profile that there was hardly a soul left on the islands who did not know that the "whale savers" had arrived. It all tied in very nicely with the legends of the native islanders who believed that when the whales disappeared, they, the people, would disappear too. The legend carried a chilling force, for native Hawaiians were now themselves almost an extinct species, and of course the whales, who had once gathered in legions around the islands, were now scarcely ever to be seen. ...

In about six months, virtually the whole learning experience that we had gone through in Vancouver was re-experienced in a telescoped fashion in Honolulu. For old dogs like Spong, it was like being trapped in a dentist's chair, having to listen to precisely the same points being raised at meetings that he had heard being raised at meetings as much as four years before, having to watch the same clumsy, groping attempts at organization, the same blind power plays and maneuvering for position, the familiar youthful ego games, the equally familiar older ego games. It was DEJA VU spelled in twelve-foot granite blocks. So while it was all very exciting, very real, very intense, it was also very boring: he had been through it all before. And sometimes it was the boredom—especially during the interminable meetings in blast-furnace heat, while new members were dickering over nickels and dimes—that was the hardest thing to cope with.

He and Korotva worked closely, drank with each other, hung around together a lot. Korotva, too, for all the kicks he was getting out of it, was feeling world-weary, jaded, exhausted. It was the same madness, only here it was all over again. New faces. New scenery. New year. Another boat, but all boats were the same. Same amount of work.

"Vat the hell ve doing Spong?" Korotva would demand after several Vodkas.

"Christ, sometimes I wonder, mate."

But if the two veterans were having trouble keeping their energy up, they at least had a mob of fresh enthusiasts who had not yet even begun to burn themselves out, who were willing to be down at the *Island Transport* at 5:00 A.M. with welding torches, and who, even though blackened with grease and soot, streaked by sweat, were still going at it by the time the sun mercifully fell behind the rich green mountains ten thousand hellish hours later.

The Greenpeace operation in Honolulu was also blessed by the fact that it was enjoying a honeymoon with the local media. The Honolulu

Star Bulletin ran an editorial with a headline: GREENPEACE MAKES VALIANT EFFORT. Typical of the kind of positive coverage they were getting was this article from the June 13 Honolulu *Advertiser*, by Bruce Benson:

> Twenty-five years ago, the workers who rushed the little ship into World War II from a Tennessee shipyard could not have sweated and labored any harder than a band of volunteers now preparing the same vessel for still another battle scheduled to begin this coming weekend.
>
> It's like watching clamoring ants down at Pier 40 as dozens of Greenpeace helpers scramble around the 175-foot Island Transport, ex-sub chaser, grooming the boat for encounters on the high seas with men who kill whales for a livelihood.
>
> Grooming is too refined a description of the work taking place on the aged, rust-scarred ship. The Island Transport, still awaiting a christening with a Hawaiian name and blessing, is in crude condition.
>
> The best it seems Greenpeace forces can hope for is to make the ship fit for sea with the most rudimentary equipment and accommodations. Capt. George Korotva, 37, who doesn't act or sound like a romantic, is optimistic.
>
> "The life-saving gear, the fire-fighting systems, the radio are all up. The hull is okay. We had one small leak and the welders repaired it. The vessel passed a complete Coast Guard inspection just last year for commercial service. I see no reason why it shouldn't pass today."

In fact, the "one small leak" was more like about two dozen leaks, which had been sprung when sandblasters went to work on the tissue-thin steel hull. The one great disadvantage of old warships was that they were mostly thrown together in a hurry with no expectation that they would last forever. The *Island Transport* was one of the last old ladies of her class to survive. By rights, she should have been bought up by the navy and placed in a maritime museum.

The biggest media break came when the ABC "American Sportsman" TV show announced it wanted to send an eight-man crew, equipped with their own helicopter, to cover the big confrontation. "American Sportsman"? Had jumping in front of harpoons become a sport already? No, the reply was that the show was trying to cover serious conservation issues, and the whales were one. Greenpeace's efforts to save them were another. ABC was willing to kick in sixty thousand dollars to cover costs. As it worked out, the offer came at a point when it was just beginning to look as though the *Island Transport* wasn't going to quite make it for lack of money. The ABC deal would just take them over the top. Of course it meant an astronomical increase in terms of pressure, because the big network-TV boys weren't going to like it if they didn't get to cover a confrontation after laying out all that cash, plus another couple of hundred thousand dollars worth of helicopter charter fees, wages, equipment costs, and so on.

"Don't worry," said Korotva, "ve get you der."

Down at the *Island Transport* the welders set to work building a helicopter landing pad at the stern. With only a twenty-three-foot beam, the ship did not have much extra space to offer a helicopter for landing. Whoever the pilot was going to be, he'd have to be damned good to try landing on the back of the old ship when she was rolling about at sea. The landing pad, made of metal, was mounted well above the level of the deck, giving it the appearance of a stage.

The workers were not entirely Hawaii residents. There were quite a few mainlanders who had come in from the other offices: John Frizell, Dalton McCarthy, and Michael Sergeant from Vancouver; Gary Zimmerman, Eddie Chavies, and Dino Pignataro from San Francisco; Dan Ebberts from Eugene, Oregon. Frizell and Zimmerman were fresh from the camp at Belle Isle: from ice to the famous blue lagoons. "Join Greenpeace," joked Frizell—or was he joking?—"and see the world." But for the most part the work crew at the boat and the staff at the office had been residents for at least six months or a year. That qualified them as "Hawaiians." When it came time to elect a board of directors, it was decided that only "Hawaiians" could be on the board—to preserve their autonomy, of course. There was a moment of embarrassed silence when someone pointed out that there wasn't a single native Hawaiian in the room, in fact not even anybody who'd been born on the islands. They were all from different parts of the States.

By the end of May, we in Vancouver were growing alarmed because the Honolulu operation was falling so far behind schedule. The whaling season was ready to start, and still there was no certainty that the *Island Transport* would even get her registration papers cleared, let alone solve all her mechanical problems.

We found ourselves starting to think about the *James Bay* again. She was sitting, still impounded pending the outcome of the suit for wages against her, in the Fraser River, and all it would take to get her going would be the posting of a bond, a couple of weeks work, a few pieces of electronic equipment—and a captain.

It just so happened that John C. Cormack had nothing in particular planned for that summer. Neither did Al Hewitt, who—we realized belatedly—could have handled the entire engineering job the year before himself, without the help of the retired chief engineer we'd eventually hired, who was now suing us for wages. In fact there were quite a few people available. It was just a question of money.

Neither Spong nor Korotva wanted us to take the *James Bay* out again. If we were going to lay out money, they wanted us to pump it toward them. For the amount of money it would take to mount another *James Bay* expedition they figured a tanker could be fueled to head out after the *Island Transport* and keep her at sea, harassing the whalers, for months. That was, of course, *if* she got to sea in the first place. And at that point, the assurances coming across the phone from Honolulu sounded hollow in Vancouver ears.

Patrick Moore and I finally went down to the *James Bay* to talk to the owner, Charlie Davis, and kick around the idea of another voyage. Time was running out. If we were to get moving, it would have to be soon. Yet, damn it, we were still in debt. How would we ever convince the bank to let us extend our line of credit? We went into a huddle with Bill Gannon, and it was not long before the accountant had his calculator in his hand, a twinkle in his eye, and a half dozen possible solutions in his head. The "money thing" could be solved. It would simply be a matter of stopping at enough ports, flogging enough merchandise, to make the *James Bay* pay for itself along the way. And if a big benefit concert could be arranged in, say, Los Angeles, there would be no problem.

The mere thought of going out on a big boat loaded with highly strung people again left me throwing up on the edge of the sidewalk. I lacked Spong's dedication and Korotva's strength. If there was to be another boat out of Vancouver in 1977, it would be up to Moore to take it. He had been through just as much as I had—more, in fact, because he'd gone along on the second seal campaign—but he was six years younger with far more physical stamina.

"It's your baby, Pat."

"Oh Gawd almighty," he said, rubbing his eyes. "I dunno."

We were suffering in Vancouver, like Korotva and Spong in Honolulu, from a feeling of saturation, as if all the wild swings in mood, the massive doses of exposure to other people, the tensions and ecstatic releases of the last couple of years had rotted away the once insatiable desire for more experiences. By then, Moore's favorite hobby was "staring at the wall." Mine was watering the plants at home, talking to them at great length. It is one thing to be seized by a vision and proceed with all your heart and soul to make it come true. It is another thing to do it again. And again. It is like a boxing match that has gone on for too many rounds. It gets harder and harder to lift your arms. In this context, what was not needed was any feeling of detachment. While the group in Honolulu was charging up the mountain toward the light for the first time, the Kitsilano tribe was shuffling about at the bottom of the same mountain, still panting from the last climb, burdened with debts, weakened by bloodletting, wondering if, indeed, we had the strength to set out again. In most great struggles, people are forced by circumstances to fight. In our case, we were under no immediate threat ourselves; it took an intellectual leap of considerable proportions to perceive the "great system of order" linking our fate to that of the whales, and in any event, the threat was still only to the distant future. The glimpses of the whales themselves had been like mystical experiences, but they were difficult to hold clearly in your minds through the weeks and months of mundane reality that were to follow. The vision was elusive: it was not something that stared us in the face every day.

"Well, I guess we've gotta do it," said Moore, somewhat glassy-eyed.

Yet once he put his mind to it, the energy started to flow out of him as though drawn by a huge, hidden suction pump. Within three weeks, the *James Bay, Greenpeace XI*, was ready to go.

Between the Honolulu boat and the Vancouver boat a keen sense of competitiveness instantly developed. Korotva and his bunch had made a point of stressing how much bigger and more effective their campaign was going to be than ours, and so—as a group—we naturally took a savage delight in getting our ship launched while the *Island Transport* still lay passively under the welding torches at Pier 40.

There could be no doubt that the launching of two former warships gave every outward appearance of growing vigor on the part of the antiwhaling movement. It gave us what political analysts called "momentum." It was a momentum that got another strong shot in June, when the International Whaling Commission held its annual meeting, for the first time in Canberra, Australia, instead of London as usual. The upshot of the meeting was a decision by the IWC to slash the whale quotas for the next year by thirty-six percent, the biggest single cut in history. Russia and Japan bitterly fought the decision, but were finally overwhelmed. Back in the scattered Greenpeace offices, the news was greeted with war whoops. The enemy seemed to be staggering and at the very time we seemed to be on the offensive. Everyone involved knew the reality of our internecine warfare, the eggshell-like fragility of our economy, and the stunning possibilities of imminent financial ruin, but the sense of momentum now was irresistible. It seemed we might really be going in for the kill. If there is truly such a thing as the "moral equivalent of war," there is also such a thing as the moral equivalent of blood-lust. To bring down the mighty *Vostok* would give us the satisfaction reserved, in ancient times, to those who had leveled a castle.

The work on the *James Bay* was in sharp contrast to the antlike chaos that surrounded the *Island Transport*. It was almost leisurely, almost routine. Most of the renovations had of course been done the year before, so it was mostly a matter of minor refitting. The biggest single problem was removing the thirty tons of sand that had been dumped in the hold for extra stability. The burlap bags containing the sand had rotted, leaving a huge pile of gritty sludge that had begun to get into the engine. All of it had to be removed by a bucket brigade, one bucket at a time. A volunteer crew came quickly and almost effortlessly together, as if the old Laws of Manifestation were starting to work again. If someone with a special skill was needed, they tended to appear suddenly and just on schedule. Bit by bit, the old sense of magic crept back into the day-to-day operation.

Without Korotva around, we needed another Russian interpreter. A twenty-year-old girl named Rusty Frank, a Russian-language student from the University of California, appeared. We needed another good heavy-duty engineer and seaman. He appeared in the form of Nelson Riley, a former biker turned tugboat operator, who wore a T-shirt with the name Smelly Nelly printed on the chest. There were a few old-timers: Moore, Cormack, Hewitt, Eileen Chivers, Rex Weyler, Mike "Generalissimo" Bailey, and even Dr. Lyle Thurston from the first Greenpeace trip, who had also sailed on the *Greenpeace VIII*—the little sailing boat *Wings II*, which we ordained in Lahaina the previous summer. There were some family additions: Moore's youngest brother, Michael Moore, and my son, Conan, thirteen, who was to serve as steward. Conan had grown fed up with watching his father go to sea every summer. He was told that if he wanted to go he'd have to prove himself to the rest of the crew by working to help prepare the boat. In his case it meant two weeks of lifting buckets of sand out of the hold, which, to his father's amazement, he did. Bailey brought his girl friend, Jackie Younge, along. Moira Farrow, a reporter from the Vancouver *Sun*—who had also covered the antisealing campaign—was taken as part of the crew. Charlie and Patsy Davis, the owners of the boat, decided to come. Michael Chechik assembled a two-man film crew. The rest of the crew was gathered together from the offices: Montreal, Toronto, London, Eugene, Edmonton, Seattle, and Los Angeles.

Good signs surrounded the launching of the *James Bay*, Sunday, July 17, as the old minesweeper set out from Vancouver's English Bay with a slick new paint job and a happy old skipper at the helm. The departure was timed to coincide with the local annual sea festival, whose symbol that year was the whale. An aerial acrobatics team looped the loop in single-engine biplanes, tailing plumes of colored smoke, as the anchor came up on schedule at 8:15 P.M. Thousands of people cheered and waved from the shore, and a flotilla of boats hooted their horns. Only three days before the launch, a baby beluga whale was born in the Vancouver aquarium, the first time such a creature had been born in captivity. It was to die not many months later, but at the time, its birth seemed auspicious, making Vancouver one of the most whale-conscious cities in the world.

Cormack was in such a good mood he was scarcely recognizable. After decades of plowing about the sea in stubby fishing boats, broken-down old freighters, tankers and tugboats, he was finally astride the bridge of a massive swordlike charger. Compared to the workhorses he'd always ridden, it was like mounting a white stallion.

There was no need to stop anywhere on Vancouver Island to do all the work that hadn't been done by launch-time. There had been such an air of professionalism and experience during the refitting period that everything was actually ready and in place. The mad scrambles that had characterized earlier voyages seemed a thing of the past. Gone too were the extravagant emotional binges that had led to early bow-cases, confusion, and recrimination. The new crew members—including Dave Taupin, Michael Earle, Peter Roscoe, Gary Cross, Jolyon Western, Pat Burke, Bob Rodvik, Bill Roxborough—were each given their assignments and fell into place with a minimum of unnecessary conflict. All that remained was a sweet, clean feeling of mission. If it was not quite as funky as before, the new efficiency at least meant that everything could be streamlined. Said Moore, as the ship was leaving: "In our first antiwhaling voyage it took us two months to find the Russians. Last year it took a month, and maybe this year it will only take two weeks." The remark was prophetic. The *James Bay* had barely cleared the Juan de Fuca Strait when Russian voices were picked up on the radio. The open sea brought a heavy swell, but there were only a half dozen seasickness cases, and by Monday night, almost everyone had recovered.

Moore had pinpoint data at his disposal, the same data that would of course be available to the *Ohana Kai* (the new Hawaiian name of the *Island Transport*) when and if it ever got away from the dock. In addition to the data, he had, from the first day at sea, definite radio-direction fixes. The ship was right on target.

The second day out, a school of thirty Pacific white-sided dolphins put on a spectacular day-long show. Wrote the *Sun* reporter: "They leaped and rolled in the water, as though deliberately performing for the Greenpeacers hanging over the ship's rail. Sometimes they turned on their backs and looked up at their audience who were hooting, hollering and yelling in excitement at the unusual sight." The magic was flowing.

On the third day, the *James Bay* passed through a fleet of draggers off the coast of northern California. She was making a steady thirteen knots, covering 215 nautical miles a day and only burning twenty-three gallons of fuel per hour. Freshwater consumption, which had been up to one hundred gallons a day—"Mostly because of all the farmers," grumbled Cormack—was brought sharply down to thirty-six gallons a day.

The week that followed developed into a stop-and-start situation. The Russians were steaming southward, away from the oncoming minesweeper, and the only radio messages they were exchanging came in short coded bursts of numbers, scarcely long enough for a radio-direction fix. Then on Sunday, July 24, a transmission between our ship and the San Francisco radio-communications center was suddenly interrupted by high-pitched maniacal laughter. "It was the kind of mechanical laughter one hears at the midway House of Horrors," reported Moira Farrow. Back in San Francisco, Dick Dillman, an experienced radio operator who was handling the communications end of things, said it was the strangest interference he had heard in twenty years in radio. It was a definite case of jamming, undoubtedly emanating from the Russian fleet. Attempts on our ship to switch to three different channels were effectively blocked: the cackling demonic laughter pursued the transmissions from channel to channel. After twenty minutes, the nerve-racking sound broke off. It recurred several times, but never for a longer period than that. It left everyone on board with the feeling that there was a devil out there somewhere, probably the same devil who was hacking the whales to death. It was as if he were mocking them.

Monday was followed by Black Tuesday. Radio signals from the Russians faded into oblivion. Al Hewitt was called upon to repeat the technical miracle he'd performed in 1975: to rerig the RDF unit.

Meanwhile, Moore ordered the engines shut down to conserve fuel. The minesweeper drifted, pitching and rolling in swells whipped to whitecaps by the wind. By the next morning, the homemade radio antenna had done its job. A new solid fix came through. Within minutes, the engines were fired and the *James Bay* was pounding along at twelve knots into four-to-six-foot swells. Water was breaking over the decks and Doc Thurston was handing out the seasickness pills like candy. Although the ship was off the coast of the Baja Peninsula, the weather seemed more like wintertime off Vancouver Island. The younger crew members were getting slightly wild with the tension of not having found the fleets—the voyage was in its twelfth day—but Cormack's only comment was: "We don't want to make it look too easy."

On the thirteenth day, some seven hundred miles southwest of Los Angeles, crew member Bob Rodvik was in the wheelhouse alone, when he saw them. "Three ships on the radar!" he yelled. Within minutes, Moore had identified them through the binoculars as Russian killer boats. "We're right among them! Whoopee!"

Another day was to pass before the weather settled enough to allow the launching of the Zodiacs, so the *James Bay* crew contented themselves with tailing the killer boats back to their mother ship. But it was not the *Vostok* whose fleet they had found: it was Russia's other great whaling factory ship, the *Vladivostok*, in outline identical to the grim hulk we'd tackled two years in a row, but its body even more battered and rust-

411

stained, more specterlike.

Back in Honolulu, the *Ohana Kai* finally broke away from Pier 40, the ABC helicopter mounted on her stern. At the last minute, most of the office staff—exhausted, punchy, half-crazed from the anxiety of trying to get their ship to sea, knowing that the *James Bay* was already closing in on a whaling fleet—had thrown themselves from the dock onto the ship, so that as it nosed out to sea, there were roughly fifty people on board, even though the crew was only supposed to include thirty-five people. The others all intended to get off before the ship cleared the last of the islands, but in the euphoria of the launch, any sense of military mission was swamped. Ross Thornwood didn't get to go along after all. Only a couple of days before, he had broken his leg. Korotva had lost close to twenty-five pounds in the months leading up to this moment and was reported actually looking skinny.

The plan of the moment was for the *Ohana Kai* to use its incredible subchaser speed to steam directly eastward to link up with the *James Bay*, so that the spectacle of two converted warships tackling a whaling fleet could be recorded by cameras in the helicopter. Moreover, while the *James Bay* only had three Zodiacs, the *Ohana Kai* had seven. Between the two of them, it would be enough to theoretically immobilize the *Vladivostok*'s fleet. From the minesweeper, Moore radioed to the subchaser: "We love you and we are waiting for you." Replied John Frizell from the *Ohana Kai*: "We're coming as fast as possible, and this ship is a runner."

A runner it might have been, but it was like both the *James Bay* and the *Phyllis Cormack* on their first voyages. The bugs weren't quite out of the system. Rather than steaming directly toward the scene of the confrontation, the Honolulu boat steamed directly toward an outlying island and put into port to straighten out all the messes and to disgorge its extra passengers, most of whom were rendered seasick by the rolling of the narrow-beamed vessel the first night out. In fact, it was the tough ABC camera crew who did most of the work keeping the ship operational that first night.

There were groans of dismay in the mainland offices when it became clear that the *Ohana Kai* wasn't going to make it to the scene where the *James Bay* and the *Vladivostok* were about to duel. It had seemed like such a fantastic opportunity: two rainbow-painted ships converging from east and west, catching the Russians in a pincer movement. On the charts we had up on the office wall back home we had already penciled in the route we expected the *Ohana Kai* to take. Likewise, we had calculated how long the *James Bay* could remain in a holding pattern, stalking the *Vladivostok*, waiting for the cavalry to arrive. Our ears were tuned for the sound of a bugle and a puff of dust on the horizon. Instead, the *Ohana Kai* went about half an inch on the charts, then stopped. The *James Bay* was too low on fuel to be able to wait around, and, besides, once the Russians started killing whales there would be no way that the boat from Vancouver could hold back from taking action. Linked by newly installed telex machines, groups in six cities—Vancouver, Portland, San Francisco, Toronto, Montreal, and London—stood around, grinding our teeth, helplessly watching the moment come and go. There were not a few souls on the *Ohana Kai* who were feeling the same frustration. But then they were still feeling giddy from the launch, frazzled, hollow-eyed, gaunt from the months of clawing desperately at the stubborn metal of the boat to get her moving. They had not had a chance to right themselves, get their bearings, or have much time to think about anything except the one great challenge in their lives: to get the damned boat out of harbor! Now that they'd done it, they were staggering around as though catatonic, like punch-drunk fighters. The fact that there was another Greenpeace ship somewhere out there thousands of miles away tackling a whaling fleet was *abstract*. Since there was no general in command of the situation—instead a committee had been set up—there was no one to order them to take an immediate bearing east-southeast. They had to mull it over themselves.

And on the *James Bay*, everybody became caught up in the whirlwind of events. The hoped-for rendezvous could always happen later.

By 11:00 A.M., July 30, two killer boats were closing in on a pod of whales, with the *James Bay* only a few hundred yards behind. There were cries from the bridge: "Whales! They're on whales!"

"Let's go, guys," yelled Moore, leaping into a Zodiac, taking cameraman Bill Roxborough with him. Into a second Zodiac leapt Mike Bailey and Rex Weyler. The third Zodiac was left behind. It had a smaller engine, and the swells were running at four to six feet. Moore didn't think it would be able to maneuver.

The sky was like a single block of granite. The waves were metallic, as though made of the same material as the distant factory ship. The wind was chill and clammy and the sound it made had an echo of the maniacal laugh that had been heard in the radio room for so many nights. A mood of mournfulness hung over the scene, a sense of lateness in the year, lateness in time, lateness in history, as though they had come to a twilight zone near the end of the world.

From the heaving deck of the *James Bay*, the crew watched as the Zodiacs zipped away toward the nearest killer boat, zigzagging across its bow. Ahead, they could see the flashes of spray and the gurgling eddies where the whales had appeared, then dived. When the Zodiacs arrived, the Russian boat's horn barked and a puff of black smoke exploded from its stacks, as though startled.

Although John Cormack had been up for thirty hours without sleep at that point, he settled himself in the padded foam seat on the flying bridge, holding the control lever lightly in his big hand, and kicked the minesweeper up to eighteen knots. Rather than hanging back to leave the Zodiacs to do their work, he brought the *James Bay* right into the fray, crashing along shoulder-to-shoulder with the killer boat across the waves.

"Hell, it's just like fishing," Cormack muttered, completely at home with the idea of jostling and butting his way into the midst of a fishing fleet. To him, the whalers were basically just other fishermen. He did not feel intimidated by them in the least, especially not when he had a boat that was bigger and faster than theirs.

Over the loudspeaker, Rusty Frank yelled in Russian at the startled whalers. Her Russian was much better than Korotva's, and this time they could understand the voice coming at them across the water perfectly. Moreover, it was a woman's voice. Whether they wanted to listen or not, they did, if for no other reason than to hear that sweet voice in the midst of a hard universe.

"Dive! Dive!" screamed Conan Hunter each time the whales surfaced.

But they were beginning to run out of breath.

And killer boats were converging from every point on the compass. Soon there were six of them in a row. Suddenly, they all stopped, and their captains communicated with each other in Morse code. They seemed confused.

The *James Bay* came to a halt too.

"They're immobilized!" Moore yelled up from his Zodiac, triumphant.

But, in fact, they were only waiting for the whales to come up again. Waiting, too, for the other boats to catch up. Ominously, they steamed in from over the horizon, then arranged themselves in a V formation—nine steel monsters, waiting, engines throbbing. The whales surfaced in a rush some distance away, and the entire fleet went into motion. There were more ships than whales. Yet the Russians did not compete with each other to be the first to the helpless prey. Rather, they maintained a stiff military formation, broken only by the presence of the floating rainbow in their midst. The two Zodiacs took up positions at the point of the V and the full procession fled in a hail of spray toward the horizon, the whales in the lead, followed by the Zodiacs, followed by nine killer boats and one minesweeper, with the factory ship pacing along some three miles away.

At the controls of the Zodiacs, Moore's and Bailey's arms felt ready to fall off. The two cameramen, who had been standing up, with a strap, around their waists, in order to get pictures, were falling to their knees repeatedly, their legs having turned to rubber. Rex Weyler was just beginning to think he couldn't take it any longer, his nerves frayed to a thread not only by the tension of waiting for the harpoon to be fired, but by the anguished cries of the whales, so close ahead that even above the roar of the outboard, he could hear them. Shrill mouselike shrieks, gagging gasps. He was desperately certain that Bailey was following too close and that one of them was going to come up any moment underneath them. The strain of being braced not only against the shock of the impact with the waves, but in anticipation of being blown to pieces or flipped into the air, left him feeling sick, at the very end of his strength, ready to turn around and beg Bailey to stop.

And then his right ear exploded. By the time he realized they'd shot, he realized also that they had missed him. His first impression was that a whale had been hit ahead—there was an erupting of huge gray bodies and flukes thrashing the air. The killer boat stopped. The Zodiac thankfully sighed to a halt. Then he realized that the harpoon had struck nothing but the water. He wanted to feel jubilant, but more than that he wanted to vomit. He could feel a scream lodged like a fish bone in his throat. He could hardly stand.

The other whalers had stopped their headlong charge as well, as though sensing some change in the situation. They closed in now in a circle, moving slowly, but confidently.

In the few minutes' reprieve, both Zodiacs roared to the side of the *James Bay* to change crews. None of the four who'd been out in the water all that time could take any more punishment. But before the crew change could occur, the entire pod of whales surfaced in the midst of the waiting Russians, perhaps knowing that they could run no farther.

A volley of cannon fire filled the air, its smoke drifting across the water where the entire pod was thrashing. From the *James Bay*, reporter Farrow could see the nearest victim of the massacre.

"It squealed and moaned like a dog run over by a car," she wrote. For

415

twenty minutes the whale fought death in a crimson sea.

By the time it finally died, most of the other corpses had been lashed to the side of the kill boats and some were being dragged toward the distant *Vladivostok*. No whales escaped this time to the horizon.

Once the bodies had all been winched up onto the flensing deck, the entire fleet turned southward and moved off rapidly. There was a mood of melancholy on the *James Bay* as the old minesweeper knifed through the waves in its wake. And since no additional fuel tanks had been added since the year before, there was the knowledge that the point of no return was coming up fast. Soon it would be time to turn around again. ...

In the press releases and statements we were putting out back in the offices, we were careful to stress the military precision with which we did everything. We were able to say, for instance, as the *James Bay* was falling back toward San Francisco, that the *Ohana Kai* was heading out to replace her. It looked like a tag-team match, as though we had planned it all along, clever tacticians that we were. The crude reality was that it had taken the Honolulu boat so much time to get going that the ball game seemed to be over already. Certainly she'd blown a glorious chance for a double-whammy opening round. But at least she was finally leaving the shelter of the islands.

We had a "war room" in the San Francisco office by then in the old Fort Mason building overlooking Alcatraz and the Golden Gate. It also served, under the brilliant guidance of Dick Dillman, as our "radio shack." He'd managed to get his hands on a giant old radio transmitter, the one that had formerly been used by the powerful KMI marine operators for linking all traffic in the eastern North Pacific. In fact, during our first voyage to the Mendocino Ridge, this transmitter was the very one through which we'd called to be "patched through" to Vancouver. The device was now Greenpeace's own communication tool. Through it we could communicate almost effortlessly with both the *Ohana Kai* near Hawaii and the *James Bay* down off the coast of Mexico. It was a very sophisticated setup. We could take calls off either boat and plug them directly into the regular telephone system, making it possible for anyone in the world to talk to either boat and for them to be able to reply almost conversationally. We had a reel-to-reel tape recorder that took "voice actualities" from the boats, which we could turn around and play directly into identical recording devices in the offices of Associated Press in New York, and United Press International and Reuters in Washington, D.C., all of whom, at that point, were taking our calls collect because we were a hot story.

The ABC network in New York was also intensely interested in what we were doing, mainly because of their investment in the *Ohana Kai*. The big question they had to ask was: *What kind of a rip-off is this?* They had signed a contract with an entity called the Greenpeace Foundation of America, Inc., which gave them exclusive rights to film the said entity's encounter with a whaling fleet in the North Pacific this summer, and had laid out good hard cash. Now, another entity, called the Greenpeace Foundation from some place called Vancouver had jumped the gun and had just had that encounter, while the *first* entity's ship was still piddling about near Hawaii. The hard-nosed executives in New York were in no mood to be told now that, actually, there were several Greenpeace entities, even though we were all part of the same organization, sort of. Nor were the ABC bosses happy to learn that there had, indeed, been cameramen on the *James Bay*, non-ABC cameramen, who proposed to release their film footage of the encounter to all the networks simultaneously. Oh, no, that would be a breach of contract. Unless the film from the *James Bay* came straight off the boat and directly into the hands of ABC, we could rest assured that lawyers would be dropping out of the sky all around us by morning, and the camera crew on the *Ohana Kai* would be ordered back immediately. Did we understand all that?

The "communications team" down at the radio shack at Fort Mason consisted at that time of Dillman, Cindy Baker, Robert O. Taunt III, and myself. We certainly understood. Taunt, who had helped to negotiate much of the original contract with ABC, understood perfectly. In exchange for our cooperation, ABC would release three-minute clips from the footage immediately after airing it nationally itself, and would allow the rights to the film to revert to us—*after* ABC's own "American Sportsman" special on the trip had run, probably within eight months. The whole situation was made that much more sticky by the fact that the original letter from ABC, expressing interest in joining an expedition, had been sent to me, personally, back when I was still president, but had been sent in care of the San Francisco office, where it had mysteriously been diverted to Honolulu. But such was the business of ecology by the summer of 1977. There was no point griping. We had to make the best of a pretty twisted situation. When I told Patrick Moore over the radio that he'd have to order his film crew to hand over the footage to ABC, otherwise we'd be up to our necks in legal actions, he almost wept.

"It sounds like highway robbery to me, Bob," he yelled.

"It's the price of greatness, Pat."

Which was not necessarily all that far from the truth. Bob Taunt was also our new Washington liaison officer and American media coordinator, and while he might not have succeeded in getting elected to office yet himself, he had family and political connections that were awesome, at least by Canadian standards. He also had a degree in communications and understood the dynamics of the U.S. mass-media system. There were certain basic rules. One of them was that you never burned off a major network with whom you had any kind of contract, otherwise your name would never see the light of video in America again. It was Taunt who demanded that we bow to the will of the New York kingpins. But it was also he who, when ABC delayed release of the footage for a day, got on the phone to New York and warned them that "a certain personage in the White House" was irritated because he had put aside all his other plans to watch the news that night especially to see the Greenpeace whale footage. Our liaison officer wasn't even kidding. Through his California Democrat connections, he'd gotten word to the President of the United States that the footage would be showing, and the President, who had appealed to environmental groups during his campaign, was only too glad to express his own personal concern about the plight of the whale by sitting down to watch the newscasts. Darkly, Taunt hinted to the ABC executives that they had come close to mingling in foreign policy matters of the highest order by not getting the footage on immediately.

It was like sticking a dime in a jukebox. Promptly at 6:00 P.M. the next night, ABC came on, in living color across America, with the scenes shot from the *James Bay*'s Zodiacs, including the convulsions of the whale who had taken so long to die, the blood splashing across the screen before millions of eyes—witnesses now, all of them, to the passing of Leviathan.

The footage was subsequently run on a "feed" to hundreds of independent TV stations, although none of the other big networks ran it nationally. Still, we had managed to arrange for the country's chief executive to see what our crew had seen out in the loneliness of the North Pacific, and that was the next best thing to bringing him right out there, along with tens of millions of other Americans. That image of a dying whale—the one who had "squealed and moaned like a dog run over by a car"—had gone out to the American mass mind, high and low, from corner pool halls and bars to the White House.

A few days later, Taunt got a phone call from Washington. The voice at the other end of the line said: "Mr. Taunt? One moment please. The

President of the United States." Then Jimmy Carter's familiar Southern drawl was coming through the earphone. Taunt was so startled he found himself coming to attention before realizing how absurd that was. Yes, Carter had seen the TV footage. He had been particularly interested in remarks made during the newscast to the effect that some of the whales that had been killed were undersized. "Definitely one of the whales we saw killed was only fifteen feet long," Moore had said. The President indicated he would like to see any photographic documentation we had to prove this, so that the U.S. delegation might present it at the next meeting of the International Whaling Commission. Preferably, the photos should be black-and-white. Taunt's main contribution to the conversation was a series of "Yes, sir. I understand. Yes, Mr. President. Thank you, sir." He promised to have the documentation to the White House within a week.

What happened from there that summer in the North Pacific was in the nature of an anticlimax.

By the time the *Ohana Kai* was finally ready for her deep-sea plunge, the *James Bay* had been forced by lack of fuel to break off contact with the *Vladivostok*, so there was no point in the ex-subchaser racing all the way over to a point some seven hundred miles off the coast of Mexico. By that time, the *Vostok* and her fleet were almost that close to the Hawaiian Islands, and so the *Ohana Kai* swung north by northeast to engage the fleet we had tackled—it seemed—so often before.

But straightaway, the *Vostok* bore northward at top speed, forcing Korotva and his crew into a long-distance chase at a hideous cost in terms of fuel.

Back in the war room in San Francisco, we groaned again—because now it looked as though the extra fuel capacity of the boat would be largely wasted before it could catch up to the fleet. Yet for the crew themselves, the mood was one of exultation. They were finally in action! Out of the thirty-five, only four had been to sea on a Greenpeace expedition before: Korotva, Spong, Zimmerman, and Kazumi Tanaka, the young Japanese photojournalist who had served as interpreter in 1976.

This was the first Greenpeace voyage that was almost completely an American exercise, and so its tone was different, its moods were, if anything, more exaggerated.

If the crew had anything in common, it was a complete absorption in the real-life movie unfolding around them, exalted by the presence of the ABC camera crew and the helicopter pilot. Certainly, they were the most heavily equipped Greenpeace expedition that had ever been launched, with a backup communication system that made our previous expeditions look pure Mickey Mouse. They had plenty of pride in what had been accomplished thus far. It didn't matter that the other offices could only moan and groan about the amount of money they'd squandered in the rush to get to sea: they had the kind of American drive and optimism that had fashioned an empire and put men on the moon. If half of them looked like psychedelic Okies, the other half looked like the kinds of guys you would have expected to see behind the scenes in Vietnam, cursing over the hardware, or at a hot rodder's rally in darkest California. The helicopter perched like a giant bug on the stern gave the *Ohana Kai*, with her ornate wheelhouse, a definite Space Age profile, and the flamboyant rainbows on the bow, along with the almost life-sized whales painted on the hull, added an almost garish, circuslike quality. The "Family of the Sea" did not entirely lack for mystics: Paul Spong was there, with his ever-present *I Ching*, Dino Pignataro radiated the kind of dervishlike ecstasy normally associated with Jesus freaks, and everyone to one degree or another was haunted by the transcendent mystery of the whales themselves and the meaning of their existence. Yet, on balance, Korotva had pulled together a crew that was weighted far more heavily on the side of turbines, generators, batteries, and diesel engines than it was toward Buddhism, the Holy Grail, or the Clear Light. It was slightly less of a religious experience than the other whale voyages had been.

As a media event, it was a nonstarter.

Only two stories were to come out of the entire eighty-five-hundred-mile voyage. The first occurred just five days after the pursuit of the *Vostok* began in earnest, taking the *Ohana Kai* to a point about one thousand miles north of the Hawaiian Islands, finally closing in on the slower-moving Russians. The skies darkened. The sea began to rise. Weather reports indicated a storm approaching from the northwest, with winds up to fifty miles an hour. Waves were soon building to sixteen feet, and the narrow-beamed subchaser was plunging and wallowing fiercely, her low decks awash. It was then that the steering mechanism broke—and it was then that Korotva's choice of a nonmystical crew paid off. Only the slightest edge of panic entered the mood. Within less than an hour, good old American know-how had triumphed, and two chain hoists had been linked to the rudder. It meant that two people had to remain below decks at the stern at all times, armed with walkie-talkies, taking their orders from the bridge and winching the rudder first one way, then the other—back-breaking, claustrophobic work, clumsy in the extreme, but just enough to keep them going. The only problem was that with such a fragile setup, Korotva was afraid to try coming about in the seas they were facing, lest the jerry-rigged steering device snap under the strain.

He opted to keep heading into the storm.

The *Vostok* and her fleet, however, turned and started heading eastward, leaving the *Ohana Kai* punching through the waves in a northwesterly direction, moving farther away from the Russians by the hour, with the wind rising.

By the time the weather calmed, the Honolulu whale savers had lost much of the ground they'd gained in the long pursuit of the whalers, and their margin of extra fuel was fading rapidly. They were able to repair the steering mechanism and turn back after the *Vostok*, but now they had to move far more slowly, conserving fuel.

When they did finally close in on the *Vostok*, everyone was gripped by battle fever. It was a fever, however, that was not to find an outlet, because, instead of attempting to chase any whales, the factory ship and her flotilla of killer boats chose to serenely keep cruising along in a straight line. All the *Ohana Kai* could do was pace them, waiting, waiting, waiting. But days passed—a week in all—and still the Russians made no attempt to hunt. Either there were no whales around, which was possible, or orders had come through from the Kremlin, or else the skipper of the *Vostok*, experienced now with Greenpeace kamikazes in a way his comrade in charge of the *Vladivostok* had not been, simply decided himself not to risk any incidents, knowing full well that the rainbow-hued protest boat could only last so long before having to turn back.

Finally, after a week of uneventful shadowing of the fleet, Paul Spong decided it was time to organize a boarding party. Taking Nancy Jack, a Honolulu organizer, and Kazumi Tanaka with him—Tanaka to speak to the Japanese observers on the *Vostok*—and an ABC cameraman, the New Zealander boldly went over to the factory ship and climbed up the slipway onto the flensing deck, there to face a crowd of very ordinary-looking men and women, few of whom spoke English. The communication was not what it might have been, since Korotva, as skipper, could not leave his post on the *Ohana Kai*. Still, the Number One whale saver was finally astride the deck of the vessel he wanted so deeply in his heart to wipe from the face of the earth. Rather than being hostile, the Russians were polite, curious, amused, and, in fact, seemed to welcome the diversion. The captain himself came down to speak to the people whose evolution he had watched, from his own peculiar vantage point on the *Vostok*'s

421

ten-story-high wheelhouse, over the past three summers—from halibut seiner to minesweeper and now to subchaser. In broken English, he allowed that he admired the stubbornness with which we had pursued him, but when Spong tried to press him to admit that the whales were vanishing, the captain just shrugged and let it be known that the visit was over. Spong and his crew were escorted, politely but firmly, back to their Zodiac. And that was the end of the *Ohana Kai*'s big confrontation with the Russian whalers.

The point inevitably came when, like all the other Greenpeace boats before her, she had to turn around and head back to port to refuel. During the entire time they were beside the *Vostok*, Spong had not seen any whales at all, had sensed, instead, an emptiness and desolation on the high seas that was that much deeper and gloomier than it had been the year before. Undoubtedly they had saved whales simply by being there, but for most of the crew—geared for a fight to the death—the feeling they had as they turned reluctantly back toward Honolulu was one of definite letdown.

The *James Bay*'s second run of the summer proved to be no more eventful than the voyage of the *Ohana Kai*.

Moore led his crew back out after the *Vladivostok* and finally tracked the factory ship and its fleet down some twelve hundred miles due west of San Francisco. This time, the second Russian fleet adopted the same policy as the more experienced *Vostok* and did no hunting so long as the *James Bay* was present. After three days of cruising side-by-side with the fleet, Moore decided to do what Spong had done a few weeks before. Taking Weyler, translator Rusty Frank, and Bob Taunt with him, the ecologist approached one of the killer boats, threw a bow line to the Russian seamen lining the rails, and the four Greenpeacers clambered on board. Rusty spoke to them in Russian and handed them buttons, which they eagerly pinned to their shirts. It was evident that at least one of the Russians promptly fell in love with the young co-ed from the University of California, but after only a few minutes of this exchange, the captain came down to the deck and ordered his men to remove the intruders. The sailors firmly shook everyone's hand.

The *James Bay* continued to pace the fleet until the fuel ran low, then turned and headed back toward Los Angeles, where a big benefit concert was being planned to raise money to cover expenses.

The *Ohana Kai*, too, set out to sea once more, but instead of actively pursuing a whaling fleet, it simply truddled across the water to San

Francisco. By then, both Korotva and Spong had left in disgust. A captain had to be hired to bring the ship across. A rendezvous between the two Greenpeace vessels in San Francisco Bay was arranged toward the end of summer, but the Honolulu boat's lack of media magic persisted to the end. Just as it was docking to meet a fairly large contingent of press people, there was a loud explosion some four blocks away. A restaurant had blown up. In an instant, the press people had all raced away to cover the disaster, and by the time the crew of the *Ohana Kai* had finally secured the lines, there was no one left on the dock to interview them.

If there was any lesson to be learned—or relearned—it was that the mass media, by and large, was not interested in diplomatic exchanges between adversaries; it was not much interested in symbolic meetings at sea; it was not interested in ships going out on the water, since ships do that every day. The only thing besides sex, politics, and sports that grabbed its attention, excited it, made its juices run, was violence. A harpoon being fired in the vicinity of human flesh was news. Dialogue was not news. And neither was the intrigue within a volunteer group by itself news. Without confrontation—the risk of life and limb—the mass media could not have cared less what we did.

The focus of the media's attention to the whaling issue now shifted to where the violence was: Australia.

Before the *James Bay* had headed out on her second run, Bobbi and I had taken off in a Qantas 747 for Sydney to lead a protest against a land-based whaling station in Western Australia. With Moore handling the *James Bay* operation and Korotva and Spong on the *Ohana Kai*, I was growing restless in my semiretirement. I hadn't particularly enjoyed being cooped up in the radio shack at Fort Mason, acting as media coordinator.

There was no money available for an expedition to Australia, but a French antiwhaling activist, Jean-Paul Fortom-Gouin, had promised to cover all the costs. He had his own group, the Whale and Dolphin Coalition, but in order to get Greenpeace expertise involved, he was willing to make it a joint campaign. All we had to do was go down to the airport and pick up our prepaid tickets. It was too good an opportunity to miss. It meant we could now open a "third front" in the whale war. The eco-navy could expand overnight into the Southern Hemisphere. Fortom-Gouin said he had a boat waiting, Zodiacs, and a crew. He had studied our tactics, knew what we needed, and had everything in readiness.

How could we resist?

Our contact was named Pat Farrington. She was an enormous woman, with dark curly hair, and she wore a great black flowing robe, which gave her the appearance of a kind of acidhead Mother Superior. She was one hundred percent American, somewhere in her early forties, a veteran of every postwar movement that had ever happened. In California, she'd organized the initiative to repeal the marijuana laws.

I thought I'd seen some funky offices in my travels as a Greenpeace organizer—but this one on New South Head Road put all the other warrens to shame. It had once been a slaughterhouse for sheep, probably back around the turn of the century, and had since been converted into an artist's studio.

If Pat Farrington looked like an old Merry Prankster, it did not set her much apart from the rest of the group that now greeted us. They all had a kind of blissed expression. There was a steady bustle, the two phones kept ringing, people crashed back and forth through the front door, carrying boxes and shouting back and forth at each other in accents so thick to our ears that they sounded as incomprehensible as Newfies. *Mate* was about the only word I could catch at first. Almost every sentence seemed to finish with the word *mate*. And then a single bark-like sound: "Eh?" It had the effect of changing each statement, at the end, into a question. One young man in particular stood out. His name was Rick. Two of his front teeth were half rotted away, so that he said "wiff" instead of "with." In California he would have been called a street freak. The others had a kind of ragtag quality, socks hanging loose around the ankles, Salvation Army clothes, and a dandruffy crumpled look having to do with the fact that several of them were using the dusty spider-webbed office as a crash pad.

It turned out that things weren't quite in the state of readiness we'd expected. The owner of the boat that had been chartered for the campaign had backed out of the deal under pressure from authorities in the Western Australia government. The Zodiacs had not yet arrived. Transportation from Sydney to Western Australia had not quite been arranged. Outboard engines, wet suits, compasses, sleeping bags, et cetera, had not been purchased. In fact, nothing was ready from an organizational point of view.

It was unnerving, to be sure, to discover that a date had been set for a demonstration at the whaling station—three thousand miles away, on the other side of the Australian landmass—and that no one really knew how we were going to get there, or what we would do when we arrived, since we didn't have a boat. But it was also refreshing. In the place of planning, there was faith. Instead of telecommunications, there seemed to be telepathy. And in the place of a board of directors, steering committee, or constitution, there was Jean-Paul Fortom-Gouin.

When he arrived in the office, the Frenchman proved to be short, dark-haired, bearded, olive-skinned, with a slightly hooked nose and an athletic grace. He smoked Gauloise cigarettes. Only thirty-five years old, he managed to look in his mid-forties. He spent most of his time living in the Bahamas—seldom visited France—and was reputed to have made his money in real estate, although nobody knew for sure. Whatever the case, he did have money. Before the campaign was over, it would cost him at least thirty thousand dollars. His only reaction was to shrug. He had already donated twenty thousand dollars to the *Ohana Kai* operation, and he was prepared to write that off, too. He had somehow arranged to get himself appointed as the delegate from Panama to the International Whaling Commission, and had made life miserable for the other commissioners. He had arranged at one point during the IWC meeting in Canberra, a few months before, to have a life-sized plastic whale balloon pumped full of air in the lobby of the hotel where most of the other delegates were staying. The inflated whale had completely blocked up the passageway, forcing reluctant Australian police to slash it with knives in front of the news cameras to get it out of the way. With his bankroll, Fortom-Gouin was rapidly becoming the godfather of the antiwhaling movement.

A demonstration had been set for August 28 at the whaling station in Albany, on the southwest coast. From the time Bobbi and I arrived, that left us a little over two weeks to get everything organized and transport the crew across the country. Fortom-Gouin, Pat Farrington, and Jonny Lewis, the main Australian organizer, would fly—the rest of us would go by ground.

It was finally decided that it didn't matter whether we had a boat or not. The whalers in Albany operated from the shore. It would be a simple matter to follow them out to sea in the Zodiacs when they set out in the morning. Of course, they were liable to travel one hundred miles or more out to sea, but that could be overcome by stuffing extra fuel tanks into the Zodiacs. It also remained that it was late winter and that the winds off the Southern Ocean were fierce and unpredictable; bad weather was the rule for the season. But—*C'est la vie*. Not to worry, mate! Eh?

When the "worldwide antiwhaling movement" finally set out from

425

Sydney, its entire "force" consisted of a beat-up 1961 Ford flatbed truck we'd bought, on the back of which we had built up a housing out of aluminum foil, and which pulled a rent-a-trailer loaded to the sky. A rented station wagon followed along behind. Crash, clank, shudder, wobble, bang. The back doors, which were tied together by string, kept flopping open, threatening to spill equipment all over the highway. The aluminum housing could be seen swaying under the weight of the Zodiacs, and the weight of the trailer made the old truck fishtail as it coughed and jerked along the road.

We drove all night and most of the day, stopping only to make meals on the propane camp stoves in the back of the truck—and to repair the truck's engine, which broke down from one cause or another at least once every day during the seven-day crossing and two or three times a day toward the end. Tom Barber was a mechanic of a very high order, even though he was, by trade, an architect. Allan Simmons was no less a native genius, but his regular occupation was fixing bicycles, not trucks. The two of them labored like Titans.

Aline Charney, an American lady who wore bracelets on her ankles under her khaki dungarees, tried to help them out by talking to the truck soothingly, as though it had a definite personality. The truck seemed to respond. Certainly, when Aline was driving, the Tinfoil Terror behaved as though slightly tranquilized. When Aline wasn't driving, it was usually Steve Jones, a thin, almost skull-faced Aussie, who had apparently always had a secret desire to be a "truckie." Either that or he was a road-junkie. He'd drive for fifteen hours at a time, grumbling and raving, dissatisfied with everything, but you couldn't pry his hands away from the wheel with crowbars. In the station wagon, it was Bobbi, a bit of a truckie herself, who did most of the driving.

A big problem we ran into was kangaroos. In Iron Bar, we'd read newspaper accounts of a "plague of roos" that was sweeping southwestern Australia. A bad drought had forced the kangaroos out of the bone-dry interior and they were moving in swarms, like giant locusts, southward. This sounded a bit exaggerated, but it proved to be close to the truth. After Iron Bar, night driving became a real hazard. The roos seemed attracted by the headlights, and since the brakes on the Tinfoil Terror weren't too good, it was only a matter of time before the first *thud*. When the roos came sailing out of the darkness, they moved slow-motion, as though in a dream. Steve Jones hardened his heart after the first half dozen roos had splattered on the front grill.

The Nullarbor Plain—twelve hundred miles of desert—almost finished our truck off for good. Twice, the engine burst into flame. The first time we managed to smother it with blankets, but the second time it got out of control. No matter what we did, the fire would not go out, and nobody could remember where the fire extinguisher was packed. Out of nowhere—which was all around—a freight truck appeared, the truckie hopped out, walked over with an extinguisher, smothered the fire in foam, grinned, and walked away. Another minute or so, and the whole rig would have blown.

"Who *was* that Masked Truckie?" Steve Jones asked.

We were all giddy from lack of sleep, dehydrated, and starting to get tense because it was beginning to look as though we weren't going to make it to Albany in time for the big demonstration.

We did, but not in time for a press conference that had been called for the day before. Toward the end, the Tinfoil Terror was only able to travel in spasms that would carry her along for three, maybe five miles, before she'd shudder to a halt again. And then one-mile spasms. And finally, it was evident she wasn't going to make it without being towed. An icy downpour obscured the land around us. We were still four hundred kilometers from our goal. There wasn't enough room in the car to carry both Zodiacs, outboards, and people, so we swept into the nearest little town and I started knocking on doors, trying to hire someone with a truck to drive the rest of our gear into Albany. Everyone was busy. I was finally directed to an elderly gentleman, who owned a repair shop.

Suspiciously, he squinted at me.

"Now listen, mate, you ain't one of them bloody whale savers, eh? Goddamn conservationists. Should all be shot. Eh?"

"Look, fifty dollars now, fifty when we get to Albany," I said.

Half an hour later, the two of us were whipping along the highway behind the rented station wagon, the back of his own car stuffed with whale protest equipment, and he was telling me stories about the war while we cranked down beer after beer, and he allowed as how all conservationists weren't necessarily bad. It was the Sierra Club that really bugged him—trying to take farmland away and make it into game reserves.

The journey across the country had set the tone for what was to follow: mechanical breakdowns, delays, last-minute bursts of speed, a nonstop jackhammer flow of events.

In Albany, Fortom-Gouin had rented several rooms in a motel

overlooking the beach at King George Sound. All we would have to do was run across the street, push the Zodiacs into the water, and zip out to sea. The whaling station itself was about four miles away on the southwest side of the sound—oddly enough, in the middle of a large state park straddling a peninsula. The moment we arrived in town, we were greeted by Fortom-Gouin, Pat Farrington, Jonny Lewis, and about a dozen Aussies and another dozen reporters and cameramen, who had been rather impatiently awaiting the arrival of the "worldwide antiwhaling movement." They had obviously expected something more romantic than this handful of bleary-eyed gypsies stepping out of a couple of cars, dragging several boxes of junk out after them. One reporter admitted afterward: "I'd read all about you guys. I thought you were gonna pull up in a convoy and all jump out in green uniforms, screaming: 'Give me whales or give me death!' Something like that."

"Come on," barked Fortom-Gouin imperiously, "get these mothers in the water!"

It was late afternoon, but the Frenchman wanted to race across the sound immediately. Within minutes, we were all struggling down to the beach with the outboards and folded-up Zodiacs, laying several bundles of equipment down on the sand, spread out like a do-it-yourself kit. "Okay Bob and Bobbi," said Fortom-Gouin, "show us what to do."

He had assumed, like everybody else, that since Bobbi and I had been on previous expeditions, we were Zodiac experts, little realizing that the most Bobbi had ever done was start an outboard engine running once, and I wasn't even sure how to do that. As for assembling one of these twelve-foot things, neither of us knew where to begin. Neither did anyone else. A small crowd of reporters and curious locals had gathered to watch the Greenpeace eco-guerrillas go into action. There was a moment or two of embarrassed silence before Fortom-Gouin and I decided to take a quick walk down the beach by ourselves to have a private conference. "I thought you knew how to do all this shit," the Frenchman hissed. "I thought *you* said you had everything all set up," I hissed back. While we were glaring at each other, Tom Barber and Allan Simmons arrived in the Tinfoil Terror—having pulled off one last mechanical miracle to get her going again. Making the best of it in front of the reporters, Fortom-Gouin said: "Ah! Here come our Zodiac experts." The architect and bicycle repairman were now elevated to the status of Zodiac-assembly specialists. Fortunately, their knack for fixing engines also applied to inflatable rubber boats. By dusk, they had figured it out.

While they were working, a cheerful, swarthy local came up to us and said: "Are you lads *serious* about taking those Rubber Duckies out in that water, eh? Listen mate, you don't leave the shore here unless you got a steel hull, mate. You go over near that whaling station and you're dead!" He laughed almost hysterically at the thought.

"Ah, forget him," said Fortom-Gouin. "Let's go."

The two of us sped across the calm of King George Sound, watching the lights of Albany coming on in the pale wintery gloom. The whaling station was easy to spot: it was the only structure on the southwest shore. As we approached, I had a moment of disorientation, a strangely powerful feeling that I had somehow been flipped back in time maybe half a century. The peninsula around the station consisted of rolling hills and spare scrub brush, with no other signs of civilization. In their isolation, the little cluster of buildings and storage tanks reminded me vividly of the first whaling station I had ever glimpsed—at Akutan Island on the edge of the Bering Sea. Whaling-station design had not changed much in fifty years, and this one looked almost exactly as the other must have looked before it had been abandoned and left to go to rust amid all the huge bones scattered on the shore. The lights of the station had come on too, and as we approached, people began to gather in the windows. We were just getting close enough to make out their faces when there was a sudden loud clank, the Zodiac bounced into the air, and the engine sputtered to a halt. We'd hit an underwater steel pipeline of some kind.

Fortom-Gouin swore in French and I thought: Oh great, we're going to have to paddle up to the station and ask those buggers to help us.

But while the propeller was dented, when Fortom-Gouin gingerly lowered the engine back in the water, it started up, albeit with a wonky edge that left you with the unnerving feeling that it might conk out any moment. We turned then and limped back across the sound to the beach beside our motel, arriving in pitch darkness, only to find that the tide was coming in, creating a phosphorescent whooshing turmoil of surf that made landing the Zodiac like trying to rein in a bull that stubbornly keeps trying to pull away.

When we finally staggered dripping and shivering into the motel, it was to discover that the reporters already knew about our accident: they had sources at the whaling station. Typewriters and phones were busy, stories were being filed quoting locals as saying that the Greenpeacers were irresponsible fanatics who were probably going to get themselves killed. The reason was not just that we had already crashed into a pipeline while roaring around in the dark. It had more to do with why the man back on the beach had been laughing so hysterically at our Rubber Duckies.

The waters around Albany were famous for the number and size of white pointer sharks who inhabited the area. The largest ever caught— only a year before—had weighed thirty-seven hundred pounds and was seventeen feet long, almost the size of an orca. For decades, the white pointers had been hanging around King George Sound, attracted by the blood spilled into the water from the whaling stations. After the whales had been killed, they were hauled back into the sound, inflated with air, and tied up to a buoy a few hundred feet offshore until the time came to process them. As a rule, the sharks would help themselves to dead-whale meat. The whaling company, in fact, employed a man to ride around the floating whales in a steel-hulled boat, armed with a high-powered rifle, to drive the sharks away. Yet still they came, day after day, year after year, maddened by the blood scent and the promise of an easy meal. "There's no point watching for their fins, mate," said one fisherman. "They come up straight from underneath you." The white pointers were the reason Albany's otherwise magnificent beaches and sand dunes had not turned the town into a major tourist center.

Not one person in town was willing to come out with us in our Rubber Duckies, even for a minute.

The next day was Sunday, August 28. Perhaps one hundred whale lovers in all, including the former Australian deputy prime minister, Dr. Jim Cairns, showed up outside the gates to the whale station, which had been reinforced with barbwire and barred to anyone except those with permits. Some fifty Commonwealth Police showed up to keep an eye on things. And about two hundred locals including the families of men who worked at the station and whose jobs were on the line. It was not a happy demonstration, even though Pat Farrington tried mightily to get everyone to join in civil rights-type hand-holding and singing. We belted out songs like "We Shall Overcome," but it was tough going against the jeering and heckling and chanting from the opposition, most noticeably about thirty members of the local "bikie" club, known as God's Garbage, most of whom worked as flensers at the whaling station. At the arrival of the bikies, their big Harley-Davidsons roaring and kicking up dust, I had almost cheered. They climbed off their bikes, wearing black leather jackets with the sleeves torn off, tatoos on their arms, some of them shaven bald, with earrings and chains hanging from their belts—

classical 1950s stuff. The Hell's Angels back in California, the prototypal bike gang of them all, had shown up several times to act as security at Greenpeace benefit concerts in San Francisco, and so naturally I made the assumption that God's Garbage would be on our side too. No such luck. It was a bit stressful to have to take an electronic loudspeaker, stand face-to-face with all thirty of them, and start giving the reasons why we thought they should have their jobs taken away.

Yet that was the lesser part of the stress of the day. The worst part had been coming across King George Sound in the Zodiacs, thinking about nothing except the white pointers that were out there somewhere around us. It was Jonny Lewis, crew member Rosie Dekanic, Bobbi, and myself who took the lead Zodiac, while Jean-Paul Fortom-Gouin remained behind to get the second one running. None of us were very happy. We were so scared that our hands were trembling. Bobbi managed to turn her fear into rage—because no one had had the foresight to check out the conditions beforehand. It was ridiculous enough to try to chase whaling boats one hundred miles out into the stormy seas in nothing but Zodiacs, but it was ridiculous to the point of insanity to zip around casually in waters that were home to some of the biggest, most savage sharks in the world. She was almost crying, she was so angry and terrified.

While the demonstration was taking place outside the barbwire barrier at the gate, work was going on as usual inside. Three dead sperm whales were moored offshore, and several others were in various states of dismemberment on the flensing platforms. Blood spread out in a slick perhaps two hundred feet from the shore, around the slipway. It was here, of course, that the greatest danger from sharks lay. Cameramen by the score had positioned themselves on the rocks and beach all around to see whether the Greenpeace kamikazes would dare to take their Rubber Duckie right into the midst of the blood.

Gritting our teeth and moving with adrenaline-fired speed, we did, whirling about in front of the station, holding up placards that said things like SHAME and STOP THE SLAUGHTER. It was a good performance— it got aired on both national TV networks and every newspaper in the country the next day, along with footage of God's Garbage, the police, and the protesters, so as a media event, it was quite worthwhile. The only thing that nobody could figure was why the white pointers hadn't showed up. Rare was the day that they had not appeared while whales were being processed.

It wasn't until late afternoon that the reason became clear.

Four dolphins had appeared, cruising up and down just off the beach beside the station, where we had landed our Zodiacs. At the first glimpse of their fins, several people had screamed: "Sharks!" The entire crowd of some three hundred people had rushed to the edge of the roadway overlooking the beach to have a look. But instead of sharks, there were only the dolphins.

The locals in the crowd were flabbergasted.

"Ain't seen dolphins hereabouts for years," said one.

"Never heard of that happening," said another. "Not when the whales were being flensed, mate. Never knew the dolphins to like blood. That's bloody weird, eh?"

The whale savers in the crowd exchanged looks. Dolphins, we knew, were the only creatures in the sea who could drive sharks away. It was the one explanation we could see for the absence of the white pointers that day. The Dolphin Patrol had arrived to protect us. Just as the demonstration was breaking up, a tremendous rainbow had flared in the sky across King George Sound. It was so spectacular that one of the TV networks started its report with footage of the rainbow hovering over the whaling station as protesters milled about at the gate. The dolphins, too, were mentioned, and the absence of sharks.

In the place of our fear that night, we felt a momentary blissful certainty that the gods were on our side.

The next morning we set out early with two Zodiacs to intercept the whaling boats as they headed out through the mouth of the sound en route to the whaling grounds. The boats proved to be almost identical in design to those used by the Russians, except that they were in better condition, freshly painted in the colors of their company, Cheynes Beach Holdings, Ltd., with black hulls and orange-brown superstructures.

We were a bulky looking lot as we clambered into the Zodiacs, wearing clothes on top of our wet suits, foul-weather gear on top of the clothes, and boots, two pairs of socks, gloves, woolen caps, and goggles to keep the spray out of our eyes. There was hardly any place to sit in the Zodiacs, since they were loaded up with six large plastic containers each full of petrol. Even if we weren't so wet, it would have been impossible to light a cigarette: we'd have blown up immediately. The Zodiacs handled heavily, clumsily. Fortom-Gouin and Jonny Lewis were in one, Allan Simmons and myself in the other. But we only got some five hundred yards away from the shore when the second Zodiac's engine broke down. Reluctantly, Lewis agreed to change places with me, so it was the Frenchman and a

Canadian who finally went catapulting out into the Southern Ocean to confront the Aussie whalers.

Because of the mechanical delays, we'd missed the passing of the first two killer boats, which had set out early from downtown Albany and had already cleared the headlands for the open sea. One boat remained. Its name was *Cheynes 2*. Just as we rounded the headlands, finding ourselves moving into an offshore swell that took us up sometimes eight feet before cresting and beginning the descent into the next trough, we spotted *Cheynes 2* coming up on our stern, throwing out a spectacular bow wave that indicated her skipper was running at top speed. We let him come abreast of us, then Fortom-Gouin said: "Okay! Let's go!" We put on a burst of speed and dove in front of the bow, then turned and spun around them in circles several times for the benefit of the half dozen reporters who were on board, courtesy of the company, whose public-relations officer had been taking media people out to sea for over a week, to show the "positive side" of whaling. The crew hurled insults at us. Fortom-Gouin gave them the finger, and I hung on at the bow, gasping because my legs were starting to buckle from the pounding already. After we'd demonstrated our ability to outrun them, we settled down in the wake of the *Cheynes 2*, prepared for the long chase out to sea. Several times, we had actually been airborne. But we'd only gotten about sixteen miles out when our engine started coughing, then abruptly died, leaving us bobbing uselessly in the swell. *Cheynes 2* doubled back, circled us, her crew jeering and laughing, then swept triumphantly away toward the horizon.

We hobbled painfully back to shore on our little six-horsepower auxiliary engine, having to go against waves all the way to the mouth of the sound. It took some five hours to make it back to the motel.

The headlines in the *West Australian* newspaper that night said: ROUND 1 TO WHALERS. The Cheynes Beach boats killed seven whales that day.

By now we had a media situation that was getting tough. On the one hand, thanks to the decision by Cheynes Beach Holdings, Ltd., to take cameramen and journalists out on the boats to record the whales being killed, Australians were for the first time being deluged with images of dying whales. Until then, most Australians had not even known that there was still one whaling station left operating in the country. At one time there had been dozens, but they had closed down, one by one, as the whales vanished. Only the Cheynes Beach operation remained, and its quota was down to seven hundred sperms a year. It was the last whaling station in the English-speaking world. And now, thanks in part to its own open-door public-relations policy, every Australian knew about it. Cheynes Beach and the protest were national news.

Three days later, Fortom-Gouin and Jonny Lewis succeeded in following a whaling boat out beyond the continental shelf. The boat—*Cheynes 2* again—led them off on a twelve-hour chase to a point some ninety miles out at sea, then stopped. It had been sent as a decoy to lure the protesters away while the other two boats went about their usual business. Still, we could also claim to have put one-third of the fleet out of commission for a day. At one point, the whaling skipper leaned over the ship of his boat and told Lewis and Fortom-Gouin that he was "lost."

"That's okay, we will take you back," replied Fortom-Gouin, even though he knew his own compass wasn't working and he was completely dependent on the whaling boat to find his way to shore.

The newspaper duly reported "some success" for the protesters.

It was not until the third confrontation—on September 5—that a shot finally rang out.

Fortom-Gouin and Tom Barber had been in the Zodiac at the time, about forty miles offshore, pulling up on the boat called *Cheynes 4*. There were no reporters around, and the whalers greeted the arrival of the whale savers with outbursts of obscenities. One man waved an iron bar. Another whipped down his pants and made gestures, yelling: "Wankers! Wankers!" Apparently an Aussie term meaning jerk-off artist. Tom Barber was stunned by their fury, but Fortom-Gouin's expression didn't change, except for his eyebrows, which went up and down. He muttered: "You bastards." Then a whale surfaced directly ahead of *Cheynes 4*, and Fortom-Gouin immediately ran the Zodiac into position in front of the bow. Unlike the deep-sea chases we'd witnessed in the Pacific, this one was over in less than twenty minutes. The harpoon cannon spat its tremendous barb so close to the Zodiac that a second after it had gone off, lodging itself in the back of the whale, the cable had twanged, rising out of the water like a knife, and snagged in the Zodiac's propellers. Desperately, Barber threw himself on the engine and pulled the pivot pin so that the still-rising cable snapped free, and the Zodiac, which had started to come out of the water, splashed heavily to the side, leaving Barber and Fortom-Gouin sitting there with a pall of white smoke hanging over their heads and a sperm whale thrashing itself to death less than thirty yards in front of them.

A week later, Barber and Jonny Lewis had a second harpoon fired at them, making it a "three-harpoon summer," counting the one that had been fired over Rex Weyler and Mike Bailey in the North Pacific.

We went back out one last time—Lewis, Allan Simmons, and myself—and got as far as the edge of the continental shelf, roughly twenty-five miles, before our thirty-five-horsepower engine broke down. In the silence, broken only by the slapping of the water and the suddenly loud whistling of the wind, Lewis and I riveted our eyes on the young bicycle repairman who now held our lives in his hands. Simmons was so seasick he could hardly crawl back over the extra fuel containers to start pawing feebly at the engine. Belatedly, we noticed that the wind had been building up all the time we were heading south and now was blasting from the land at some twenty knots. Unless Simmons could get the big engine going, we'd have to hope that the six-horsepower job could push us back against the rising sea. There was no land at all to be seen. Under the circumstances, it was no surprise that we all got a bit silly. Jonny Lewis started singing "Lavender Blue," and I burst, absurdly, into "O Canada." When, after five or six agonizing moments, Simmons got the engine coughing and spluttering, the three of us began to sing "Rub a Dub Dub, Three Men in a Tub." The ride back took from around noon to almost dark. The engine kept conking out if it was made to go any faster than half throttle, so we putted slowly, tensely, up and down the waves. From time to time, the bow of the air-filled Zodiac would be caught by a gust of wind as it came up over the crest of a wave, and we would hover, as though astride a kite, three-quarters out of the water. When Lewis went to dig out the "tucker," it was to find that the plastic bag full of sandwiches had been reduced to a salty soup that he had to cup in his hands and drink, rather than eat.

By the time the church spires of Albany finally came into sight, we were numb, shivering, and bruised. It seemed to me that we'd used up every drop of luck that remained. In Perth, sitting on the doorstep of the company that owned the whaling station, Pat Farrington had started a hunger strike that was to go on for a month before she finally had to give up.

When we left Albany, we felt depressed, letdown, and exhausted. It wasn't until months later that the federal government bowed to the pressure that had been generated by all the media coverage of the event, and ordered an inquiry into whaling in Australia, the first such inquiry ever called. Public-opinion polls had shown that feeling against whaling was running at something like seventy percent nationwide. We had not directly saved a single whale—yet, indirectly, we had precipitated a strong movement in that direction.

There was one particular scene that occurred just before we left town that lingers in my mind. We had gone down to the bar in the motel when a gentle-spoken man approached us and asked if he could sit down. We recognized him from his pictures in the newspaper. His name was Kase van der Gaag, and he was the skipper of *Cheynes 2*. A Dutchman, he had been living in Australia, making his living as a whaler, for seven years. He was married and had three children. A good-looking man with sensitive features, he was forty-six years old. The stories in the newspapers had described him as a man who did not enjoy killing whales, and who became agitated if the whale didn't die quickly after the first shot. He loved being at sea, enjoyed the hunt, but detested the kill. Yet it was his "job."

"You don't have to remind me," he said softly, "that that's what the Nazis said, too."

He had a pained expression and was obviously tormented.

Ironically, what he wanted to know was why we were wasting our time harassing him and his fellow workers when there was another hunt going on in the world that he considered thoroughly detestable and unnecessary. He was referring to the harp-seal slaughter off the coast of Labrador. Here we were, sitting in Albany, Western Australia, almost precisely on the opposite side of the globe from St. Anthony, Newfoundland, being told by a whaler that we should stop the sealers, just as we had been told in Newfoundland that we should lay off the sealing issue and concentrate on stopping whaling, something most of the Newfies we'd talked to were against. There was some yin-yang principle at work here that defied ordinary logic and perhaps pointed to the real crunch faced by the conservation movement: what was close to home was sacred, what was distant was profane. It was easy to win agreement that what the "other guy" was doing was bad, impossible to win agreement that what one was doing oneself was equally bad. Yet the arguments that Captain van der Gaag was using to defend the whale hunt—that the whales were not endangered, that the hunt was an economic necessity, that jobs depended on it—were precisely the arguments used in Newfoundland to defend the seal hunt, even though, there, they saw the whale hunt as a thing to be ended.

The captain and I shook hands on a private deal. If I could get the seal hunt stopped in Newfoundland, he would quit whaling.

My wife and I left Albany not knowing whether to laugh or cry.

From there we went to Perth, and from Perth flew to London, England,

and went by boat from there to Amsterdam, where I had been invited by a Dutch television station to appear on a show to appeal for funds to be used by the Greenpeace offices in London and Paris to purchase a 165-foot trawler, the *Sir William Hardy*, which would be sent into the North Atlantic to intercept the Icelandic whaling fleet in the summer of 1978.

The boat would be renamed the *Rainbow Warrior*.

Epilogue

When we got back to Vancouver it was to find that we had indeed come a full circle.

The emphasis was once again on plutonium and nuclear military hardware.

In early June 1977, three members of Greenpeace Toronto had "invaded" a nuclear power plant at Douglas Point, Ontario. In a predawn raid, the three paddled ashore from Lake Huron and walked unopposed into the Bruce nuclear power station. Seeking to prove how vulnerable these plants are to terrorism or sabotage, the three—John Bennett, Doug Saunders, and Rick Curry—wandered through the buildings for about an hour, placing Greenpeace stickers on the doorways to sensitive installations.

Bennett, the coordinator for the Toronto group, said later: "Ontario Hydro has spent the last twenty years persuading the citizens of Ontario that its operation of nuclear power plants is safe, clean, and inexpensive. This action demonstrates the vulnerability of Ontario Hydro nuclear plants to organized attack by international terrorist groups with resulting loss of life and release of radioactivity."

After their hour-long stroll through the plant, the three were finally caught by a startled guard, an unarmed man in his mid-fifties. He started to call the police, but when he discovered they were from Greenpeace Foundation, he decided against it and released them instead.

A day later, Ontario's energy minister tried to brush the incident off as the work of "a few environmental streakers." But the fact was that the incident made national news, embarrassing the power utility.

In early October, Greenpeace Toronto struck again. This time, twelve members were arrested when they attempted to set up camp on a site where Ontario Hydro planned to build yet another nuclear reactor. The provincial government had allowed the corporation to proceed with construction without having to submit an environmental-impact

438

statement. Accompanied by close to one hundred local residents, the Greenpeacers penetrated the site and remained for several hours, attempting to set up camp, when the Ontario Provincial Police moved in to cart them away. All twelve were charged with trespassing. The protests and arrests were peaceful. No one was hurt.

Back on the West Coast, the summer saw a total of fifty persons being arrested at another type of nuclear site at the little Washington State resort town of Bangor, just south of the Canadian border. Construction had begun on a "Trident" nuclear-submarine base. The Trident submarine, the largest ever built, would be four stories high, 560 feet long—almost the length of two football fields—weighing 18,700 tons and equipped to carry 408 independently targeted warheads mounted on twenty-four missiles, each warhead with the capacity to annihilate an entire city. Thirty of these machines—described as the most powerful weapons of destruction ever devised—were to be based at the Bangor site, once it was finished.

The actions against the Trident base were mainly organized by a group called the Pacific Life Community and Greenpeace, with Walrus Oakenbough in the role of spokesman and coordinator. For the most part, leaflets and buttons were handed out at the gates of the site on a day-to-day basis, culminating in a demonstration involving some two thousand people on August 14. An attempt by seven Canadians to march 160 miles from Vancouver to Bangor was turned back at the border. At various times through the summer, people clipped barbwire fences, rowed ashore onto the site, and tried to block the path of construction workers.

By the end of the year, most of the energy in the various mainland North American offices was directed to mounting a third anti-sealing campaign, while the European offices concentrated on preparing the *Rainbow Warrior* for its assault on the Icelandic whaling fleet. The *Ohana Kai* remained tied up to a wharf in San Francisco while several Greenpeace boards of directors pondered its fate. The 1978 seal campaign proved to be different than the previous expeditions out to the Front ice, partially because of a decision to attempt to bring our new American political muscle to bear and partially because the Canadian government passed yet another order-in-council, making it illegal for anyone except a sealer to so much as put a foot down on the ice in the vicinity of the hunt, without special permission from the federal fisheries minister.

The "muscle" part of the exercise involved utilizing political connections that had been established mostly by our San Francisco liaison officer, Bob Taunt, who had succeeded earlier in getting California

Congressman Leo Ryan to put up a resolution before the U.S. House of Representatives calling on Canada to "reassess its present policy" of permitting the harp-seal-pup slaughter. The Canadian government had been furious about the House resolution. Back home, we had chortled. When was the last time a Canadian-based group managed to swing the weight of America against a Canadian operation? Moreover, for a group that had begun its career in opposition to American militarism—and remained opposed to such militarism—it was one of the more convoluted political exercises to oppose U.S. imperialism on the West Coast, yet bring it to bear against Canada on the other side of the continent.

When the expedition was finally launched, in March, it included Congressman Ryan, a Democrat, and a Republican from Vermont, Jim Jeffords. Until that point, the Canadian government thought it had closed off every avenue leading to the ice and that it could simply sit back, knowing that there was nothing we could do to reach the seals. The other antisealing groups had already abandoned the field.

The one contingency for which Ottawa had not prepared itself was a demand from two U.S. congressmen to be granted permits to inspect the slaughter—"at the invitation of Greenpeace."

Greenpeace's forte had always been international incidents, or the threat of them, and now the same specter rose to haunt the bureaucrats of the federal fisheries department.

The expedition included fifteen Greenpeacers, the two congressmen, with three legislative assistants and two actresses, Pamela Sue Martin of Hollywood and Monique van der Ven of Holland. The Greenpeacers came from San Francisco, Vancouver, Portland, and Los Angeles, with Patrick Moore as expedition leader. Lawyer Peter Ballem came along to fend off the dozens of fisheries officers who would be on hand with coils of red tape, ready to lasso us at every point. I was yanked out of semiretirement again to help with the "media coordination." The veterans included Eileen Chivers—now Mrs. Eileen Moore—Rex Weyler, Eddie Chavies, who had been on the *Ohana Kai*, film maker Michael Chechik, cameraman Ron Orieux, who had served briefly on the *Phyllis Cormack*'s first run at the whalers, and Nelson Riley. The others were Steve Bowerman, Bob Taunt, Gary Young, Debbie Imoson, Edwina Ruzette, Jon Sargent, and Gil Wald, all of them Americans.

By the time we arrived in St. Anthony—determined this year not to avoid face-to-face contact with the Newfoundlanders—the anti-sealing campaign had already begun. In Halifax, a blockade of the sealing ships as

440

they left harbor had already been attempted, led by the Greenpeace groups in Toronto and Montreal. Toronto's John Bennett found himself being flipped out of a Zodiac into icy water as one of the sealing ships ignored the blockade and backed up right into one of the Zodiacs. No one was hurt, although the sealers themselves were so outraged they threw rocks and bottles and even a folded table, at the Greenpeacers in the water. On the other side of the Atlantic, in Alesund, Norway, five Greenpeacers, from London, Paris, and Norway, were arrested when they chained themselves to a departing sealing boat, delaying it for several hours.

In St. Anthony, it was Ballem who led the assault on the bureaucratic barricades that had been thrown up in our path. Even with the congressmen present—assured by Ottawa that there would be no difficulties—the federal fisheries officers on hand did everything they could to stall, delay, and deflect the drive to the ice.

The tactics included referring to "regulations" that did not exist. Ballem was the only man in all of Newfoundland who actually possessed a copy of the new regulations, and when he was able to point out that certain rules did not, in fact, exist, he was told by the chief fisheries spokesman: "Well, it's department policy." The Americans on the expedition, including the congressmen and their aides, were dumbfounded by the legal atmosphere in Canada, where lower-echelon bureaucrats could rearrange the laws of the land at whim, one moment saying that the "regulations" demanded that each Greenpeace helicopter be accompanied by an RCMP officer—at Greenpeace's expense—and the next moment blithely admitting that there really wasn't such a regulation, only a policy. Yet policies seemed to have the force of law—a law emanating from Ottawa like an all-pervading fog. The more players Ottawa sent into the field, the more bulldoglike became Ballem's determination to batter them aside. And in the end, he did.

Within one day of our arrival by aircraft in St. Anthony, two helicopters were lifting from the ice just outside Decker's Boardinghouse, with the two congressmen, their aides, and actress Pamela Sue Martin, led by Patrick Moore. Landing directly beside the sealing ships, the ecologist took his party on a tour of the open-air slaughterhouse, trailed along by a small army of fisheries officers, Mounties, cameramen, and reporters. A Newfoundland cabinet minister even put in an appearance. Dressing himself up as a sealer, he ranted at the U.S. congressmen that they were just up to "cheap political tricks." A day later, the cunning cabinet minister flew back to St. John's and took advantage of his instant television fame as the province's champion and defender, to defect from his party and cross to the opposition, where he had been promised a better ministry after the next election.

The trip to the Front ice had a powerful and visible effect on the congressmen and the young actress. They returned as from a horror film. Unfocused eyes stared out from pale faces. The actress couldn't speak even though a crowd of newsmen and TV interviewers had arrived in the boardinghouse, eager for statements. All Leo Ryan could manage to say was: "I'm in a state of shock. I just want to say enough ... enough! Stop!" Jim Jeffords described the walk across the ice: "We saw three seal pups clubbed, and in each case the mother defended them as best she could. I can't relate to the experience in what you might call a rational manner right now. The experience of witnessing the slaughter transcends all figures, arguments, or rationality."

The next day, both men left, along with their aides, heading back to Washington. It seemed clear at the time that the seals now had friends in high places, and considering the attitude of the Canadian government, the seals needed all the friends they could get. Jeffords himself was chairman of a powerful "invisible lobby" of political figures in the U.S. Congress, which included the men who had pushed through the Marine Mammal Protection Act, the loftiest piece of environmental legislation in the world. According to that act, the killing of any "nursing marine mammal" is forbidden.

With the disappearance of our American political musclemen, we were left in St. Anthony on our own hook, faced by fisheries officials whose only comment was: "That's it. You're finished. There's no way you'll get out on the ice now."

When Moore went down to the fisheries officer's hotel room the next morning to ask for permission to get airborne again, he was refused an answer. "Get out of this office," he was told. When he insisted that he had a right to get an answer to his questions, he was promptly arrested by the RCMP, along with Rex Weyler, who likewise refused to leave the room. They were charged with "loitering," then released.

The following day was Tuesday, March 14. Our entire energy was channeled into prying loose more permits to get out to the ice. We had agreed, as a condition of getting the congressmen to come, that we would perform no acts of civil disobedience when they were there, lest they be implicated and touch off a major diplomatic row—something we would have favored, but not the politicians. The task now became one of returning to the scene to make our point.

442

Sensing this, the government men were determined not to give us our chance.

It now became a purely political battle, with Ballem on the phone to contacts in Ottawa, right up to the prime minister's office, and ourselves mobilizing support back in Vancouver, trying to bring pressure to bear on British Columbian politicians in whose ridings our members lived. Telegrams began to flow from Kitsilano. It took three days before the pressure built up enough to force a change of heart in Ottawa, but the word finally came down that we would be allowed eight permits to go out once, but only to an area where there were some seals. We would still not be allowed into the "area of the hunt."

The offer was a sop. We rejected it. Our job was to go to the hunt itself, not simply to behave as tourists.

A last burst of threats, phone calls, telex messages, and press releases decried the creation of a "Gulag Archipelago" on the ice floes. "The government of Canada is turning the coast of Labrador into a police state," raged Moore. "All civil rights have been suspended, freedom of the press has been suspended, and minor public officials are passing laws in beer parlors." All of which was true: the press had been refused further permits, too. But the challenge was to make the truth widely enough known to embarrass federal government-party politicians, forcing them to make concessions, if only to avoid the charge of "Gestapo tactics."

Two hours later, the federal officials came back with their "final" offer: four permits to go to the hunt. There was a new Catch-22. The permits would only be valid for twenty-four hours, beginning at 2:00 P.M., March 17. There was to be no flexibility to allow for weather. That left us roughly an hour to get into the air, and a major storm was predicted for the area. Its front was sweeping directly toward us from out on the Atlantic. Obviously, the fisheries department had decided to let us have permits right away, since the storm was sure to ground us. Afterward, they would be able to say: "We were good guys. We gave them permits. Too bad they had bad luck with the weather." And if we refused the permits, they could always shrug and say: "We gave them their chance." Tactically, it was a brilliant move on their part.

So now, after all the nights of political bargaining and arm wrestling, we were left with a situation that depended entirely on our "luck." It was the one door open; if we failed to leap through it, we knew we'd never get another chance. If we leapt for it and it slammed shut in our face, that would be the end, too.

443

Within minutes, Rex Weyler and I were up in my room, nervously throwing the *I Ching* coins. Had the hard-nosed politicians of the federal government known that it was to an ancient Chinese oracle that we would turn, they would have either fallen out of their chairs, laughing, or gone weeping to their rooms.

The *I Ching* reading noted the "awakening of a consciousness of the common origin of all creatures." It was important that "the clouds should be dispersed before they have brought storm." Its judgment was:

> DISPERSION. Success.
> The king approaches his temple.
> It furthers one to cross the great water.

So it was "go" for a run out to the ice.

"You're crazy," snapped one experienced maritime reporter. "You'll be grounded by morning." The weather reports were firm: there would definitely be a storm. It was on its way.

By 2:30 P.M., the helicopters were off the ground, heading northward for the Labrador community of Cartwright, where they would stay overnight before attempting to leap across the water and ice to the sealing ships, operating closer to the arctic than usual. The crew included Moore and Weyler, Bob Taunt, and cameraman Steve Bowerman. At the last minute they were joined by Peter Ballem, who seemed confident he could arrange a permit for himself, as legal advisor, by morning.

Saturday, March 18, the skies dawned blue and clear, leaving the experienced maritime reporter shaking his head, fisheries officers groaning with dismay, and veteran Greenpeacers smiling knowingly. The storm had "inexplicably" changed course. By 9:00 A.M., with less than five hours to go on the permits, the two choppers put down on the ice in the midst of the seal slaughter, followed by a flotilla of fisheries and RCMP choppers and a small mob—fifteen in all—of nervous officials, all there to protect the seals from Greenpeace.

But for the blood staining the ice in great mile-long streaks and the high-pitched screaming of seal pups and mothers alike, it would have been a Keystone Kops farce.

Moore approached a group of sealers standing around a pile of pelts. The combative mood of the year before had all but vanished there on the ice, just as it had back in St. Anthony. The locals seemed resigned to our recurring presence. Only a few threats had been made. Mostly, there had seemed to be a mood—reflected by the surprising results of a radio poll conducted in the provincial capital, showing some forty percent of the callers actually opposed to the hunt—that indicated enthusiasm for the hunt was waning. Certainly, the men Moore now found himself confronting on the ice were less than righteous about their work.

One man said: "We's getting used as bad as the whitecoats, b'ye." The hunters themselves shared only twenty-four percent of the profits, he said. The rest—seventy-six percent—went to the owners of the boats, back in Halifax and Norway.

"Doesn't sound like a very good deal to me," said Moore. No one argued.

Toward noon, Moore made his move—striding over to a seal pup which was next in line to be killed. He calmly climbed on top of the animal, holding its flippers firmly, making it impossible for anyone to kill it so long as he was there. Once again, human flesh had come between the hunters and their prey.

Frantic fisheries officials ran about in circles, going into huddles; whispering furiously to one another, trying to figure out what to do.

"I believe I have as much right to give this pup life as you do to kill it," Moore said. "I claim this pup in the name of Greenpeace and ask you to please let this one live." For twenty minutes he remained on top of the animal—"a tough little bugger," he called it—before the senior fisheries officer marched forward and ordered him off the seal. When Moore shook his head, the officer tried again. A third time he warned the stubborn ecologist to get up, only to be refused with a shake of the head again. Finally, the officer said: "You are under arrest for breach of the seal-protection regulations."

As they pulled Moore to his feet, the seal pup waddled frantically away.

"Kill her, b'ye," an older nearby sealer said to his youthful companion. *Whack!* The seal pup lay dead and bleeding on the ice.

Moore was taken into Cartwright by helicopter and put into the local jail. Within two hours, Ballem had him out and the party returned to St. Anthony. Weyler's photos of the arrest were on the wire service within half an hour.

By early the next morning, most of the fisheries officers in St. Anthony packed their bags and left town. Several of them were seen screaming obscenities at one another, trying to shift the blame for what had turned into a political disaster. The gods of wind had worked against them, and

Moore's willingness to stand up to their master web of rules, regulations, and threats of imprisonment had left them in a position where, as one weary fisheries officer put it, "the shit's already coming down the hill." For his part, Moore faced up to a year in jail and a five-thousand-dollar fine. Shortly after getting back to Vancouver, lawyer Ballem also found himself being charged with unspecified violations of the seal-protection rules. The Canadian government was obviously not amused.

Out on the ice, the blood-stained ships were winding a fruitless course through broken pans that were mostly empty. They had killed every seal pup they could find, but it still did not add up to even half of the quota they had been assigned. Somewhere between the "population models" conjured out of banks of computers in Ottawa and the reality of the remains of the life cycle in the North Atlantic, there was a discrepancy. The ocean had not brought forth pups in anything like the abundance the experts had predicted. And so the tired, unhappy men on the boats went home in a rage of frustration, having made far less money than they had expected, and the ice itself, where it was not already streaked with blood, was strangely silent: there were few sounds except for the tinkle of melting crystals and a cold lonely wind from the north.

Index

Acknowledgements

This second edition of what many would consider my husband's best known work would not have been possible without the support and help of many. You know who you are. I will name some but forgive me if not all are listed.

I recall vividly the months Bob spent writing this, much of the time standing as he typed, trying to ignore signals from an injured back. He chuckled and cried and pushed through the pain. And in the end the work was finished.

Thank you to all who journeyed with Bob in those early years from that first voyage in 1971 to the time he handed over leadership to others. Bob remembered you all and always wished you the best. Many were dear friends then and remain so. To mention just a few: John Bennett, Hamish Bruce and Moira, Aline and Tom Barber, Kimiko Bruce, Michael Chechik, Al Clapp, Bill Darnell, Anne Dingwall, Fred Easton, Pat Farrington, Jean-Paul Gouin, Don and Lily Francks, David Garrick, Bill Gannon and Linda, Zoe Hunter, Will Jackson, Al Johnson, Bob Keziere, George Korotva, Barry Lavender, Johnny Lewis, Myron MacDonald, Dan McDermott, Rod Marining, Michael Manolson, Ron Precious, Nelson Riley, Peter and Mary Speck, Paul Spong, Marvin Storrow, Alan Thornton, Paul Watson, Rex Weyler, David Weiss. Deceased: Dave Birmingham, Jim Bohlen, John and Phyllis Cormack, Bob Cummings, Bree Drummond, Garry Gallon, Davey Gibbons, Ben Metcalfe, Dorothy and Irving Stowe, Lyle Thurston and Don Webb. These friends, and many more who were quite possibly missed with a mention, inspired and invigorated Bob through the years. But, his great love was for his children Conan, Justine, Will and Emily, and his grandchildren Alex, Chaz, Dexter and Rhys, and one grandchild he did not get to welcome with a kiss, Gwynn. To all his friends and family, mentioned or not, he cherished you all and thought of you as his gurus.

Bob would have been amazed and humbled to know that *Warriors of the Rainbow* was to be republished and would again, decades after it was written, influence another generation to protect the planet.

The story of the book's reissue on the 40th anniversary of Greenpeace would not be complete without thanking those who worked on the project: Kumi Naidoo, his predecessor Gerd Leipold, and the global organization that is Greenpeace; Steve Shallhorn, Bob's friend from later years, who made this project a personal cause; Jane Fraser, Clive Newman and the team at Fremantle Press, a small quality publisher in Australia; the communications folks at Greenpeace International, Martin Lloyd, Nicolette Bouwman, Karen Guy and Bernhard Drummel; Greenpeace International's Legal department and Kumi's assistant Tinu Otoki; and journalist Chris Pash, author of *The Last Whale* (Fremantle Press 2008), who met Bob in Albany, Western Australia, in 1977 and is writing Bob's biography.

Finally, thank you to everyone who has ever sailed with, volunteered for, knocked on doors for, stood up for or helped in any way the organization Bob founded. Without their support, this story would not have been possible.

For more information on Greenpeace and its campaigns please visit www.greenpeace.org.

Bobbi Hunter
Toronto, Canada
2011

Covers of the original U.S., U.K. and Japanese editions.

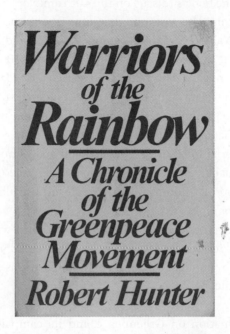

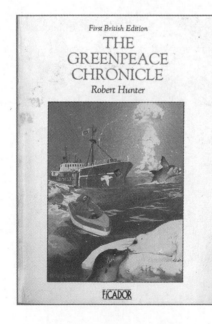